CODATA KEY VALUES FOR THERMODYNAMICS

CODATA Series on Thermodynamic Properties

CODATA Thermodynamic Tables: Selections for Some Compounds of Calcium and
Related Mixtures: A Prototype Set of Tables
Edited by **D. Garvin, V. B. Parker, and H. J. White, Jr.**

CODATA Key Values for Thermodynamics
Edited by **J. D. Cox, D. D. Wagman, and V. A. Medvedev**

CODATA KEY VALUES FOR THERMODYNAMICS

Edited by

J. D. Cox
National Physical Laboratory, Teddington, UK

D. D. Wagman
National Bureau of Standards, Gaithersburg, Maryland, USA

V. A. Medvedev
Institute for High Temperatures, Moscow, USSR

Final Report of the CODATA Task Group
on Key Values for Thermodynamics

⬤ **HEMISPHERE PUBLISHING CORPORATION**
A member of the Taylor & Francis Group

New York   Washington   Philadelphia   London

CODATA KEY VALUES FOR THERMODYNAMICS

Copyright © 1989 by CODATA. All rights reserved. Printed in the United States of America. Except as permitted under the United States Copyright Act of 1976, no part of this publication may be reproduced or distributed in any form or by any means, or stored in a data base or retrieval system, without the prior written permission of the publisher.

1 2 3 4 5 6 7 8 9 0   E B E B   8 9 8 7 6 5 4 3 2 1 0 9 8

Library of Congress Cataloging-in-Publication Data

CODATA key values for thermodynamics.

  Bibliography: p.
  Includes index.
  1. Thermodynamics—Tables I. Cox, J. D. II. Wagman,
  Donald D. III. Medvedev, V. A. (Vadim Andreevich)
QD504.C62 1989   541.3'69   88-24743
ISBN 0-89116-758-7

On 24 February 1988, Vadim A. Medvedev, a long-time friend, associate, colleague, and co-author of this work, died suddenly of a pulmonary embolism. He was associated with the Key Values Task Group from the beginning and provided invaluable assistance, especially in the evaluation of thermochemical and solution data. He was also a Consultant to the CODATA Task Group on Geothermodynamic Data, thus providing an invaluable liaison between the two groups. He worked tirelessly in support of the CODATA programs, and his death is a great loss to all of us. The members of the Task Group on Key Values for Thermodynamics dedicate this book to his memory.

Contents

FOREWORD

Because thermodynamics—with its breadth of utility in technology, chemistry, physics, and geosciences—spans a wide range of scientific disciplines, it is particularly embraced within the CODATA mandate of "improving the quality, reliability, processing, management, and accessibility of data of importance to science and technology." The provision of key values for chemical thermodynamics tables is an endeavor of fundamental importance since it provides the foundation upon which the compilation and evaluation of Gibbs energy, entropy, and enthalpy data may be elaborated. This was recognized early within CODATA. This volume represents the conclusion of the first stage of this endeavor in readily accessible, more comprehensive formulation than was possible in the previous aperiodic reports prepared in CODATA Bulletin format.

Although these data will find much application in current endeavors in subsequent updating of thermodynamics tables they provide also an archival record of the embodiment of newer techniques of evaluation. The current endeavors of the CODATA Task Groups on Chemical Thermodynamic Tables and on Geothermodynamic Tables are firmly anchored by these initiatory selections and evaluation achievements. The production of CODATA's internationally generated comprehensive thermodynamic tables depends upon the essential foundation provided by these key values.

As the 20th anniversary of the founding of the original Task Group on Key Values approaches, this rich harvest attests to the reliability and thoroughness of their endeavors. On the other hand, the essential expansion of these endeavors will probably continue to be a high priority aspect of CODATA's program.

E. F. Westrum, Jr.
(formerly Secretary
General of CODATA)

PREFACE

This work chronicles the production of an internationally agreed data set for some basic thermodynamic properties of 151 key chemical species. The project was among the first to be launched by the Committee on Data for Science and Technology (CODATA) of the International Council of Scientific Unions, following its creation in 1966. When we joined the project we little realized that we would still be involved with it 20 years later! During the intervening years many scientists have helped the enterprise along, and we would like to emphasize that the work of critical selection of data, followed by further critical assessment of the initial selections, has been a cooperative, participatory process. We hope, therefore, that the results of the project presented here will find widespread acceptance among thermodynamicists, and that others will build our recommended data into more comprehensive data tables and thence into calculations in many branches of science and technology. The use of internationally agreed key values for thermodynamics should go far in assuring harmony among thermodynamicists worldwide.

J. D. Cox
D. D. Wagman
V. A. Medvedev

ACKNOWLEDGMENTS

The participation of successive members of the CODATA Task Group on Key Values for Thermodynamics (TGKVT) is recorded in Chapter 1 and Annex I. Beyond these colleagues, who bore the brunt of the original work of data selection, we would like to thank those who contributed to the final stages of checking and revision after the formal end of TGKVT, namely the late W. H. Evans, G. A. Bergman, D. Garvin, L. V. Gurvich, I. L. Khodakovsky, V. B. Parker, C. E. Vanderzee and V. S. Yungman.

The editors wish to acknowledge the work of the late Stig Sunner, first Chairman of the Task Group on Key Values for Thermodynamics, who was instrumental in organizing the work and providing scientific guidance during the early years of the Task Group.

The Task Group expresses its appreciation to the Chemical Thermodynamics Data Center of the U.S. National Bureau of Standards and to the Department of Chemical Thermodynamics of the U.S.S.R. Institute for High Temperatures for their continued assistance and support during the project.

TGKVT was fortunate to enjoy cordial relationships with the IUPAC Commission on Thermodynamics and with CODATA's Task Groups on Fundamental Constants and on Internationalization and Systematization of Thermodynamic Tables (later renamed Task Group on Chemical Thermodynamic Tables). Edgar F. Westrum, Jr., former Secretary General of CODATA, provided much encouragement for the scientific work and help in the solution of organizational problems; he also was responsible for the preparation of the final camera-ready version of the manuscript. For all of this assistance we are truly grateful.

Finally, our warm thanks are offered to those who helped produce this document: to Sheena Smart (National Physical Laboratory) for typing the manuscript of text and bibliography, to Dorothea Megow, Ann Cecila Wood, and Roey Shaviv (University of Michigan) for copyediting and typescript production, to Ralph L. Nuttall (National Bureau of Standards) for computer analyses of the aqueous ion data, and to Dorothy M. Bickham, David B. Neumann, Nancy Young and Geraldine R. Dalton (all of NBS) for producing Annexes II and III.

J. D. Cox
D. D. Wagman
V. A. Medvedev

Chapter 1

OVERVIEW

1.1 General Introduction

1.1.1 CODATA and its objectives

CODATA is an acronym for the Committee on Data for Science and Technology, established in 1966 by the International Council of Science Unions. The constitution of CODATA states the organization's objectives as follows: "CODATA, working on an inter-disciplinary basis, shall seek to improve the quality, reliability and accessibility of data of importance to science and technology, including not only quantitative information on the properties and behavior of matter, but also other experimental and observational data".

The scientific program of CODATA is pursued through Task Groups and ad hoc Panels. The body to which Task Groups and Panels report is the CODATA General Assembly, which meets biennially. Coordination, communication, and publishing activities are undertaken by the CODATA Secretariat.

1.1.2 Origin of the Task Group on Key Values for Thermodynamics

The setting up of a Task Group on Key Values for Thermodynamics (abbreviated to TGKVT) was one of the earliest actions of the founders of CODATA. In particular, Professor F.D. Rossini (first President of CODATA), Dr Guy Waddington (first Director of CODATA's Central Office), and Sir Gordon Sutherland, Professor W. Klemm, Academician M.A. Styrikovich, Professor M. Kotani, and Professor B. Vodar (members of the first Executive Committee) all had a keen appreciation of the scientific and technical importance of thermodynamic data and of the need for a coordinated effort to create internationally agreed thermodynamic tables. As a result of their guidance and prompting, TGKVT was set up at the First International CODATA Conference, held in Arnoldshain, Federal Republic of Germany, in 1968.

TGKVT sustained an active program for the compilation and assessment of thermodynamic data from 1968 to 1979; the subsequent years were devoted to the preparation of this final report.

The history of TGKVT is recorded in Annex I.

1.1.3   Objectives of TGKVT

At its foundation meeting in 1968, TGKVT resolved that the key thermodynamic properties with which the Task Group would concern itself were standard enthalpies of formation at 298.15 K, standard entropies at 298.15 K, and increments in standard enthalpy between zero thermodynamic temperature and 298.15 K, and that the key chemical species with which the Task Group would concern itself were all the elements in their standard reference states and standard monatomic gas states and "such additional compounds as the Task Group may decide".

Hence the original objectives of TGKVT were to compile literature data pertaining to key thermodynamic data for key chemical species, to assess the data for accuracy, and to publish recommended values based on the assessment. These objectives remained essentially unchanged during the years of TGKVT's existence.

1.1.4   Membership of the TGKVT

The membership changed during the course of the Task Group's work. At one time or another, the following persons have served as members of, or consultants to, the Task Group.

C.B. Alcock	(Canada)
J.D. Cox	(UK), Chairman 1974 - 1980
J. Drowart	(Belgium)
W.H. Evans	(USA)
L.V. Gurvich	(USSR)
L.G. Hepler	(Canada)
V.A. Medvedev	(USSR)
D.M. Newitt	(UK)
J.B. Pedley	(UK)
S. Sunner	(Sweden), Chairman 1968-1974
I. Wadsö	(Sweden)
D.D. Wagman	(USA)
L. Waldmann	(FRG)

1.1.5   Relationship of TGKVT with other bodies

The relationship of TGKVT to CODATA as a whole, as represented by the General Assembly, is stated in Sub-section 1.1.1. TGKVT has benefited from its relationship with two particular CODATA Task Groups, namely those on Fundamental Constants and on Internationalization and Systematization of Thermodynamic Tables; it is a responsibility of the latter Task Group to promote the production of wide-ranging tabulations based on the key values recommended by TGKVT.

TGKVT has enjoyed a special relationship with the International Union of Pure and Applied Chemistry (IUPAC). Thus, advice and encouragement on thermodynamic matters have been received from its Commission on Thermodynamics, and the latest data on relative atomic masses have been communicated by its Commission on Atomic Weights.

1.2 The Task Group's Method of Working

1.2.1 The work plan

The following plan was agreed upon at TGKVT's earliest meetings.

The task would be broken down into "blocks", each of which would be studied in depth over a period of about one year; a "block" would consist of the thermodynamic data for a group of elements and their derived compounds, based on the Periodic Table of the Elements.

The sequence in which blocks would be studied, and the sequence of elements within a given block, would follow from the "Standard Order of Arrangement". This procedure had been successfully used by the National Bureau of Standards (NBS)[1] and the Institute for High Temperatures (IHT)[2] in their preparation of thermodynamic tables. The concept is illustrated in Figure 1, where the continuous

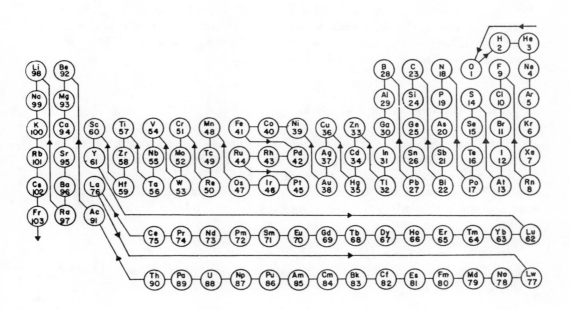

Standard Order of Arrangement of the Elements and Compounds
based on the
Periodic Classification of the Elements

line, marked with arrowheads, shows the progress through the Periodic Table, start-
ing with oxygen and ending with francium. Only the elements whose symbols are
encircled in Figure 1 were actually included in TGKVT's program.

Relevant data for a given block of elements and their simple compounds would
be abstracted from the literature and assessed critically (see Section 1.4) by staff of
NBS and IHT; the assessments, made independently in the two institutes, would be
compared by correspondence or during exchange visits, then a combined viewpoint
would be offered for comment to the whole Task Group.

At an ensuing meeting of the Task Group an attempt would be made to achieve a
consensus on each item of data within a given block. On achievement of consensus,
the data items would be incorporated in a report headed "Tentative Set of Key
Values for Thermodynamics. Part ...", which would be published under CODATA
auspices. Where a consensus view could not be achieved, the data item would
be reserved for discussion the following year, or dropped from the program if the
prospect of achieving agreement seemed unlikely.

The text of the tentative report would invite comments from experts, or in
cases where the extant data seemed poor would draw attention to the need for new
measurements.

About one year after publication of a tentative report, the comments sent in
by readers plus any pertinent new data would be considered at a meeting of the
Task Group. An attempt would then be made to affirm or modify values, according
to the particular circumstances. Values needing major modification would remain
"tentative", but affirmed or slightly modified values would be incorporated in a
report entitled "Recommended Key Values for Thermodynamics. Part ...", to be
published by CODATA.

While assessment work on one block of data was going forward, as described
previously, work on the next block of data would be initiated, until the whole work
program had been encompassed. Then an overall revision of data would be undertaken and a final report prepared.

1.2.2   The selection of data blocks

The general plan for data assessment described in Sub-section 1.2.1 left open
the question of how many elements, ions and compounds a block should contain.
No *a priori* decisions on this matter were made by TGKVT. Rather, the question
was resolved at each of our meetings in respect of the following year's program. The
species actually treated were derived from the following blocks of elements:*

O, H, He, Ne, Ar, Kr, Xe, Cl, Br, I, N, C

S, Li, Na, K, Rb, Cs

---

* The early inclusion of Group I species. as aqueous ions. formed an exception to the use of the Standard
Order of Arrangement as the basis of the Task Group's work-plan: this was necessitated by the ubiquity
of the alkali metal ions in the thermochemistry of the other species.

N, Zn, Ag

F, S, Si, B, Al, Zn, Cu

S, P, Al, Cu, U, Th, Be, Mg, Ca

S, N, P, Ge, Sn, Pb, Cd, Hg, U, Ca

Cl, S, P, C, Ge, Pb, B, Al, Cd, Ti, U

N, Cl, S, P, C, Al, Li, Na, K, Rb, Cs

Some repetition of elements between blocks is to be noted. The block-wise approach had the merits of economy of effort (by treating together species of like properties) and of providing data users with published output throughout the program. However, it involved the continual build-up of uncertainties, since the uncertainty attaching to the experimental value for a given species had to be compounded with uncertainties from connected species treated earlier. To avoid such a build-up of uncertainties and to facilitate revision of data, consideration was given to methods of assessment in which very large sets of thermodynamic data for chemical and physical processes are arranged in the form of linear equations which are then solved by use of digital computers. This approach, first described by Pedley and co-workers[3] for thermochemical data, was progressively improved during the period of TGKVT's work; in general, such sets of equations are over-determined, so best fits are sought, with the use of weighting factors.[3,4] The situation can be expressed mathematically in terms of finding a set of values for the thermodynamic quantities (x_i) from data for a network of interconnecting chemical reactions or phase changes, by solving the equations:

$$\sum_i a_{ij} x_i = y_j \pm u_j$$

in which $i = 1 \ldots n, j = 1 \ldots m, m \geq n$. The a_{ij} values are (known) stoichiometric coefficients for the processes in question; the y_j values are measured values for thermodynamic processes; and the u_j values are the corresponding uncertainties. Computer fitting does not absolve the assessor from detailed consideration of the basic experimental data, but once all corrections and adjustments to the input data have been made and the weighting factors assigned, the subsequent output should be free from any subjective judgment.

In practice, TGKVT took the view that best fits from computer solution of data sets should be kept subordinate to the element-by-element and block-by-block assessment of data by experienced thermodynamicists. Thus, the computer method was used as an occasional tool during the establishment of tentative and recommended values, for example to process the data for boron compounds and for certain aqueous ions. The main use of the technique was in the final revision of data for aqueous ions in the preparation of this report. (Annex II).

1.2.3   The scope of the work program

Within its overall objectives (Section 1.1.3) the Task Group had a considerable amount of freedom in determining which species were "key" and which were not. The choice was made on grounds reflecting the frequency of occurrence of the species in thermochemical processes for which good data had been published. The economic importance of candidate species was also taken into consideration, but this factor was sometimes overridden by the importance, or lack of it, of the species in thermodynamic networks.

After ten years of endeavor, it became clear to members of TGKVT that achievement of the objective of assessing data on all the elements in their reference and monatomic-gas states still lay far in the future; it was then decided to abandon the attempt to encompass all the elements, in the belief that a foreshortened report by 1985 would be of more use to the scientific community than a longer report by 1995. Regretfully, therefore, it has not been possible to include many economically important elements, like iron and chromium, in the program. On the other hand, it proved possible to include more ionic species than were envisaged in the early stages. In all 151 species have been evaluated.

Readers who wish to follow the chronology of the program will find the necessary information in Annex I.

1.3   Symbols, definitions, and units

1.3.1   Glossary of symbols

The symbols used mostly correspond to the recommendations of the International Union of Pure and Applied Chemistry (IUPAC):[5]

g	the symbol for a gas or vapor.
l	the symbol for a liquid.
s	the symbol for a solid.
cr	the symbol for a crystalline solid; extra words or symbols are added to identify a particular crystal modification.
am	the symbol for an amorphous solid.
aq	the symbol for an aqueous solution, in general.
aq, undissoc.	the symbol for an aqueous solution in which a weak electrolyte is considered to be undissociated.
$\Delta_f H^o(T)$	the standard molar* enthalpy of formation of a species from its constituent elements in their standard reference states at temperature T.

$S^o(T)$	the standard molar* entropy of a species at temperature T.
$H^o(T)-H^o(0)$	the difference in the standard molar* enthalpy of a species between temperature T and zero thermodynamic temperature.
$\Delta_f G^o(T)$	the standard molar* Gibbs energy of formation of a species from its constituent elements in their standard reference states at temperature T.
$\Delta_r H^o(T)$	the change in standard molar* enthalpy for an indicated chemical reaction at temperature T with the amount of each species being expressed in moles.
$\Delta_r S^o(T)$	the change in standard molar* entropy for an indicated chemical reaction at temperature T with the amount of each species being expressed in moles.
$\Delta_r G^o(T)$	the change in standard molar* Gibbs energy for an indicated chemical reaction at temperature T with the amount of each species being expressed in moles.
$D_o(0)(X_n)$	the standard molar* energy of dissociation of a species X_n, in its ground vibrational state, into n moles of $X(g)$, all at zero thermodynamic temperature.
C_p	the standard molar* isobaric heat capacity.
$E^o(T)$	the standard potential of an electrochemical cell, for a stated chemical reaction at temperature T.
$K(T)$	the thermodynamic equilibrium constant for a stated chemical reaction at temperature T. Subscripts to K are used to symbolize the type of reaction.
a_B	the activity of a species B.
$\gamma_\pm$	the stoichiometric mean molal activity coefficient of an electrolyte.
ν_B	the stoichiometric coefficient of a species B.
A_r	the relative atomic mass.

1.3.2 Definitions of standard states

We have followed the definitions published by IUPAC:[5]
- the standard state for a pure gaseous substance is that of the substance as a (hypothetical) ideal gas at the standard-state pressure;
- the standard state for a pure liquid substance is that of the liquid under the standard-state pressure;
- the standard state for a pure crystalline substance is that of the crystalline

---

* All extensive thermodynamic properties in this report are molar quantities. i.e., divided by amount of substance; therefore. it has been possible to omit the subscript m for "molar" in the symbols for extensive quantities.

substance under the standard-state pressure;

- the standard state for an undissociated substance in aqueous solution is the hypothetical state of the pure substance dissolved in pure water without dissociation at unit molality to give an ideal solution, under the standard-state pressure; [6]

- the standard state for an ionized electrolyte in aqueous solution is the hypothetical state of the pure ions dissolved in pure water to give an ideal solution at unit mean ionic molality, under the standard-state pressure. [6]

The following points are to be carefully noted.

- During the period of the Task Group's data-assessment program, the conventional value for the standard-state pressure, p^o, was 1 standard atmosphere, equal to 101 325 Pa, and this value of the standard-state pressure was used for all calculations in this report. It should be noted, however, that IUPAC has recently recommended [5] 10^5 Pa as the value for the standard-state pressure. Within the accuracy of the data tabulated in Chapter 2, the only values that are altered when p^o is taken as 10^5 Pa, rather than 101 325 Pa, are those for S^o of gaseous species as follows: S^o (g, $p^o = 10^5$ Pa)/(J·K^{-1}·mol^{-1}) = S^o(g, $p^o =$ 101 325 Pa)/(J·K^{-1}·mol^{-1}) + 0.109. The recommended values in Table 5 and in Annex IV are given for both values of the standard state pressure. All values quoted in the Notes to Table 5 and those in Annexes II and III are based on $p^o = 1$ atm only.

- For aqueous ions, values of $\Delta_f H^o$ depend on the conventions that $\Delta_f H^o$ (ionized solute B, aq) $= \Sigma \nu_{anion} \Delta_f H^o$ (anion, aq) $+ \Sigma \nu_{cation} \Delta_f H^o$ (cation, aq), and that $\Delta_f H^o$ (H$^+$, aq) = 0.

- For aqueous ions, values of S^o depend on the conventions that S^o (ionized solute B, aq) $= \Sigma \nu_{anion} S^o$(anion, aq) $+ \Sigma \nu_{cation} S^o$(cation, aq), and that S^o(H$^+$, aq) = 0.

All thermodynamic data in this report relate to natural isotopic compositions, with no allowance for mixing of different isotopes.

No allowance has been made for entropies arising from nuclear spins.

1.3.3   Reference states

In general, thermodynamic data quoted here for reactions involving elements, including formation of a substance from its elements, assume that the reference states for elements are their thermodynamically stable states, at the quoted reference temperature. However, some deviations from this generality should be mentioned:

Values of S^o and H^o(298.15 K) - H^o(0) for all species whose reference state at 298.15 K is the gas take the ideal-gas state as reference state at zero temperature also.

Values of D_o(0) take, by definition, the ideal gas as the reference state at zero temperature for the products of dissociation of the species in question.

For elemental phosphorus at 298.15 K and at all temperatures down to zero, a metastable allotrope ("white") is taken as the reference state, because most experimental work relates to this metastable allotrope.

For tin, the allotrope which is stable at 298.15 K (the "white" form) is adopted as the reference state at all temperatures down to zero, despite the fact that it becomes metastable below 287 K.

To avoid any doubts, the reference states for the elements, as adopted in this report, are listed in Table 1, in the Standard Order of Arrangement.

Table 1. Reference States for the Elements, as Adopted in this Report

	Reference State			Reference State	
	at 298.15 K	at zero K		at 298.15 K	at zero K
O_2	Gas	Gas	Pb	Cubic crystal	Cubic crystal
H_2	Gas	Gas	B	Hexagonal crystal	Hexagonal crystal
He	Gas	Gas	Al	Cubic crystal	Cubic crystal
Ne	Gas	Gas	Zn	Hexagonal crystal	Hexagonal crystal
Ar	Gas	Gas	Cd	Hexagonal crystal	Hexagonal crystal
Kr	Gas	Gas	Hg	Liquid	Tetragonal crystal
Xe	Gas	Gas	Cu	Cubic crystal	Cubic crystal
F_2	Gas	Gas	Ag	Cubic crystal	Cubic crystal
Cl_2	Gas	Gas	Ti	Hexagonal crystal	Hexagonal crystal
Br_2	Liquid	Rhombic crystal	U	Orthorhombic crystal	Orthorhombic crystal
I_2	Rhombic crystal	Rhombic crystal	Th	Cubic crystal	Cubic crystal
S	Rhombic crystal	Rhombic crystal	Be	Hexagonal crystal	Hexagonal crystal
N_2	Gas	Gas	Mg	Hexagonal crystal	Hexagonal crystal
P	Cubic crystal	Cubic crystal	Ca	Cubic crystal	Cubic crystal
C	Hexagonal crystal (graphite)*	Hexagonal crystal (graphite)*	Li	Cubic crystal	Hexagonal crystal
Si	Cubic crystal	Cubic crystal	Na	Cubic crystal	Hexagonal crystal
Ge	Cubic crystal	Cubic crystal	K	Cubic crystal	Cubic crystal
Sn	Tetragonal crystal	Tetragonal crystal	Rb	Cubic crystal	Cubic crystal
			Cs	Cubic crystal	Cubic crystal

---

* As manufactured by the Acheson Company

1.3.4   Units of measurement

TGKVT began its program of data assessment in 1968, before the International System of Units (SI) had been widely adopted by generators and users of thermodynamic data; in particular, the calorie held sway as the most common unit for expression of energy and cognate quantities. However, the majority of thermodynamicists now use SI units, so these (or their multiples and submultiples) are utilized in this report. Our interim reports employed both joules (J) and thermochemical calories (cal_{th}) as units; the two units are interrelated by 1 cal_{th} = 4.184 J (exactly). Problems connected with earlier definitions of various types of calorie (the "steam-table", the "15 degree", and the "thermochemical" calorie) and of the joule (the "international" and the "absolute" joule) have been taken into account during critical assessment of literature data. In principle, the joule used here is that defined by the International Committee of Weights and Measures (CIPM),[7] i.e. J = N·m = $m^2 \cdot kg \cdot s^{-2}$, where the base units for length (meter, m), mass (kilogram, kg), time (second, s) are to be realized in the way prescribed by the International Committee of Weights and Measures.[7] Similar remarks apply to the other SI units employed in this report, viz. those for temperature (kelvin, K), electric current (ampere, A), amount of substance (mole, mol), electric potential (volt, V), wave number (reciprocal length, m^{-1}), pressure (pascal, Pa). One non-SI unit employed, the standard atmosphere (atm), is defined by 1 atm = 101 325 Pa.

1.3.5   Relative atomic masses

Since the present key values are expressed on a molar basis, TGKVT required values of relative atomic masses, A_r, of the elements in their natural isotopic compositions, for calculation of the molar masses of the species whose thermodynamic data we have assessed. The values of A_r used initially in the derivation of key values for thermodynamics were those provided in 1969 by IUPAC's Commission on Atomic Weights;[8] they are scaled to A_r (^{12}C) = 12. In the revision of key values for inclusion in this final report, IUPAC's 1981 values of A_r were adopted.[9]

Uncertainties in relative atomic masses are propagated into the uncertainties of our key values. IUPAC has drawn attention to special problems about the uncertainties of A_r for some elements in their naturally occurring states or as purchased from chemical suppliers. For example, the IUPAC Commission on Atomic Weights added the following footnotes to its 1981 tables:[9]

g   Geologically exceptional specimens are known in which the element has an isotopic composition outside the limits for normal material. The difference between the atomic weight of the element in such specimens and that given in the Table may exceed considerably the implied uncertainty.

m   Modified isotopic compositions may be found in commercially available material, because it has been subjected to an undisclosed or inadvertent isotopic separation. Substantial deviations in atomic weight of the element from that given in the Table can occur.

r   Range in isotopic composition of normal terrestrial material prevents a
           more precise atomic weight being given; the tabulated A_r value should
           be applicable to any normal material.

L   Longest half-life isotope mass is chosen for the tabulated A_r value.

Table 2 lists the elements treated in this report to which notes g, m, r and L
apply. Inasmuch as the values of molar thermodynamic quantities in this report are
traceable to IUPAC's values of A_r, the notes g, m, r and L will have a bearing on key
values for thermodynamics of species whose constituent elements appear in Table 2.
Thus experimenters should be aware of the possibility that the materials they use
may not be "normal"; the use of isotopically abnormal materials will obviously call
for adjustment of the relevant key values for thermodynamics.

Table 2.   Elements whose Relative Atomic Masses are Subject to
            Special Considerations

Element	Note	Element	Note	Element	Note
O	g,r	S	r	U	g,m
H	g,m,r	C	r	Th	g,L
He	g	Pb	g,r	Mg	g
Ne	g,m	B	m,r	Ca	g
Ar	g,r	Cd	g	Li	g,m,r
Kr	g,m	Cu	r	Rb	g
Xe	g,m	Ag	g		

1.3.6   Fundamental constants

The principal constants needed in calculations relating to key values for thermodynamics are the molar gas constant (R), the Faraday constant (F), the Avogadro
constant (N_A), the speed of light in a vacuum (c), and the Planck constant (h). The
last three named constants are used chiefly in the form of the product $N_A hc$ to convert energies expressed in wave numbers per molecule to molar energies. The values
of the constants used in TGKVT's early publications were in principle made obso-
lete in 1973 on publication of "Recommended consistent values of the fundamental
physical constants" by the CODATA Task Group on Fundamental Constants.[10]
After much discussion within TGKVT, the decision was taken to persevere with our
original set of constants until a final readjustment of our key values was called for,
when a switch to the then extant fundamental constants would be made. Such a
switch became possible in 1981 when the Chairman of the CODATA Task Group
on Fundamental Constants, Dr. E. R. Cohen, sent us that Task Group's estimates

of the best values of R, F, and $N_A\ hc$ that were then available, prior to completion of a least-squares adjustment of the values of all the fundamental constants. These new estimates, shown in the third column of Table 3, were utilized by TGKVT in the final round of data revision in the preparation of the present report. For comparison, values of the fundamental constants used in TGKVT's previous reports are shown in the second column of Table 3. Differences between the two sets of constants are at the margin of significance for even the best established of key values for thermodynamics.

Table 3.   Fundamental Constants

Constant	Value used hitherto by TGKVT	Value used in this report
$R/\mathrm{J\cdot K^{-1}\cdot mol^{-1}}$	$8.314\ 33 \pm 0.001\ 20$[a]	$8.314\ 48 \pm 0.000\ 14$[a]
$F/\mathrm{A\cdot s\cdot mol^{-1}}$	$96\ 487.0 \pm 1.0$[a]	$96\ 484.6 \pm 0.5$[a]
$N_A\ hc/\mathrm{J\cdot m\cdot mol^{-1}}$	$0.119\ 625\ 6$[b] $\pm 0.000\ 002\ 6$[a]	$0.119\ 626\ 09$[b] $\pm 0.000\ 000\ 62$[a]

[a]   The $\pm$ terms are estimated uncertainties at the 95% confidence level.
[b]   For conversion of energies (per molecule) expressed in wave numbers ($\mathrm{cm^{-1}}$) to molar energies expressed in $\mathrm{kJ\cdot mol^{-1}}$, these values are to be multiplied by 10^{-1}.

1.4   Principles of Critical Assessment

1.4.1   Critical perusal of published thermodynamic measurements

The approach adopted by staff of the National Bureau of Standards and the Institute for High Temperatures can be described in outline only, as so many subtleties are involved. Typical points which the assessor considers in a published paper are the following:

Was the experimental method sufficiently sensitive and reliable? Were all necessary corrections made to the raw data? Have the measurements been reported in sufficient detail to permit checking of the calculational procedures? Have appropriate physical constants and auxiliary data been used? Was the chemical purity of the material well established? Was the physical state of the material well characterized? If a chemical reaction was studied, were the products satisfactorily analyzed? Was the equipment reliably calibrated? Was the performance of the equipment or operator tested in some way, e.g. by measurements on a certified reference material? Were sufficient replications performed in the calibration and "unknown" experiments? Was there a proper discussion of random and systematic uncertainties?

It is a rare publication that fully satisfies the assessor. If he is dissatisfied with a few aspects of a piece of published work, it may be possible to firm up the results by consideration of other publications from the same institute or by direct contact with the author(s). Where the author's auxiliary data or calibration constants can be improved upon, then recalculation of the results may be feasible. However, some published results may not be amenable to revision. Two courses are then open: the assessor may reject the result altogether, or he may accept it but with an enhanced uncertainty, i.e. he will use a low weighting factor in the overall assessment.

1.4.2 Thermodynamic data and their consistency

Once the perusal stage has been passed, individual values must be appraised for their consistency one with another and with the laws of thermodynamics. The most straightforward case is that in which a given datum has been determined by the same method by two or more experimenters. It often happens, however, that data have been obtained by different methods; inconsistencies in values obtained by different routes indicate the presence of systematic errors. By consideration of sufficient case histories, the assessor may be able to deduce that a particular method is prone to systematic error and will then give that method a low weight in his overall assessments.

Section 2.3 describes the actual problems encountered in data assessment, and the judgments made, during TGKVT's program. Data were accepted from many different experimental or calculational routes, the chief of which are summarized in Table 4; it should be noted that many auxiliary data (e.g. heat capacities, enthalpies of dilution, activity coefficients, virial coefficients) are needed to realize the chief routes to key thermodynamic quantities.

1.4.3 The uncertainty of thermodynamic data

Detailed recommendations were put forward by Rossini[11] on how to derive and present the uncertainties of measured energies of combustion; many, but not all, combustion calorimetrists have followed Rossini's recommendations. Some experimenters have adapted his approach to their own branches of thermodynamics. Recently IUPAC has issued a guide to the assignment of uncertainties to the results of thermodynamic measurements,[12] which may lead to more uniformity of practice in the future.

Table 4.   Chief Methods for Obtaining Key Thermodynamic Quantities

Type of species	Sources of data[a]		
	$\Delta_f H^\circ(298.15 \text{ K})$	$S^\circ(298.15 \text{ K})$	$H^\circ(298.15 \text{ K}) - H^\circ(0)$
Elements in their standard reference states	The quantity is zero, by definition	Statistical mechanics (gas-state only), C_p and $\Delta_\alpha^\beta H^\circ$	Statistical mechanics (gas-state only), C_p and $\Delta_\alpha^\beta H^\circ$
Elements as monatomic gases, where this is not the reference state	Vapor pressure(T), $D_o(0)$ from spectroscopy, $K_r(T)$ for dissociation to atoms	Statistical mechanics	Statistical mechanics
Compounds	$\Delta_r H^\circ$ for reactions involving elements or other compounds in the program, $K(T), E(T)$	C_p and $\Delta_\alpha^\beta H^\circ$, statistical mechanics (gas-state only) $K(T), E(T)$	C_p and $\Delta_\alpha^\beta H^\circ$, statistical mechanics (gas-state only)
Aqueous ions	$\Delta_r H^\circ$ for reactions involving solutions, $\Delta_{sol} H^\circ, E(T)$	E or solubility and $\Delta_{sol} H^\circ$	The quantity cannot be evaluated

[a]   In this table subscript r refers to chemical reactions involving a key species; subscript sol refers to the dissolution of a solute in a solvent; superscript β and subscript α represent states of matter for which a change of state is measured; i.e. from solid I to solid II, from solid to liquid, from solid to gas, from liquid to gas.

Rossini advocated use of the term "uncertainty interval", which was defined as twice the "overall standard deviation of the mean". The latter quantity is defined in an exact way [11] and can be realized in many instances by replication of relevant measurements. However, the *significance* of "uncertainty interval" is less exact than is sometimes supposed. The significance was stated by Rossini in the following way: 'When two investigations yield, for a given thermochemical constant, values which differ by more than the sum of the two assigned uncertainty intervals, it is probable, but not at all certain, that a systematic error or combination of errors exists in one or both of the investigations. Conversely, when the two values are in accord within their assigned uncertainty intervals, it is probable, but not at all certain, that systematic errors are absent."

Experience shows that the results of thermochemical investigations often do differ by more than the sum of the respective uncertainty intervals. This may imply that systematic errors are fairly prevalent in experimental thermochemistry; inadequate purity of the reactants and uncertainty in the stoichiometry of the actual reactions may be the causes in many cases. Another reason for the occasional lack of agreement may be the rather small number of replicate experiments made: this has the consequence that a normal distribution of results cannot be assured, and that an estimated standard deviation may be far from the true standard deviation for a normally distributed population of results. If there are problems with deriving the uncertainty of a rather small number of thermochemical results, then the problem is exacerbated when a single result only is available, as sometimes happens in the determination of S^o (298.15 K) and H^o(298.15 K) - H^o(0) by low-temperature calorimetry, albeit many experimental values of heat capacity will have contributed to the determination. Experimenters often place an estimated overall uncertainty on such determinations, by consideration of possible sources of error. If the experimenter pitches his estimate at the level of one standard deviation, it may reasonably be combined in quadrature with standard deviations, statistically derived, for other parts of the experiment or pertaining to other data.[13]

The assessor of thermodynamic data has a difficult job to do in attaching a measure of uncertainty to his assessed value. Though he may start with the original experimenters' estimates of uncertainty, he must allow for systematic errors and for uncertainties in auxiliary data. There is no way of proceeding which will satisfy all scientists and will produce a measure of uncertainty which is itself exact. Indeed, all measures of uncertainty are themselves uncertain. Our endeavor in this report has been to attach a ± term to each recommended value whose significance, paraphrasing Rossini, is as follows: "It is probable, but not at all certain, that the true values of the thermodynamic quantities differ from the recommended values given in this report by no more than twice the ± terms attached to the recommended values."

1.4.4   Harmonization and revision of recommended values

As explained in section 1.2, TGKVT broke down the work of data assessment and recommendation of key values in terms of blocks of elements. Because that phase of the program occupied twelve years, some unevenness of treatment inevitably occurred over such a long time-span. At the end of it, the need to harmonize all recommendations to a uniform standard (so far as the data themselves permitted) had become evident. In fulfilling this task, TGKVT also effected the following forms of revision:

- Reconsideration of the "key value" accolade: this led to the deletion of a few less important species and the incorporation of a few extra species whose thermodynamic values had previously been regarded as auxiliary.
- Finalising and solving a network of thermodynamic equations involving reactions/processes for the aqueous ions OH^-, F^-, Cl^-, Br^-, I^-, ClO_4^-, SO_4^{2-}, NO_3^-, NH_4^+, Hg_2^{2+}, Ag^+, Li^+, Na^+, K^+, Rb^+, Cs^+ (See Section 2.1 and Annex II).
- Consideration of relevant thermodynamic data that had been published since our original recommendations had been made; publications up to 1983 were included.

- Re-appraisal of our earlier data selections in the light of the new information on ions and other species, and of the newly available values for the fundamental constants (Sub-section 1.3.6).
- Preparation of new tables of CODATA Recommended Key Values for Thermodynamics (Section 2.1) and of new explanatory Notes to the Tables (Section 2.3).

1.5 A Forward Look

The CODATA Recommended Key Values for Thermodynamics presented in Chapter 2 constitute a data set on which others can build tabulations that are more comprehensive in their coverage of species, phases, properties, and variables like temperature, pressure and composition. Whereas it will be open to any individual or group to prepare such comprehensive tables, we would emphasize the need to achieve consistency between the respective tabulations. Our own experience has shown that an international consensus among thermodynamic data assessors can be achieved, at least on a comparatively narrow front, and already another CODATA Task Group, now called the Task Group on Chemical Thermodynamic Tables, has revealed plans [14] for building much larger databases by international co-operation: the "Babel" effect can thereby be avoided, as has been achieved for relative atomic masses and fundamental constants.

As to future work on key values for thermodynamics, a glance at Chapter 2 will show the uneven quality of the current values. As time goes by, new experiments and calculations will undoubtedly yield values of higher quality than those recorded here. In due course, therefore, revision and expansion of the present key values will be called for: we hope that an international group, like the CODATA Task Group on Chemical Thermodynamic Tables, will carry forward the work that we now leave off.

1.6   References

1. Rossini, F.D.; Wagman, D.D.; Evans, W.H.; Levine, S.; Jaffe, I. *Selected values of chemical thermodynamic properties, NBS Circular 500,* US Government Printing Office, Washington DC, **1952.**

2. Glushko, V.P.; Medvedev, V.A.; Bergman, G.A.; Gurvich, L.V.; Yungman, V.S.; Vorob'ev, A.F.; Kolesov, V.P.; Reznitskii, L.A.; Mikhailov, V.V.; Gal'chenko, G.L.; Karapet'yants, M. Kh., *Termicheskie konstanty veshchestv,* Vol. I, Academy of Sciences, Moscow, **1965.**

3. Guest, M.F.; Pedley, J.B.; Horn, H.*J. Chem. Thermodynamics.* **1969,** 1, 345.

4. Garvin, D.; Parker, V.B.; Wagman, D.D.; Evans, W.H. *A combined least sums and least squares approach to the evaluation of thermodynamic data networks. Proc. Fifth Biennial International CODATA Conf., B. Dreyfus, ed., Pergamon Press, Oxford,* **1977,** pgs. 567 − 575.

5. *Pure App. Chem.* **1982,** 54, 1239.

6. Wagman, D.D.; Evans, W.H.; Parker, V.B.; Schumm, R.H.; Halow, I.; Bailey, S.M.; Churney, K.L.; Nuttall, R.L. *J. Phys. Chem. Ref. Data,* **1982,** 11, Supplement No. 2.

7. *Le Système International d'Unités, OFFILIB,* Paris, **1970.**

8. *Pure App. Chem.* **1970,** 21, 95.

9. *Pure App. Chem.* **1983,** 55, 1101.

10. *CODATA Bulletin,* No. 11, **1974.**

11. Rossini, F.D. (Editor), *Experimental Thermochemistry,* Interscience, New York, **1956,** p.297.

12. *Pure App. Chem.* **1981,** 53, 1805.

13. *Metrologia,* **1981,** 17, 73.

14. *CODATA Bulletin,* No. 47, **1982.**

Chapter 2

RECOMMENDED KEY VALUES
FOR THERMODYNAMICS

2.1 Tabulated Values

Table 5 lists CODATA-recommended values for all the 151 chemical species treated in this report; the ordering system used is the Standard Order of Arrangement. (See Sub-section 1.2.1.)

The column headed 'Substance' gives the chemical formula of each species treated; the column headed 'State' gives the corresponding state of aggregation at 298.15 K to which the thermodynamic data refer. (See Sub-section 1.3.1 for an explanation of the symbols used.)

The column headed 'Relative molecular mass', which is taken to subsume 'Relative atomic mass' in the case of atomic species, gives values of this ratio as derived from the IUPAC 1981 table, (see Sub-section 1.3.5). For compounds in which the relative atomic masses of the constituent elements are quoted by IUPAC with differing numbers of decimals, the relative molecular mass is given here to one decimal place more than the least; the greatest number of decimal places given for compounds is four. The relative atomic mass of the electron is taken as 0.00055. The purpose of the column is to permit readers to inter-convert molar and specific values of thermodynamic quantities by means of the same values of relative molecular mass as the Task Group utilized.

The third, fourth and fifth columns of Table 5 list recommended values of $\Delta_f H^\circ$(298.15 K), S°(298.15 K) and H°(298.15 K)$-H^\circ$(0), respectively. Readers are referred to Sub-sections 1.3.1, 1.3.2 and 1.3.3 for explanations of assumptions, conventions and other details involved in the derivation of values for these thermodynamic quantities. Values of S° for gases are given for standard state pressures of 101 325 Pa (1 atm) and 100 000 Pa (1 bar).

The last column in Table 5, headed 'Note', assigns a unique numerical code to each species treated. In section 2.3, immediately after Table 5, these numerical codes head up Notes which explain how the various recommended values were obtained. Thus the Notes cite literature references from which the basic data were drawn and describe the pathways by which recommended values were derived from basic data. Values cited in these Notes are based on a standard state pressure of 1 atm. Full references, in alphabetic order of the first author's surname, are given in Chapter 3. In general, only lead references are cited. For more comprehensive bibliographies, readers are referred to the following sources:

Gurvich, L.V.; Bergman, G.A.; Veits, I.V.; Medvedev, V.A.; Khachkuryzov, G.A.; Yungman, V.S., *Termodinamicheskie svoistva individualnykh veshchestv*, Nauka, Moscow, Vol. I, **1978**, Vol. II, **1979**, Vol. III, **1981**, Vol. IV, **1982**. This work contains bibliographies on most of the elements and compounds covered in the present report, save for Zn, Cd, Cu, Hg and Ag. It describes methods for calculating thermal functions for mono-, di-, and poly-atomic gases and gives tables of selected molecular constants needed for this purpose. Tables showing the results of Second Law and Third Law evaluations of vapor-pressure data are given for many substances.

Hultgren, R.; Desai, P.D.; Hawkins, D.T.; Gleiser, M.; Kelley, K.K.; Wagman, D.D., *Selected values of the thermodynamic properties of the elements*, **1973,** American Society of Metals, Metals Park, Ohio. This work contains complete bibliographies, as well as values, for the heat capacity, vapor pressure, entropy and enthalpy of vaporization for all the elements.

Stull, D.R.; Prophet, H., (Project Directors), *JANAF Thermochemical Tables,* **1971,** Second edition, NSRDF-NBS 37, with supplements in *J. Phys. Chem. Reference Data,* **1974,** 3, 311; **1975,** 4,1; **1978,** 7, 793; **1982,** 11, 695. This work contains bibliographies and critically selected values of enthalpies of formation, entropies and thermal functions for many of the elements, oxides and halides treated in the present report.

In Annex II, basic data are given for the ions OH^- (aq), F^- (aq), Cl^- (aq), Br^- (aq), I^- (aq), ClO_4^- (aq), SO_4^{2-} (aq), NO_3^- (aq), NH_4^+ (aq), Hg_2^{2+} (aq), Ag^+ (aq), Li^+ (aq), Na^+ (aq), K^+ (aq), Rb^+ (aq), and Cs^+ (aq), for which a least-squares simultaneous treatment of Gibbs energy, enthalpy and entropy data was carried out, according to the method of Garvin et al., NBSIR 76-1147, **1976.**

In Annex III, thermal functions at 1 atm (where available) for species treated in this report are tabulated, permitting readers to convert the present recommended values for 298.15 K to those for other temperatures.

In Annex IV, recommended values from Table 5 have been reformatted for the benefit of readers with particular interests. Thus Table IV-1 contains recommended values for monatomic gaseous elements, whilst Table IV-2 contains recommended values for aqueous ions. Both tables are ordered alphabetically in respect of element symbols.

2.2   Literature References in the Notes to Table 5

In keeping with the highly condensed style of the Notes to Table 5, citations of literature references are given in abbreviated form; the full references are given in Chapter 3, in alphabetical order by the first author's surname. In the Notes, the following conventions for references are used:

Surnames are quoted in full, together with the year of publication, wherever a publication has either one or two authors; the surname of the first author followed by 'et al.', together with the year of publication, is quoted wherever a publication has three or more authors; wherever these conventions do not correspond to a unique reference in Chapter 3 (e.g., there were two publications by Ahrland and Kullberg in 1971), then the reference year is annotated by a,b ...

Table 5. CODATA Recommended Key Values for Chemical Species at 101 325 Pa (1 atm) and 100 000 Pa (1 bar).[a]

Substance	State	Relative molecular mass	$\Delta_f H°(298.15\ K)$ $kJ \cdot mol^{-1}$	$S°(298.15\ K)$ $J \cdot K^{-1} \cdot mol^{-1}$	$[H°(298.15\ K)$ $-H°(0)]$ $kJ \cdot mol^{-1}$	Note
O	g	15.9994	249.18 ± 0.10	160.950 ± 0.003 (161.059 ± 0.003)	6.725 ± 0.001	1
O_2	g	31.9988	0	205.043 ± 0.005 (205.152 ± 0.005)	8.680 ± 0.002	2
H	g	1.00794	217.998 ± 0.006	114.608 ± 0.002 (114.717 ± 0.002)	6.197 ± 0.001	3
H^+	aq	1.0074	0	0	—	4
H_2	g	2.0159	0	130.571 ± 0.003 (130.680 ± 0.003)	8.468 ± 0.001	5
OH^-	aq	17.0079	-230.015 ± 0.040	-10.90 ± 0.20	—	6
H_2O	l	18.0153	-285.830 ± 0.040	69.95 ± 0.03	13.273 ± 0.020	7
H_2O	g	18.0153	-241.826 ± 0.040	188.726 ± 0.010 (188.835 ± 0.010)	9.905 ± 0.005	8
He	g	4.00260	0	126.044 ± 0.002 (126.153 ± 0.002)	6.197 ± 0.001	9
Ne	g	20.179	0	146.219 ± 0.003 (146.328 ± 0.003)	6.197 ± 0.001	10
Ar	g	39.948	0	154.737 ± 0.003 (154.846 ± 0.003)	6.197 ± 0.001	11
Kr	g	83.80	0	163.976 ± 0.003 (164.085 ± 0.003)	6.197 ± 0.001	12
Xe	g	131.29	0	169.576 ± 0.003 (169.685 ± 0.003)	6.197 ± 0.001	13
F	g	18.99840	79.38 ± 0.30	158.642 ± 0.004 (158.751 ± 0.004)	6.518 ± 0.001	14
F^-	aq	18.9989	-335.35 ± 0.65	-13.8 ± 0.8	—	15
F_2	g	37.9968	0	202.682 ± 0.005 (202.791 ± 0.005)	8.825 ± 0.001	16
HF	g	20.0063	-273.30 ± 0.70	173.670 ± 0.003 (173.779 ± 0.003)	8.599 ± 0.001	17
Cl	g	35.453	121.301 ± 0.008	165.081 ± 0.004 (165.190 ± 0.004)	6.272 ± 0.001	18
Cl^-	aq	35.4535	-167.080 ± 0.10	56.60 ± 0.20	—	19
Cl_2	g	70.906	0	222.972 ± 0.010 (223.081 ± 0.010)	9.181 ± 0.001	20

Table 5. (continued)

Sub-stance	State	Relative molecular mass	$\Delta_f H^o(298.15$ K$)$ kJ·mol^{-1}	$S^o(298.15$ K$)$ J·K^{-1}·mol^{-1}	$[H^o(298.15$ K$)$ -$H^o(0)]$ kJ·mol^{-1}	Note
ClO_4^-	aq	99.4511	-128.10 ± 0.40	184.0 ± 1.5	—	21
HCl	g	36.4609	-92.31 ± 0.10	186.793 ± 0.005 (186.902 ± 0.005)	8.640 ± 0.001	22
Br	g	79.904	111.87 ± 0.12	174.909 ± 0.004 (175.018 ± 0.004)	6.197 ± 0.001	23
Br$^-$	aq	79.9045	-121.41 ± 0.15	82.55 ± 0.20	—	24
Br$_2$	l	159.808	0	152.21 ± 0.30	24.52 ± 0.01	25
Br$_2$	g	159.808	30.91 ± 0.11	245.359 ± 0.005 (245.468 ± 0.005)	9.725 ± 0.001	26
HBr	g	80.9119	-36.29 ± 0.16	198.591 ± 0.004 (198.700 ± 0.004)	8.648 ± 0.001	27
I	g	126.9045	106.76 ± 0.04	180.678 ± 0.004 (180.787 ± 0.004)	6.197 ± 0.001	28
I$^-$	aq	126.9050	-56.78 ± 0.05	106.45 ± 0.30	—	29
I$_2$	cr	253.8090	0	116.14 ± 0.30	13.196 ± 0.040	30
I$_2$	g	253.8090	62.42 ± 0.08	260.578 ± 0.005 (260.687 ± 0.005)	10.116 ± 0.001	31
HI	g	127.9124	26.50 ± 0.10	206.481 ± 0.004 (206.590 ± 0.004)	8.657 ± 0.001	32
S	cr, rhombic	32.06	0	32.054 ± 0.050	4.412 ± 0.006	33
S	g	32.06	277.17 ± 0.15	167.720 ± 0.006 (167.829 ± 0.006)	6.657 ± 0.001	34
S$_2$	g	64.12	128.60 ± 0.30	228.058 ± 0.010 (228.167 ± 0.010)	9.132 ± 0.002	35
SO$_2$	g	64.059	-296.81 ± 0.20	248.114 ± 0.050 (248.223 ± 0.050)	10.549 ± 0.010	36
SO$_4^{2-}$	aq	96.059	-909.34 ± 0.40	18.50 ± 0.40	—	37
HS$^-$	aq	33.068	-16.3 ± 1.5	67 ± 5	—	38
H$_2$S	g	34.076	-20.6 ± 0.5	205.70 ± 0.05 (205.81 ± 0.05)	9.957 ± 0.010	39
H$_2$S	aq, undissoc.	34.076	-38.6 ± 1.5	126 ± 5	—	40
HSO$_4^-$	aq	97.066	-886.9 ± 1.0	131.7 ± 3.0	—	41
N	g	14.0067	472.68 ± 0.40	153.192 ± 0.003 (153.301 ± 0.003)	6.197 ± 0.001	42

Table 5. (continued)

Sub-stance	State	Relative molecular mass	$\Delta_f H^o(298.15\ K)$ kJ·mol^{-1}	$S^o(298.15\ K)$ J·K^{-1}·mol^{-1}	$[H^o(298.15\ K) -H^o(0)]$ kJ·mol^{-1}	Note
N_2	g	28.0134	0	191.500 ± 0.004 (191.609 ± 0.004)	8.670 ± 0.001	43
NO_3^-	aq	62.0054	-206.85 ± 0.40	146.70 ± 0.40	—	44
NH_3	g	17.0305	-45.94 ± 0.35	192.66 ± 0.05 (192.77 ± 0.05)	10.043 ± 0.010	45
NH_4^+	aq	18.0379	-133.26 ± 0.25	111.17 ± 0.40	—	46
P	cr, white	30.97376	0	41.09 ± 0.25	5.360 ± 0.015	47
P	g	30.97376	316.5 ± 1.0	163.090 ± 0.003 (163.199 ± 0.003)	6.197 ± 0.001	48
P_2	g	61.9475	144.0 ± 2.0	218.014 ± 0.004 (218.123 ± 0.004)	8.904 ± 0.001	49
P_4	g	123.8950	58.9 ± 0.3	279.90 ± 0.50 (280.01 ± 0.50)	14.10 ± 0.20	50
HPO_4^{2-}	aq	95.9804	-1299.0 ± 1.5	-33.5 ± 1.5	—	51
$H_2PO_4^-$	aq	96.9878	-1302.6 ± 1.5	92.5 ± 1.5	—	52
C	cr, graphite	12.011	0	5.74 ± 0.10	1.050 ± 0.020	53
C	g	12.011	716.68 ± 0.45	157.991 ± 0.003 (158.100 ± 0.003)	6.536 ± 0.001	54
CO	g	28.0104	-110.53 ± 0.17	197.551 ± 0.004 (197.660 ± 0.004)	8.671 ± 0.001	55
CO_2	g	44.0098	-393.51 ± 0.13	213.676 ± 0.010 (213.785 ± 0.010)	9.365 ± 0.003	56
CO_2	aq, undissoc.	44.0098	-413.26 ± 0.20	119.36 ± 0.60	—	57
CO_3^{2-}	aq	60.0103	-675.23 ± 0.25	-50.0 ± 1.0	—	58
HCO_3^-	aq	61.0177	-689.93 ± 0.20	98.4 ± 0.5	—	59
Si	cr	28.0855	0	18.81 ± 0.08	3.217 ± 0.008	60
Si	g	28.0855	450 ± 8	167.872 ± 0.004 (167.981 ± 0.004)	7.550 ± 0.001	61
SiO_2	cr, α-quartz	60.0843	-910.7 ± 1.0	41.46 ± 0.20	6.916 ± 0.020	62
SiF_4	g	104.0791	-1615.0 ± 0.8	282.65 ± 0.50 (282.76 ± 0.50)	15.36 ± 0.05	63
Ge	cr	72.59	0	31.09 ± 0.15	4.636 ± 0.020	64

Table 5. (continued)

Sub-stance	State	Relative molecular mass	$\Delta_f H^o$(298.15 K) kJ·mol^{-1}	S^o(298.15 K) J·K^{-1}·mol^{-1}	$[H^o$(298.15 K) $-H^o(0)]$ kJ·mol^{-1}	Note
Ge	g	72.59	372 ± 3	167.795 ± 0.005 (167.904 ± 0.005)	7.398 ± 0.001	65
GeO$_2$	cr, tetragonal	104.589	-580.0 ± 1.0	39.71 ± 0.15	7.230 ± 0.020	66
GeF$_4$	g	148.584	-1190.20 ± 0.50	301.8 ± 1.0 (301.9 ± 1.0)	17.29 ± 0.10	67
Sn	cr, white	118.69	0	51.18 ± 0.08	6.323 ± 0.008	68
Sn	g	118.69	301.2 ± 1.5	168.383 ± 0.004 (168.492 ± 0.004)	6.215 ± 0.001	69
Sn^{2+}	aq	118.689	-8.9 ± 1.0	-16.7 ± 4.0	—	70
SnO	cr, tetragonal	134.689	-280.71 ± 0.20	57.17 ± 0.30	8.736 ± 0.020	71
SnO$_2$	cr, tetragonal	150.689	-577.63 ± 0.20	49.04 ± 0.10	8.384 ± 0.020	72
Pb	cr	207.2	0	64.80 ± 0.30	6.870 ± 0.030	73
Pb	g	207.2	195.2 ± 0.8	175.266 ± 0.005 (175.375 ± 0.005)	6.197 ± 0.001	74
Pb^{2+}	aq	207.2	0.92 ± 0.25	18.5 ± 1.0	—	75
PbSO$_4$	cr	303.26	-919.97 ± 0.40	148.50 ± 0.60	20.050 ± 0.040	76
B	cr, rhombic	10.81	0	5.90 ± 0.08	1.222 ± 0.008	77
B	g	10.81	565 ± 5	153.327 ± 0.015 (153.436 ± 0.015)	6.316 ± 0.002	78
B$_2$O$_3$	cr	69.618	-1273.5 ± 1.4	53.97 ± 0.30	9.301 ± 0.040	79
H$_3$BO$_3$	cr	61.832	-1094.8 ± 0.8	89.95 ± 0.60	13.52 ± 0.04	80
H$_3$BO$_3$	aq, undissoc.	61.832	-1072.8 ± 0.8	162.4 ± 0.6	—	81
BF$_3$	g	67.805	-1136.0 ± 0.8	254.31 ± 0.20 (254.42 ± 0.20)	11.650 ± 0.020	82
Al	cr	26.98154	0	28.30 ± 0.10	4.540 ± 0.020	83
Al	g	26.98154	330.0 ± 4.0	164.445 ± 0.004 (164.554 ± 0.004)	6.919 ± 0.001	84
Al^{3+}	aq	26.9799	-538.4 ± 1.5	-325 ± 10	—	85
Al$_2$O$_3$	cr, corundum	101.9613	-1675.7 ± 1.3	50.92 ± 0.10	10.016 ± 0.020	86

Table 5. (continued)

Sub-stance	State	Relative molecular mass	$\Delta_f H°(298.15 \text{ K})$ kJ·mol^{-1}	$S°(298.15 \text{ K})$ J·K^{-1}·mol^{-1}	$[H°(298.15 \text{ K})$ $-H°(0)]$ kJ·mol^{-1}	Note
AlF$_3$	cr	83.9767	-1510.4 ± 1.3	66.5 ± 0.5	11.62 ± 0.04	87
Zn	cr	65.38	0	41.63 ± 0.15	5.657 ± 0.020	88
Zn	g	65.38	130.40 ± 0.40	160.881 ± 0.004 (160.990 ± 0.004)	6.197 ± 0.001	89
Zn^{2+}	aq	65.38	-153.39 ± 0.20	-109.8 ± 0.5	—	90
ZnO	cr	81.379	-350.46 ± 0.27	43.65 ± 0.40	6.933 ± 0.040	91
Cd	cr	112.41	0	51.80 ± 0.15	6.247 ± 0.015	92
Cd	g	112.41	111.80 ± 0.20	167.640 ± 0.004 (167.749 ± 0.004)	6.197 ± 0.001	93
Cd^{2+}	aq	112.41	-75.92 ± 0.60	-72.8 ± 1.5	—	94
CdO	cr	128.409	-258.35 ± 0.40	54.8 ± 1.5	8.41 ± 0.08	95
CdSO$_4$ ·8/3H$_2$O	cr	256.508	-1729.30 ± 0.80	229.65 ± 0.40	35.56 ± 0.04	96
Hg	l	200.59	0	75.90 ± 0.12	9.342 ± 0.008	97
Hg	g	200.59	61.38 ± 0.04	174.862 ± 0.005 (174.971 ± 0.005)	6.197 ± 0.001	98
Hg^{2+}	aq	200.59	170.21 ± 0.20	-36.19 ± 0.80	—	99
Hg$_2^{2+}$	aq	401.18	166.87 ± 0.50	65.74 ± 0.80	—	100
HgO	cr, red	216.589	-90.79 ± 0.12	70.25 ± 0.30	9.117 ± 0.025	101
Hg$_2$Cl$_2$	cr	472.086	-265.37 ± 0.40	191.6 ± 0.8	23.35 ± 0.20	102
Hg$_2$SO$_4$	cr	497.238	-743.09 ± 0.40	200.70 ± 0.20	26.070 ± 0.030	103
Cu	cr	63.546	0	33.15 ± 0.08	5.004 ± 0.008	104
Cu	g	63.546	337.4 ± 1.2	166.289 ± 0.004 (166.398 ± 0.004)	6.197 ± 0.001	105
Cu^{2+}	aq	63.545	64.9 ± 1.0	-98 ± 4	—	106
CuSO$_4$	cr	159.604	-771.4 ± 1.2	109.2 ± 0.4	16.86 ± 0.08	107
Ag	cr	107.8682	0	42.55 ± 0.20	5.745 ± 0.020	108
Ag	g	107.8682	284.9 ± 0.8	172.888 ± 0.004 (172.997 ± 0.004)	6.197 ± 0.001	109
Ag$^+$	aq	107.8677	105.79 ± 0.08	73.45 ± 0.40	—	110
AgCl	cr	143.321	-127.01 ± 0.05	96.25 ± 0.20	12.033 ± 0.020	111
Ti	cr	47.88	0	30.72 ± 0.10	4.824 ± 0.015	112
Ti	g	47.88	473 ± 3	180.189 ± 0.010 (180.298 ± 0.010)	7.539 ± 0.002	113

Table 5. (continued)

Sub-stance	State	Relative molecular mass	$\Delta_f H^o$(298.15 K) kJ·mol^{-1}	S^o(298.15 K) J·K^{-1}·mol^{-1}	$[H^o$(298.15 K) $-H^o(0)]$ kJ·mol^{-1}	Note
TiO$_2$	cr, rutile	79.879	-944.0 ± 0.8	50.62 ± 0.30	8.68 ± 0.05	114
TiCl$_4$	g	189.692	-763.2 ± 3.0	353.1 ± 4.0 (353.2 ± 4.0)	21.5 ± 0.5	115
U	cr	238.0289	0	50.20 ± 0.20	6.364 ± 0.020	116
U	g	238.0289	533 ± 8	199.68 ± 0.10 (199.79 ± 0.10)	6.499 ± 0.020	117
UO$_2$	cr	270.0277	-1085.0 ± 1.0	77.03 ± 0.20	11.280 ± 0.020	118
UO$_2^{2+}$	aq	270.0266	-1019.0 ± 1.5	-98.2 ± 3.0	—	119
UO$_3$	cr, gamma	286.0271	-1223.8 ± 1.2	96.11 ± 0.40	14.585 ± 0.050	120
U$_3$O$_8$	cr	842.0819	-3574.8 ± 2.5	282.55 ± 0.50	42.74 ± 0.10	121
Th	cr	232.0381	0	51.8 ± 0.5	6.35 ± 0.05	122
Th	g	232.0381	602 ± 6	190.06 ± 0.05 (190.17 ± 0.05)	6.197 ± 0.003	123
ThO$_2$	cr	264.0369	-1226.4 ± 3.5	65.23 ± 0.20	10.560 ± 0.020	124
Be	cr	9.01218	0	9.50 ± 0.08	1.950 ± 0.020	125
Be	g	9.01218	324 ± 5	136.166 ± 0.003 (136.275 ± 0.003)	6.197 ± 0.001	126
BeO	cr	25.0116	-609.4 ± 2.5	13.77 ± 0.04	2.837 ± 0.008	127
Mg	cr	24.305	0	32.67 ± 0.10	4.998 ± 0.030	128
Mg	g	24.305	147.1 ± 0.8	148.539 ± 0.003 (148.648 ± 0.003)	6.197 ± 0.001	129
Mg^{2+}	aq	24.304	-467.0 ± 0.6	-137 ± 4	—	130
MgO	cr	40.3044	-601.60 ± 0.30	26.95 ± 0.15	5.160 ± 0.020	131
MgF$_2$	cr	62.3018	-1124.2 ± 1.2	57.2 ± 0.5	9.91 ± 0.06	132
Ca	cr	40.08	0	41.59 ± 0.40	5.736 ± 0.040	133
Ca	g	40.08	177.8 ± 0.8	154.778 ± 0.004 (154.887 ± 0.004)	6.197 ± 0.001	134
Ca^{2+}	aq	40.08	-543.0 ± 1.0	-56.2 ± 1.0	—	135
CaO	cr	56.079	-634.92 ± 0.90	38.1 ± 0.4	6.75 ± 0.06	136
Li	cr	6.941	0	29.12 ± 0.20	4.632 ± 0.040	137
Li	g	6.941	159.3 ± 1.0	138.673 ± 0.010 (138.782 ± 0.010)	6.197 ± 0.001	138
Li$^+$	aq	6.9405	-278.47 ± 0.08	12.24 ± 0.15	—	139
Na	cr	22.98977	0	51.30 ± 0.20	6.460 ± 0.020	140

Table 5. (continued)

Sub-stance	State	Relative molecular mass	$\Delta_f H°(298.15\ K)$ kJ·mol^{-1}	$S°(298.15\ K)$ J·K^{-1}·mol^{-1}	$[H°(298.15\ K)$ $-H°(0)]$ kJ·mol^{-1}	Note
Na	g	22.98977	107.5 ± 0.7	153.609 ± 0.003 (153.718 ± 0.003)	6.197 ± 0.001	141
Na$^+$	aq	22.9892	-240.34 ± 0.06	58.45 ± 0.15	—	142
K	cr	39.0983	0	64.68 ± 0.20	7.088 ± 0.020	143
K	g	39.0983	89.0 ± 0.8	160.232 ± 0.003 (160.341 ± 0.003)	6.197 ± 0.001	144
K$^+$	aq	39.0978	-252.14 ± 0.08	101.20 ± 0.20	—	145
Rb	cr	85.4678	0	76.78 ± 0.30	7.489 ± 0.020	146
Rb	g	85.4678	80.9 ± 0.8	169.985 ± 0.003 (170.094 ± 0.003)	6.197 ± 0.001	147
Rb$^+$	aq	85.4673	-251.12 ± 0.10	121.75 ± 0.25	—	148
Cs	cr	132.9054	0	85.23 ± 0.40	7.711 ± 0.020	149
Cs	g	132.9054	76.5 ± 1.0	175.492 ± 0.003 (175.601 ± 0.003)	6.197 ± 0.001	150
Cs$^+$	aq	132.9049	-258.00 ± 0.50	132.1 ± 0.5	—	151

[a] Only the values of $S°$ for gases are affected by the change in pressure. The values at 1 bar are in parentheses.

2.3   Notes to Table 5

1. O(g). $\Delta_f H^o$ was calculated using $D_o(0)(O_2) = (41260 \pm 15)$ cm^{-1} or (493.58 ± 0.18) kJ·mol^{-1} obtained by Brix and Herzberg (1954), assuming that the products of dissociation are O atoms in the ground state $^3P_2$.

The entropy and the enthalpy increment were calculated using the energy levels of the O atom given by Moore (1970).

2. O$_2$(g).   Defined reference state.

The entropy and the enthalpy increment were calculated by direct summation over the electronic vibrational-rotational levels. The uncertainty in S^o takes into account the slight inaccuracy in the calculation of the effect of splitting in the lowest rotational levels. See Krupenie (1972) and Gurvich et al. (1978) part 1, pages 92-96.

3. H(g). $\Delta_f H^o$ was calculated using $D_o(0)(H_2) = (36118.3 \pm 1)$ cm^{-1} or (432.069 ± 0.012) kJ·mol^{-1} from Herzberg (1970). The entropy and the enthalpy increment were calculated assuming the electronic partition function to be 2.

4. H$^+$(aq). $\Delta_f H^o$ and S^o are zero by convention.

5. H$_2$(g).   Defined reference state.

The entropy and the enthalpy increment were calculated by direct summation over the vibrational-rotational levels of the ground electronic state of the H$_2$ molecule assuming an equilibrium mixture of ortho and para forms; the contribution of nuclear spin ($R \ln 4$) is not included in the selected value. See Gurvich et al. (1978) part 1, pages 103-107.

6. OH$^-$(aq). $\Delta_f H^o$ and S^o of the ion were derived from an analysis of a set of measurements of entropies, enthalpies, and Gibbs energies for a large number of reactions involving many substances and their aqueous ions, using a weighted least-sums, least-squares procedure. See Annex II. References to additional measurements on ΔH^o and ΔG^o for ionization of H$_2$O(l) may be found in Parker (1965), Cerutti et al. (1978), Hepler and Hopkins (1979), and in Medvedev et al. (1965, 1971).

7. H$_2$O(l). $\Delta_f H^o$ was obtained from measurements of the combustion of H$_2$(g) and O$_2$(g) by Rossini (1931a, 1939) and by King and Armstrong (1968).

The entropy was calculated from the value selected for H$_2$O(g), (see Note 8), and the data for H$_2$O(g) and H$_2$O(l) given by Haar et al. (1983). From the latter data we calculated for the vaporization of H$_2$O(l) at 298.15 K, $\Delta_{vap} H^o = (44.004 \pm 0.002)$ kJ·mol^{-1} and $\Delta_{vap} G^o = (8.591 \pm 0.005)$ kJ·mol^{-1}. This value of $\Delta_{vap} G^o$ is in excellent agreement with the value calculated from the vapor pressure of H$_2$O(l) reported by Stimson (1969) and appropriate equation of state data. Direct calculation of S^o from low-temperature heat-capacity data fails to include the contribution of the "frozen-in" zero-point entropy.

The enthalpy increment was calculated from the heat-capacity data of Giauque and Stout (1936), (16-268 K), Flubacher et al. (1960), (2-27 K), and Osborne et al. (1939), (273-373 K).

8. H$_2$O(g). $\Delta_f H^o$ was obtained from the enthalpies of formation and vaporization of H$_2$O(l), (see Note 7). The entropy and the enthalpy increment were calculated by direct summation over the vibrational levels of the ground electronic state of the H$_2$O molecule, with integration over the rotational levels for each vibrational state. See Gurvich et al. (1978), part 1, pages 118-120.

9. He(g). Defined reference state.
The entropy and the enthalpy increment, which include only translational effects, were calculated taking the electronic partition function equal to 1.

10. Ne(g). See note 9.

11. Ar(g). See note 9.

12. Kr(g). See note 9.

13. Xe(g). See note 9.

14. F(g). $\Delta_f H^o$ was calculated using $D_o(0)(F_2) = (12920 \pm 50)$ cm^{-1} or (154.56 ± 0.60) kJ·mol^{-1} obtained by Colbourn et al. (1976). It is in good agreement with the data for $\Delta_f H^o$(HF) and $D_o(0)$(HF) (see Note 17). The entropy and the enthalpy increment were calculated using the energy of the ^{2}P$_{\frac{1}{2}}$ level equivalent to 404.1 cm^{-1}, as recommended by Moore (1970).

15. F$^-$(aq). $\Delta_f H^o$ was calculated from the following sets of measurements:
a. Johnson et al. (1973) reported the enthalpy of formation and neutralization of HF(l), yielding $\Delta_f H^o = -(335.68 \pm 0.30)$ kJ·mol^{-1}.
b. Johnson et al. (1973) measured the enthalpies of solution and dilution of HF(l) to HF·5500H$_2$O. Extrapolation to infinite dilution yielded $\Delta_f H^o = -(335.35 \pm 0.30)$ kJ·mol^{-1}.
c. Vanderzee and Rodenburg (1971) measured the enthalpy of neutralization of HF(g). Extrapolation to infinite dilution yielded $\Delta_f H^o = -(335.01 \pm 0.54)$ kJ·mol^{-1}.
d. King and Armstrong (1968) measured the enthalpy of reaction of H$_2$(g) and F$_2$(g), followed by addition to H$_2$O(l) to obtain HF·50H$_2$O. Extrapolation to infinite dilution yielded $\Delta_f H^o = -(333.9 \pm 0.4)$ kJ·mol^{-1}.
e. The measurements by Good and Månsson (1966) yielded for the reaction B(cr) + 0.75 O$_2$(g) + 3 HF(aq) = BF$_3$(g) + 1.5 H$_2$O(l), $\Delta_r H^o = -(558.67 \pm 0.80)$ kJ·mol^{-1}. Combination with values for BF$_3$(g), (see Note 79), and H$_2$O(l) yielded $\Delta_f H^o = -(335.3 \pm 1.7)$ kJ·mol^{-1}.
f. The measurements by Kilday and Prosen (1973) yielded for the reaction SiO$_2$ (cr, α-quartz) + 4 HF(aq) = SiF$_4$(g) + 2 H$_2$O(l), $\Delta_r H^o = (65.29 \pm 0.80)$ kJ·mol^{-1}. Combination with values for SiF$_4$(g), (see Note 63), and SiO$_2$ (cr, α-quartz), (see Note 62), and H$_2$O(l) yielded $\Delta_f H^o = -(335.3 \pm 1.7)$ kJ·mol^{-1}.

Values of the enthalpy of dilution for HF(aq) were calculated from the solution data of Johnson et al. (1973). Extrapolation of the values to infinite dilution was made using for the reaction H$^+$(aq) + F$^-$(aq) = HF(aq, undissoc.), $\Delta_r H^o = (13.30 \pm 0.10)$ kJ. mol^{-1}, $\Delta_r G^o = -(18.10 \pm 0.10)$ kJ·mol^{-1}, and for the reaction H$^+$(aq) + 2 F$^-$(aq) = HF$_2^-$ (aq), $\Delta_r H^o = (16.3 \pm 0.8)$ kJ. mol^{-1}, $\Delta_r G^o = -(21.2 \pm 0.4)$ kJ·mol^{-1}.

These values were based on data taken from Hepler et al. (1953), Ahrland (1967), Baumann (1969), Broene and DeVries (1947), Ellis (1963), and Kleboth (1970). See also Hamer and Wu (1970).

See Annex II for the derivation of S^o for the ion.

16. $F_2(g)$. Defined reference state.

The entropy and the enthalpy increment were calculated by direct summation over the vibrational-rotational levels of the ground electronic state. The uncertainty in S^o takes into account the inaccuracy in the value of the rotational constant of F_2 ($\pm$ 0.0002 cm^{-1}).

17. HF(g). $\Delta_f H^o$ was calculated from the measurements by Johnson et al. (1973) on the enthalpy of formation of HF(l), -(303.55 $\pm$ 0.25) kJ·mol^{-1}, and the enthalpy of vaporization to monomeric HF(g), (30.26 $\pm$ 0.10) kJ·mol^{-1}, obtained by Vanderzee and Rodenburg (1971). The spectroscopic value for the dissociation energy of HF(g), $D_o(0) = (566.2 \pm 0.7)$ kJ·mol^{-1}, measured by Di Lonardo and Douglas (1973), is in excellent agreement with the calculated value $D_o(0) = 566.5$ kJ·mol^{-1}. See also Johns and Barrow (1959) and Berkowitz et al. (1971).

The entropy and the enthalpy increment were calculated by direct summation over the vibrational-rotational levels of the ground electronic state of the HF molecule.

18. Cl(g). $\Delta_f H^o$ was calculated for the natural isotopic mixture of chlorine atoms using $D_o(0)(^{35}Cl_2) = (19997.25 \pm 0.3)$ cm^{-1} obtained by LeRoy and Bernstein (1971) from the measurements of Douglas et al. (1963). This yields the value $D_o(0)(^{35.453}Cl_2) = (19999 \pm 1)$ cm^{-1} or (239.240 $\pm$ 0.012) kJ·mol^{-1}. The entropy and the enthalpy increment were calculated using for the $^2P_{\frac{1}{2}}$ sublevel of the ground state the energy equivalent to 882.36 cm^{-1}, as recommended by Moore (1970).

19. Cl$^-$(aq). $\Delta_f H^o$ was calculated from the selected value for $\Delta_f H^o$ for HCl(g), (see Note 22), and measurements of the enthalpy of solution by Gunn and Green (1963), (see Gunn (1964)), Vanderzee and Nutter (1963), and Roth and Bertram (1937). From these measurements we selected $\Delta_{sol} H^o$ = -(74.770 $\pm$ 0.050) kJ·mol^{-1}. Enthalpies of dilution were taken from Parker (1965).

See Annex II for the derivation of S^o.

20. Cl$_2$(g). Defined reference state.

The entropy and the enthalpy increment were calculated for the normal isotopic mixture by direct summation over the vibrational-rotational levels of the ground state of the Cl$_2$ molecule, using averaged molecular constants. The uncertainty in the entropy value includes the contribution from the effect of using averaged molecular constants.

21. ClO$_4^-$(aq). See Annex II for the derivation of $\Delta_f H^o$ and S^o.

22. HCl(g). $\Delta_f H^o$ was measured by Rossini (1932), -(92.312 $\pm$ 0.150) kJ·mol^{-1}, King and Armstrong (1970), -(92.47 $\pm$ 0.30) kJ·mol^{-1}, Lacher et al.(1956), -(92.38 $\pm$ 0.15) kJ·mol^{-1}, Lacher et al. (1962), -(92.60 $\pm$ 0.50) kJ·mol^{-1}, Davies and Benson

(1965), -(92.36 ± 1.50) kJ·mol^{-1}, and Cerquetti et al. (1968), -(96.10 ± 0.20) kJ·mol^{-1}. See also Faita et al. (1967).

The entropy and the enthalpy increment were calculated by direct summation over the vibrational-rotational levels of the ground state of the HCl molecule. The uncertainty in the entropy value includes the contribution from the effect of using averaged molecular constants.

23. Br(g). $\Delta_f H^\circ$ was calculated using the value for $\Delta_f H^\circ$(Br_2, g), (see Note 26), together with $D_o(0)(^{79}Br^{81}Br) = (15895.64 ± 0.05)$ cm^{-1}, or (190.153 ± 0.001) kJ·mol^{-1} obtained by Barrow et al. (1974) from the convergence limit of the vibrational levels of the Br_2 molecule in the $B^3\Pi_0^+$ state and from the rotational predissociation in this state. The value (15895.5 ± 0.5) cm^{-1} was given by LeRoy and Bernstein (1971). The entropy and the enthalpy increment were calculated using for the ground state sublevel $^2P_{\frac{1}{2}}$ the energy equivalent to 3685.24 cm^{-1}, as recommended by Moore (1970).

24. Br$^-$(aq). See Annex II for the derivation of $\Delta_f H^\circ$ and S°.

25. Br$_2$(l). Defined reference state.

The entropy was calculated using the value for S°(Br_2, g), (see Note 26), and $\Delta_{vap}S^\circ$(Br_2, l, 298.15 K) equal to (93.15 ± 0.30) J·K^{-1}·mol^{-1}, obtained from the calorimetric and vapor-pressure data of Hildenbrand et al. (1958). Low-temperature heat capacity measurements (15-310 K) by Hildenbrand et al. (1958) give a slightly less accurate value of the entropy, (152.2 ± 1.0) J·K^{-1}·mol^{-1}, the contribution below 15 K being 2.7 J·mol^{-1}·K^{-1}. The enthalpy increment was calculated from the heat-capacity data of Hildenbrand et al. (1958).

26. Br$_2$(g). $\Delta_f H^\circ$ was obtained from the data of Hildenbrand et al. (1958) on the vapor pressure of Br_2 and calorimetric measurements of $\Delta_{vap}H$. The entropy and the enthalpy increment of the natural isotopic mixture were calculated by direct summation over the vibrational-rotational levels of the ground and four excited electronic states of the molecule; see Gurvich et al. (1978), part 1, pages 183-185. The uncertainty in S° takes into account the contribution from the effect of using averaged molecular constants.

27. HBr(g). $\Delta_f H^\circ$ was calculated from the value of $\Delta_f H^\circ$ for Br$^-$(aq), (see Note 24), and of $\Delta_{sol}H^\circ$, -(85.12 ± 0.04) kJ·mol^{-1}, obtained from the measurements by Vanderzee and Nutter (1963), Roth and Bertram (1937), and Thomsen (1882).

The entropy and the enthalpy increment were calculated by direct summation over the vibrational-rotational levels of the ground state, using molecular constants calculated from data for the H^{81}Br molecule obtained by Barrow and Stamper (1961), Rank et al. (1965a), Ginter and Tilford (1971), and Jones and Gordy (1964).

28. I(g). $\Delta_f H^\circ$ was calculated using the value for $\Delta_f H^\circ$(I_2, g), (see Note 31), and $D_o(0)(I_2) = (12440.06 ± 0.05)$ cm^{-1} or (148.816 ± 0.001) kJ·mol^{-1} obtained by Barrow and Yee (1973) by a short extrapolation of the vibrational levels of the $B^3\Pi_0^+$ state. LeRoy and Bernstein (1971) gave essentially the same value (12440.9 cm^{-1}) using data from Brown (1931).

The entropy and the enthalpy increment were calculated using for the ground state sublevel $^2P_{\frac{1}{2}}$ the energy equivalent to 7603.15 cm^{-1}, as recommended by Moore (1970).

29. I$^-$(aq).   See Annex II for the derivation of $\Delta_f H^o$ and S^o.

30. I$_2$(cr).   Defined reference state.

The entropy and the enthalpy increment were calculated from the heat-capacity data obtained by Shirley and Giauque (1959), (13-322 K), the entropy contribution below 13 K being 2.50 J·K^{-1}·mol^{-1}. Calculation from the value of S^o for I$_2$(g) and $\Delta_{sub}S^o$(I$_2$, cr) from the Second Law treatment of the vapor-pressure data for I$_2$(cr), (see Note 31), gave S^o(I$_2$, cr) = 116.05 J·mol^{-1}·K^{-1}.

31. I$_2$(g). $\Delta_f H^o$ was calculated from vapor-pressure data measured by Baxter et al. (1907), Baxter and Grose (1915), Gillespie and Fraser (1936), and Kokovin et al. (1970). The entropy and the enthalpy increment were calculated by direct summation over the vibrational-rotational levels of the X$^1\Sigma_g^+$, a$^3\Pi_{2u}$, A$^3\Pi_{1u}$, and B$^3\Pi_{0u}^+$ electronic states. See Gurvich et al. (1978), part 1, pages 200-203.

32. HI(g). $\Delta_f H^o$ was calculated from the measurements of $\Delta_{sol}H^o$ for HI(g) by Vanderzee and Gier (1974), -(83.283 ± 0.085) kJ·mol^{-1}, combined with the selected value for $\Delta_f H^o$(I$^-$, aq), (see Note 29). Results in good agreement with the selected value were obtained from equilibrium measurements on H$_2$(g) + I$_2$(g) = 2 HI(g), by Taylor and Crist (1941), (26.23 ± 0.10) kJ·mol^{-1}, Rittenberg and Urey (1934b), (26.69 ± 0.30) kJ·mol^{-1}, Bright and Hagerty (1947), (26.57 ± 0.30) kJ·mol^{-1}, and Bodenstein (1894), (26.55 ± 0.20) kJ·mol^{-1}. See Murphy (1936).

The entropy and the enthalpy increment were calculated by direct summation over the vibrational–rotational levels of the ground state, using data from Boyd and Thompson (1952), Haeusler et al. (1963), and Cowan and Gordy (1956).

33. S(cr, rhomb.).   Defined reference state.

The entropy and the enthalpy increment were calculated from heat-capacity data obtained by Berezovskii and Paukov (1978), (5-306 K), Eastman and McGavock (1937), (15-361 K), and Montgomery (1975), (12.3-361 K).

34. S(g). $\Delta_f H^o$ was calculated from $\Delta_f H^o$(S$_2$, g, 298.15 K), (see Note 35), and D_o(0)(S$_2$) = (35240.2 ± 2.5) cm^{-1} or (421.565 ± 0.030) kJ·mol^{-1}, obtained from the rotational predissociation of the molecules ^{32}S$_2$, ^{34}S$_2$ and ^{32}S ^{34}S in the state B$^3\Sigma_u^-$, as reported by Ricks and Barrow (1969). See Huber and Herzberg (1979), page 563. The entropy and the enthalpy increment were calculated using for sublevels $^3P_1$ and $^3P_0$ the energies recommended by Moore (1970). The uncertainty in S^o includes a contribution of 0.002 J·mol^{-1}·K^{-1} from the uncertainty in the relative atomic mass. See also Grønvold et al. (1984).

35. S$_2$(g). $\Delta_f H^o$ was calculated as a weighted average of the equilibrium measurements on the system 2H$_2$(g) + S$_2$(g) = 2H$_2$S(g), made by Randall and Bichowsky (1918), (128.4 ± 1.3) kJ·mol^{-1}, and Preuner and Schupp (1909), (128.4 ± 1.0) kJ·mol^{-1}, and also from measurements of the vapor pressure of sulfur by West and Menzies (1929) coupled with the density measurements of Braune et al.

(1951), (128.9 ± 1.2) kJ·mol^{-1}, and measurements of the partial pressure of $S_2(g)$ by Drowart et al. (1968), (128.61 ± 0.20) kJ·mol^{-1}. For data on $H_2S(g)$ see Note 39.

The entropy and the enthalpy increment were calculated by direct summation over the vibrational-rotational levels of the $X^3 \Sigma_g^-$, $a^1\Delta_g$, and $b^1\Sigma_g^+$ electronic states, taking into account the multiplet splitting in the triplet state. See also Grønvold et al. (1984).

36. $SO_2(g)$. $\Delta_f H^o$ was calculated from the combustion measurements of Eckman and Rossini (1929). Earlier measurements were less accurate because of uncertainties in the composition of the combustion products; an analysis of these measurements is given in Eckman and Rossini (1929).

The entropy and the enthalpy increment were calculated using the anharmonic-oscillator, non-rigid-rotator approximation. See Gurvich et al. (1978), part 1, pages 257-258.

37. SO_4^{2-} (aq). $\Delta_f H^o$ was obtained from measurements of $\Delta_c H^o$ of S(cr, rhomb.) by McCullough et al. (1953), $-(908.53 \pm 0.80)$ kJ·mol^{-1}, Good et al. (1960), $-(909.03 \pm 0.50)$ kJ·mol^{-1}, Månsson and Sunner (1963), $-(909.45 \pm 0.35)$ kJ·mol^{-1}. Values were also obtained from the enthalpy of oxidation of $SO_2(g)$ or $SO_2(aq)$ by Roth et al. (1930), $-(909.22 \pm 1.60)$ kJ·mol^{-1}, Johnson and Ambrose (1963), $-(909.26 \pm 0.60)$ kJ·mol^{-1}, and Johnson and Sunner (1963), $-(909.34 \pm 0.45)$ kJ·mol^{-1}. Values of $\Delta_{dil} H^o$ used in these calculations were obtained from the measurements by Lange et al. (1933), Giauque et al. (1960), Gunn (1970), and Wu and Young (1980).

See Annex II for the derivation of S^o.

38. HS^- (aq). The enthalpy of ionization of $H_2S(aq, undissoc.)$ = $HS^-(aq)$ + $H^+(aq)$, (22.2 kJ·mol^{-1}) was calculated in a review of the literature by Rao and Hepler (1977) from the temperature coefficient of the ionization constant. The measurements of Kury et al. (1953) on the enthalpy of reaction of NaOH(aq) with $H_2S(aq)$ yielded $\Delta_r H^o = (22.3 \pm 0.4)$ kJ·mol^{-1} for the ionization reaction. The value $\Delta_r H^o = (22.3 \pm 0.4)$ kJ·mol^{-1} was selected. The Gibbs energy of ionization was taken from the review by Rao and Hepler (1977), (39.90 ± 0.05) kJ·mol^{-1}, Khodakovsky et al. (1965), (39.84 ± 0.05) kJ·mol^{-1}, Goldhaber and Kaplan (1975), (39.84 ± 0.05) kJ·mol^{-1}, Pavlyuk and Kryukov (1976), (40.13 ± 0.10) kJ·mol^{-1}, and Sretenskaya (1974), (39.97 ± 0.05) kJ·mol^{-1}. The value $\Delta_r G^o = (39.90 \pm 0.15)$ kJ·mol^{-1} was selected. The values selected here were combined to give $\Delta_r S^o = -(59.0 \pm 1.2)$ J·K^{-1}·mol^{-1}.

Evaluation of the data from Kury et al. (1953) and from Stephens and Cobble (1971) yielded $\Delta_r H^o = (53 \pm 3)$ kJ·mol^{-1} for the reaction $HS^-(aq) = S^{2-}(aq)$ + $H^+(aq)$. There is considerable uncertainty in the value of $\Delta_r G^o$; see Rao and Hepler (1977). The value $\Delta_r G^o = (74 \pm 8)$ kJ·mol^{-1} was used in the present calculations.

39. $H_2S(g)$. $\Delta_f H^o$ was calculated from the combustion measurements of Kapustinskii and Kankovskii (1958), $-(20.57 \pm 0.50)$ kJ·mol^{-1}, and Zeumer and Roth (1934), $-(20.7 \pm 2.0)$ kJ·mol^{-1}. The measurements by Pollitzer (1909) on the equilibrium reaction $H_2S(g) + I_2(cr) = 2 HI(g) + S(cr)$ yielded $\Delta_f H^o = -(19.4 \pm 1.1)$ kJ·mol^{-1}.

The entropy and the enthalpy increment were derived using the anharmonic-oscillator, non-rigid-rotator approximation. See Gurvich et al. (1978), part 1, page 265.

40. H_2S(aq, undissoc.).   Calorimetric measurements of $\Delta_{sol}H^\circ$ of H_2S(g) in H_2O(l) were made by Sunner and Wadsö (1957), -(19.7 ± 0.8) $kJ \cdot mol^{-1}$, Zeumer and Roth (1934), -(18.2 ± 0.5) $kJ \cdot mol^{-1}$, and Thomsen (1882), -(18.7 ± 0.6) $kJ \cdot mol^{-1}$. Measurements of the temperature dependence of the solubility of H_2S(g) in H_2O(l), yielding values of $\Delta_{sol}H^\circ$, were reported by Wright and Maass (1932), -(18.3 ± 0.5) $kJ \cdot mol^{-1}$, Helgeson (1967), -(17.1) $kJ \cdot mol^{-1}$, and Clarke and Glew (1971), -(17.48 ± 0.25) $kJ \cdot mol^{-1}$. $\Delta_{sol}H^\circ$ was selected as -(18.0 ± 1.5) $kJ \cdot mol^{-1}$.

$\Delta_{sol}G^\circ$ of H_2S(g) in H_2O(l), (5.66 ± 0.05) $kJ \cdot mol^{-1}$, was calculated from the solubility measurements of Wright and Maass (1932), (5.64 ± 0.03) $kJ \cdot mol^{-1}$, Kendall and Andrews (1921), (5.58 ± 0.05) $kJ \cdot mol^{-1}$, Kapustinskii and Anvaer (1941), (5.60 ± 0.10) $kJ \cdot mol^{-1}$, Winkler (1906), (5.76 ± 0.20) $kJ \cdot mol^{-1}$, Clarke and Glew (1971), (5.66 ± 0.05) $kJ \cdot mol^{-1}$, and Gamsjäger et al. (1967), (5.66 ± 0.05) $kJ \cdot mol^{-1}$. Combination of these selected values yielded $\Delta_{sol}S^\circ$ = -(79.3 ± 5.0) $J \cdot K^{-1} \cdot mol^{-1}$.

41. HSO_4^-(aq). Δ_rH° for HSO_4^-(aq) = H^+(aq) + SO_4^{2-}(aq) has been measured by Izatt et al. (1969a), -(20.50 ± 0.80) $kJ \cdot mol^{-1}$, Austin and Mair (1962), -(24.02 ± 0.80) $kJ \cdot mol^{-1}$, Zielen (1959), -(19.79 ± 0.10) $kJ \cdot mol^{-1}$, Pitzer (1937), -(21.76 ± 0.40) $kJ \cdot mol^{-1}$, and Ahrland and Kullberg (1971b), -(24.60 ± 0.50) $kJ \cdot mol^{-1}$. Values calculated from the temperature dependence of the equilibrium constant were obtained by Lietzke et al. (1961), -(20.54 ± 0.80) $kJ \cdot mol^{-1}$, Nair and Nancollas (1958), -(23.43 ± 0.80) $kJ \cdot mol^{-1}$, Vdovenko et al. (1967), -(22.55 ± 0.40) $kJ \cdot mol^{-1}$, and Fletcher (1964), -(22.25 ± 0.40) $kJ \cdot mol^{-1}$. Pitzer et al. (1977) obtained Δ_rH° = -(23.47 ± 0.20) $kJ \cdot mol^{-1}$ from a correlation of data.

Δ_rG° for the above reaction was measured by Marshall and Jones (1966), (11.348 ± 0.050) $kJ \cdot mol^{-1}$, Evans and Monk (1971), (11.32 ± 0.10) $kJ \cdot mol^{-1}$, Izatt et al. (1969a), (11.24 ± 0.20) $kJ \cdot mol^{-1}$, Nair and Nancollas (1958), (11.19 ± 0.10) $kJ \cdot mol^{-1}$, Vdovenko et al. (1967), (11.34 ± 0.05) $kJ \cdot mol^{-1}$, Lietzke et al. (1961), (11.340 ± 0.050) $kJ \cdot mol^{-1}$, Monk and Amira (1978), (11.353 ± 0.050) $kJ \cdot mol^{-1}$, Covington et al. (1973), (11.387 ± 0.050) $kJ \cdot mol^{-1}$, and Covington et al. (1965), (11.273 ± 0.050) $kJ \cdot mol^{-1}$. Pitzer et al. (1977) obtained (11.30 ± 0.10) $kJ \cdot mol^{-1}$ from a correlation of data.

We have selected Δ_rH° = -(22.4 ± 1.0) $kJ \cdot mol^{-1}$ and Δ_rG° = (11.340 ± 0.050) $kJ \cdot mol^{-1}$, from which Δ_fH° = -(886.9 ± 1.0) $kJ \cdot mol^{-1}$ and S° = (131.7 ± 3.0) $J \cdot K^{-1} \cdot mol^{-1}$ for HSO_4^- (aq).

42. N(g). Δ_fH° was calculated using $D_o(0)(N_2)$ = (78715 ± 50) cm^{-1} or (941.64 ± 0.60) $kJ \cdot mol^{-1}$. This value was obtained from the data of Büttenbender and Herzberg (1935) on predissociation in the $C^3\Pi_u$ state; see Gaydon (1968), page 184. The entropy and the enthalpy increment were calculated taking the electronic partition function equal to 4.

43. N_2(g).   Defined reference state.

The entropy and the enthalpy increment were calculated by direct summation over the vibrational–rotational levels of the ground electronic state. See Gurvich et al. (1978), part 1, pages 280-284.

44. NO_3^- (aq). Δ_fH° was calculated from values for the enthalpy of decomposition and solution of NH_4NO_3(cr, IV). Δ_rH° for NH_4NO_3(cr, IV) = N_2(g) + 0.5 O_2(g) + 2 H_2O(l) was determined by Becker and Roth (1934), -(206.7 ± 0.4) $kJ \cdot mol^{-1}$,

Médard and Thomas (1953), -(206.1 ± 1.0) kJ·mol^{-1}, and Cox et al. (1979), -(206.05 ± 0.31) kJ·mol^{-1}. The value $\Delta_r H^\circ$ = -(206.05 ± 0.31) kJ·mol^{-1} was selected.

$\Delta_{sol} H^\circ$ of NH_4NO_3(cr, IV) was determined by Medvedev et al. (1978), (25.418 ± 0.025) kJ·mol^{-1}, and by Vanderzee et al. (1980b), (25.544 ± 0.030) kJ·mol.$^{-1}$. The value (25.50 ± 0.05) kJ·mol^{-1} was selected. Combination of the two selected values with the value of $\Delta_f H^\circ$ for NH_4^+(aq), (see Note 46), yielded $\Delta_f H^\circ(NO_3^-$, aq) = -(206.85 ± 0.40) kJ·mol^{-1}.

Values in good agreement were obtained from other calorimetric and equilibrium studies. The decomposition measurements of Becker and Roth (1934) may be combined with their measurements on the enthalpy of solution of NH_4NO_3(cr) and the neutralization of HNO_3(aq) by NH_3(g) to yield $\Delta_f H^\circ(NO_3^-$, aq) = -(206.4 ± 0.6) kJ·mol^{-1}. Equilibrium measurements of reactions involving NO_2(g) to yield HNO_3(g) have been reported by Jones (1943), Abel et al. (1930), and Chambers and Sherwood (1937). Third Law treatment of these data yielded for HNO_3(g), $\Delta_f H^\circ$ = -(133.8 ± 0.6), -(134.1 ± 0.5), and -(134.5 ± 0.4) kJ·mol^{-1}, respectively. For NO_2(g), $\Delta_f H^\circ$ = (34.2 ± 0.4) kJ·mol^{-1}, based on the review by Giauque and Kemp (1938); for NO(g), $\Delta_f H^\circ$ = (91.20 ± 0.10) kJ·mol^{-1}, based on the data from Callear and Pilling (1970), Dingle et al. (1975), and Frisch and Margrave (1965). The equilibrium decomposition of NH_4NO_3(cr) as a function of temperature was studied by Feick (1954), (463-543 K), and by Brandner et al. (1962), (349-513 K). A Third Law treatment of their data yielded for HNO_3(g), $\Delta_f H^\circ$ = -(134.36 ± 0.45) and -(134.47 ± 0.60) kJ·mol^{-1}, respectively. We have selected for HNO_3(g), $\Delta_f H^\circ$ = -(134.2 ± 0.4) kJ·mol^{-1}. $\Delta_{vap} H^\circ$ for HNO_3(l) = (39.16 ± 0.13) kJ·mol^{-1} was taken from the data of Wilson and Miles (1940), and $\Delta_{sol} H^\circ$ for HNO_3(l) = -(33.280 ± 0.080) kJ·mol^{-1} was taken from Forsythe and Giauque (1942). Combination of these values yielded for NO_3^-(aq), $\Delta_f H^\circ$ = -(206.7 ± 0.6) kJ·mol^{-1}.

We have selected $\Delta_f H^\circ$ = -(206.85 ± 0.40) kJ·mol^{-1} based on the decomposition measurements by Cox et al., because of the attention paid to characterizing the crystal form of NH_4NO_3 studied.

See Annex II for the derivation of S°.

45. NH_3(g). $\Delta_f H^\circ$ was calculated from calorimetric measurements of the decomposition of NH_3(g) by Haber et al. (1915), -(46.3 ± 0.6) kJ·mol^{-1}, Haber and Tamaru (1915), -(45.9 ± 0.6) kJ·mol^{-1}, and Wittig and Schmatz (1959), -(45.94 ± 0.42) kJ·mol^{-1}. Measurements of the enthalpy of combustion and solution of oxalic acid dihydrate and ammonium oxalate hydrate by Becker and Roth (1934) yielded $\Delta_f H^\circ$ (NH_3, g) = -(46.10 ± 0.50) kJ·mol^{-1}. The data on the $\frac{1}{2} N_2$(g) + $\frac{3}{2} H_2$(g) = NH_3(g) equilibrium have been reviewed by Vanderzee and King (1972) and by Gurvich et al. (1978), part 1, page 311. These data yielded $\Delta_f H^\circ(NH_3$, g) = -(45.56 ± 0.25) kJ·mol.$^{-1}$. The data on the enthalpy of solution of NH_3(g) and ionization of NH_3(aq), (see Note 46), confirm the value selected here.

The entropy and the enthalpy increment were calculated using the anharmonic-oscillator, non-rigid-rotator approximation, including the contribution due to the inversion doubling; see Gurvich et al. (1978), part 1, pages 309-10. Calculations by Haar (1968) are in agreement with these values.

46. NH_4^+(aq). Values of $\Delta_f H^\circ$, as calculated from measurements of $\Delta_c H^\circ$ of mono- and di-ammonium salts of succinic acid, of $\Delta_{sol} H^\circ$ of these salts and of succinic acid by Vanderzee et al. (1972c), and of $\Delta_c H^\circ$ of succinic acid by Vanderzee et al.

(1972b), were -(133.53 $\pm$ 0.60) kJ·mol^{-1} and -(133.15 $\pm$ 0.25) kJ·mol^{-1}, respectively. The value calculated from $\Delta_f H^o$(NH$_3$, g), (see Note 45), the enthalpy of solution of NH$_3$(g) measured by Vanderzee and King (1972), -(35.346 $\pm$ 0.060) kJ·mol^{-1}, and the enthalpy of ionization of NH$_3$(aq) measured by Vanderzee et al, (1972a), (3.849 $\pm$ 0.025) kJ·mol^{-1}, yielded -(133.25 $\pm$ 0.40) kJ. mol^{-1}. Earlier measurements are discussed by these authors.

See Annex II for the derivation of S^o of the ion.

47. P(cr, white). Defined reference state. Note that this is not the thermodynamically stable state at 298.15 K.

The entropy and the enthalpy increment were obtained from the heat capacity data of Stephenson et al. (1969), (15-320 K). The entropy contribution below 15 K is 1.13 J·K^{-1}·mol^{-1}.

48. P(g). $\Delta_f H^o$ was calculated using the value of $\Delta_f H^o$(P$_2$, g, 298.15 K), (see Note 49), and D_o(0)(P$_2$) = (4059 $\pm$ 30) cm^{-1} or (48.56 $\pm$ 0.36) kJ·mol^{-1} obtained from measurements by Herzberg (1932) on predissociation in the state C$^1\Sigma_u^+$.

The entropy and the enthalpy increment were calculated taking the electronic partition function equal to 4.

49. P$_2$(g). $\Delta_f H^o$ for P$_2$(g) was calculated from the value of $\Delta_f H^o$ for P$_4$(g), (see Note 50), and the value for $\Delta_r H^o$ for the reaction P$_4$(g) = 2 P$_2$(g) calculated by the Third Law method from the vapor pressure and vapor density measurements of phosphorus vapor by Stock et al. (1912), (229 $\pm$ 3) kJ·mol^{-1}, and by Rau (1975), (228.3 $\pm$ 1.0) kJ·mol^{-1}. Mass spectrometric studies of the phosphorus equilibrium in phosphide vapors by Drowart and Goldfinger (1958) and photoionization studies by Smets et al. (1977) yielded $\Delta_r H^o$ = (233 $\pm$ 8) kJ·mol^{-1}. See Gurvich et al. (1978), part 1, pages 349-351, for a summary of other measurements.

The entropy and the enthalpy increment were calculated by direct summation over the vibrational-rotational levels of the ground X$^1\Sigma_g^+$ state, using the molecular constants obtained by Douglas and Rao (1958).

50. P$_4$(g). $\Delta_f H^o$ was calculated by the Third Law method from the vapor pressure measurements on P(cr, white) by Centnerszwer (1913), Fischbeck and Eich (1937), and Dainton and Kimberly (1950). The uncertainty derives mainly from uncertainties in the thermal functions of P$_4$(g) and P(l).

The entropy and the enthalpy increment were calculated in the harmonic-oscillator, rigid-rotator approximation using vibrational frequencies ν_1(606.5 cm^{-1}) and ν_2(367.2 cm^{-1}) obtained by Bosworth et al. (1973) from the Raman spectrum of the vapor and ν_3(464.5 cm^{-1}) obtained by Gutowsky and Hoffman (1950) from the infrared spectrum of the vapor.

51. HPO$_4^{2-}$(aq). $\Delta_f H^o$ was calculated from the following measurements: Irving and McKerrell (1967b) measured $\Delta_r H^o$ for the reaction of H$_3$PO$_4$·0.6416 H$_2$O with 2 NaOH(aq), and $\Delta_{sol} H^o$ of Na$_2$HPO$_4$(cr) in the final mixture. These data were combined with appropriate dilution data to obtain $\Delta_r H^o$ = -(89.358 $\pm$ 0.095) kJ·mol^{-1} for the reaction H$_3$PO$_4$ · 100H$_2$O + 2 NaOH(aq) = Na$_2$HPO$_4$(cr) + 2 H$_2$O(l). Using the value $\Delta_f H$(H$_3$PO$_4$ · 100H$_2$O) = -(1296.2 $\pm$ 1.0) kJ·mol^{-1} (see below) we obtained $\Delta_f H^\circ(Na_2HPO_4, cr) = -(1754.61 \pm 1.15)$ kJ·mol^{-1}. The enthalpy of solution measurements by Mitchell (1965) and Waterfield and Staveley (1967), when corrected for hydrolysis and dilution, (see below and Note 52), yielded for the reaction $Na_2HPO_4(cr) = 2 Na^+(aq) + HPO_4^{2-}(aq)$, $\Delta_r H^\circ = -(25.1 \pm 0.5)$ kJ·mol^{-1} and $\Delta_f H^\circ(HPO_4^{2-}, aq) = -(1299.0 \pm 1.3)$ kJ·mol^{-1}. de Passille and Séon (1934) measured $\Delta_r H$ for the reaction of $H_3PO_4 \cdot 300H_2O$ with 2 moles of aqueous NH_3 and for the solution of $(NH_4)_2HPO_4(cr)$ at 288 K; combination with estimated temperature and dilution corrections yielded for the reaction $H_3PO_4 \cdot 100H_2O + 2 NH_3(300H_2O) = (NH_4)_2HPO_4(cr)$, $\Delta_r H = -(118.4 \pm 1.5)$ kJ·mol^{-1}. Taking $\Delta_f H^\circ$ for $NH_3(300H_2O) = -(81.21 \pm 0.30)$ kJ·mol^{-1}, we obtained $\Delta_f H^\circ((NH_4)_2HPO_4, cr) = -(1577.0 \pm 2.0)$ kJ·mol^{-1}. The enthalpy of solution of $(NH_4)_2HPO_4(cr)$, (12.8 ± 0.5) kJ·mol^{-1}, was calculated from measurements made by de Passille and Séon (1934), (12.4 ± 1.0) kJ·mol^{-1}, and by Hardesty and Ross (1937), (12.9 ± 0.5) kJ·mol^{-1}. By this means we obtained $\Delta_f H^\circ(HPO_4^{2-}, aq) = -(1297.8 \pm 2.5)$ kJ·mol^{-1}. Because of the larger uncertainties in the latter value (cf. above) we selected $\Delta_f H^\circ = -(1299.0 \pm 1.5)$ kJ·mol^{-1}.

S° of the $HPO_4^{2-}(aq)$ ion has been calculated from the following sets of measurements: Waterfield and Staveley (1967) measured $S^\circ(Na_2HPO_4 \cdot 7H_2O, cr)$, (434.59 ± 0.80) J·K^{-1}·mol^{-1}, the enthalpies of solution of $Na_2HPO_4 \cdot 7H_2O(cr)$, (48.85 ± 0.05) kJ·mol^{-1}, and of $Na_2HPO_4 \cdot 12H_2O(cr)$, (95.82 ± 0.05) kJ·mol^{-1}, and the equilibrium water-vapor pressure over the heptahydrate + dodecahydrate mixture. From these data we obtained $S^\circ(Na_2HPO_4 \cdot 12H_2O, cr) = (634.6 \pm 1.0)$ J·K^{-1}·mol^{-1}. From the solubility (0.812 M) obtained by Wendrow and Kobe (1952) and the activity data $(\gamma_\pm = 0.223, a_w = 0.9732)$ from Scatchard and Breckenridge (1954) we calculated $\Delta_{sol} G^\circ = (10.080 \pm 0.050)$ kJ·mol^{-1} and $S^\circ(HPO_4^{2-}, aq) = -(34.1 \pm 1.2)$ J·K^{-1}·mol^{-1}.

For the reaction $H_2PO_4^-(aq) = HPO_4^{2-}(aq) + H^+(aq)$, $\Delta_r H^\circ = (3.55 \pm 0.30)$ kJ·mol^{-1} based on measurements by Pitzer (1938), (3.35 ± 0.20) kJ·mol^{-1}, and by Christensen and Izatt (1962), (3.77 ± 0.45) kJ·mol^{-1} (see Note 52). $\Delta_r G^\circ$ was measured by Bates and Acree (1943), 41.084 kJ·mol^{-1}, Grzybowski (1958b), 41.100 kJ·mol^{-1}, Ender et al. (1957), 41.091 kJ·mol^{-1}, and Nims (1933), 41.126 kJ·mol^{-1}. We have selected $\Delta_r G^\circ = (41.100 \pm 0.030)$ kJ·mol^{-1}. From the $\Delta_r H^\circ$ and $\Delta_r G^\circ$ values and from the selected value of S° for $H_2PO_4^-(aq)$, (see Note 52), we obtained $S^\circ(HPO_4^{2-}, aq), = -(33.4 \pm 1.0)$ J·K^{-1}·mol^{-1}. Corrections for hydrolysis in the phosphate solutions were made using for the reaction $HPO_4^{2-}(aq) = PO_4^{3-}(aq) + H^+(aq)$, $\Delta_r H^\circ = (16.1 \pm 1.5)$ kJ·mol^{-1}, based on measurements by Pitzer (1937) and Christensen et al. (1966), and $\Delta_r G^\circ = (70.55 \pm 0.40)$ kJ·mol^{-1} from Vanderzee and Quist (1961).

The value of $\Delta_f H^\circ$ for $H_3PO_4 \cdot 100H_2O$ was calculated from the measurements by Schumm et al. (1974) on $\Delta_f H^\circ$ of $PCl_5(cr)$ and Birley and Skinner (1968) on the enthalpy of hydrolysis of $PCl_5(cr)$, and on measurements by Head and Lewis (1970) on $\Delta_c H^\circ$ of $P(cr)$ to form aqueous H_3PO_4. The combustion data of Egan and Luff (1963b) on $P(cr)$ to form $P_4O_{10}(cr)$, combined with the hydrolysis measurements of Irving and McKerrell (1967a) and Holmes (1962), yielded an agreeing, but less precise, value.

52. $H_2PO_4^-$(aq). $\Delta_f H^\circ$ was obtained from the following data sets: Irving and McKerrell (1967b) measured the enthalpy of reaction of $H_3PO_4 \cdot 0.6416H_2O$ with NaOH(aq) and the enthalpy of solution of $NaH_2PO_4(cr)$ in the final mixture.

These data were combined with appropriate dilution data to obtain for the reaction $H_3PO_4 \cdot 100H_2O + NaOH(aq) = NaH_2PO_4(cr) + H_2O(l)$, $\Delta_r H^\circ = -(62.919 \pm 0.065)$ kJ·mol^{-1}. Using $\Delta_f H^\circ(H_3PO_4 \cdot 100H_2O, l) = -(1296.2 \pm 1.0)$ kJ·mol^{-1}, (see Note 51), we obtained $\Delta_f H^\circ(NaH_2PO_4, cr) = -(1543.6 \pm 1.1)$ kJ·mol^{-1}. The data of Mitchell (1965), corrected for hydrolysis (see Note 51), yielded for the reaction $NaH_2PO_4(cr) = Na^+(aq) + H_2PO_4^-(aq)$, $\Delta_r H^\circ = (0.69 \pm 0.25)$ kJ·mol^{-1}, and for $H_2PO_4^-(aq)$, $\Delta_f H^\circ = -(1302.6 \pm 1.2)$ kJ·mol^{-1}.

de Passille and Séon (1934) measured $\Delta_r H^\circ$ for the reaction of $H_3PO_4 \cdot 300H_2O$ with one mole of aqueous NH_3 and also $\Delta_{sol} H^\circ$ of $NH_4H_2PO_4(cr)$ at 288 K; combination with estimated temperature and dilution corrections yielded for the reaction $H_3PO_4 \cdot 300H_2O + NH_3(300H_2O) = NH_4H_2PO_4(cr)$, $\Delta_r H^\circ = -(78.7 \pm 1.5)$ kJ·mol^{-1} and for $NH_4H_2PO_4(cr)$, $\Delta_f H^\circ = -(1456.1 \pm 2.0)$ kJ·mol^{-1}. The enthalpy of solution of $NH_4H_2PO_4(cr)$, (16.27 ± 0.05) kJ·mol^{-1}, was derived from the work of Egan and Luff (1963a). From these data we obtained $\Delta_f H^\circ(H_2PO_4^-, aq) = -(1306.6 \pm 2.5)$ kJ· mol^{-1}.

$\Delta_r H^\circ$ of the reaction $H_2PO_4^-(aq) = HPO_4^{2-}(aq) + H^+(aq)$ was measured by Pitzer (1937), (3.35 ± 0.20) kJ·mol^{-1}, Irani and Taulli (1966), (2.51 ± 1.20) kJ·mol^{-1}, and Christensen and Izatt (1962), (3.77 ± 0.45) kJ·mol^{-1}. Values of $\Delta_r H^\circ$ were also calculated from measurements of the temperature coefficient of the ionization constant by Bates and Acree (1943), (3.5 ± 0.3) kJ·mol^{-1}, Grzybowski (1958b), (3.75 ± 0.65) kJ·mol^{-1}, Mesmer and Baes (1974), (3.19 ± 0.20) kJ·mol^{-1}, and Ender et al. (1957), (4.08 ± 0.20) kJ·mol^{-1}. We have selected $\Delta_r H^\circ = (3.55 \pm 0.30)$ kJ·mol^{-1}. Combining this value with $\Delta_f H^\circ$ for $HPO_4^{2-}(aq)$, (see Note 51), we obtained for $H_2PO_4^-(aq)$, $\Delta_f H^\circ = -(1302.6 \pm 1.6)$ kJ·mol^{-1}.

Because the corrections to 298.15 K and to standard conditions for the de Passille and Séon measurements are rather uncertain, we have selected $\Delta_f H^\circ = -(1302.6 \pm 1.5)$ kJ·mol^{-1}.

S° of the ion was calculated from the following data sets:

Heat-capacity measurements, 16-300 K, on $KH_2PO_4(cr)$ by Stephenson and Hooley (1944), $S^\circ(298.15 \text{ K}) = (134.85 \pm 0.45)$ J·K^{-1}·mol^{-1}, the standard enthalpy of solution, (19.69 ± 0.12) kJ·mol^{-1}, measured by Egan and Luff (1963), and the standard Gibbs energy of solution, (2.50 ± 0.20) kJ·mol^{-1}, from the measurements by Childs et al. (1973) (solubility = 1.828 M, $\gamma_\pm = 0.3306$) yielded $S^\circ(H_2PO_4^-(aq)) = (91.3 \pm 1.0)$ J·K^{-1}·mol^{-1}.

Heat-capacity measurements on $NH_4H_2PO_4(cr)$ reported by Stephenson and Zettelmoyer (1944), $S^\circ(298.15 \text{ K}) = (151.96 \pm 0.45)$ J·K^{-1}·mol^{-1}, the standard enthalpy of solution, (16.20 ± 0.10) kJ·mol^{-1}, selected from the measurements by Egan and Luff (1963), de Passille (1936), and de Passille and Séon (1934), and the standard Gibbs energy of solution, (0.57 ± 0.35) kJ·mol.$^{-1}$, from the solubility given in Linke (1958), with $\gamma_\pm$ estimated to be 0.25, yielded the value $S^\circ(H_2PO_4^-, aq) = (93.2 \pm 1.5)$ J·K^{-1}· mol^{-1}.

53. C(cr, graphite). Graphite is selected as the defined reference state for carbon. However, there is considerable variation among measured values of the low-temperature heat capacity and entropy, depending upon the source and nature of the graphite samples. See DeSorbo (1955b), DeSorbo and Nichols (1958), and DeSorbo and Tyler (1953). The entropy and the enthalpy increment were calculated from the low-temperature measurements by DeSorbo and Nichols (1-20 K) and DeSorbo and Tyler (13-300 K) on Acheson spectroscopic graphite, corresponding to the material used for the determination of $\Delta_f H^\circ$ of $CO_2(g)$. (See Note 56).

54. C(g). $\Delta_f H^\circ$ was calculated using the value for $\Delta_f H^\circ(CO, g)$, (see Note 55), and $D_o(0)(CO) = (89595 \pm 30)$ cm^{-1} or (1071.79 ± 0.36) kJ·mol^{-1} from Douglas and Møller (1955).

The entropy and the enthalpy increment were calculated taking the energies of the $^3P_1$ and $^3P_2$ sublevels equivalent to 16.4 and 43.4 cm^{-1}, from Moore (1970).

55. CO(g). $\Delta_f H^\circ$ was calculated from combustion measurements by Rossini (1931b, 1939), $\Delta_f H^\circ = -(110.52 \pm 0.17)$ kJ·mol^{-1}, Awbery and Griffiths (1933), $\Delta_f H^\circ = -(110.63 \pm 0.40)$ kJ·mol^{-1}, and Fenning and Cotton (1933), $\Delta_f H^\circ = -(110.36 \pm 0.25)$ kJ·mol^{-1}.

The high-temperature equilibrium $CO(g) + 0.5\ O_2(g) = CO_2(g)$ was measured by Peters and Möbius (1958), (1017-1624 K), and Caneiro et al. (1981), (970-1270 K). Third Law evaluation of their results yielded $\Delta_r H^\circ = -(281.5 \pm 1.5)$ and $-(283.05 \pm 0.50)$ kJ·mol^{-1}, respectively. Similar evaluation of the measurements on the equilibrium $CO(g) + H_2O(g) = CO_2(g) + H_2(g)$ by Neumann and Köhler (1928), (635-1259 K), and Meyer and Scheffer (1938), (895-1086 K), yielded $\Delta_r H^\circ = -(41.15 \pm 0.20)$ and $-(41.95 \pm 0.50)$ kJ·mol^{-1}, respectively. The equilibrium $C(cr, graph.) + CO_2(g) = 2\ CO(g)$ was studied by Peters and Möbius (1958), (1051-1463 K), and Smith (1946), (1073-1273 K). Third Law evaluation of these data yielded $\Delta_r H^\circ = (171.9 \pm 1.5)$ and (172.13 ± 0.40) kJ·mol^{-1}, respectively. All of these results are in good agreement with the values selected here.

The entropy and the enthalpy increment were calculated by direct summation over the vibrational-rotational levels of the ground electronic state $X^1\Sigma^+$ using the molecular constants of the $^{12}C^{16}O$ molecule recommended by Todd et al. (1976). The assigned uncertainties arise mainly from the use of averaged molecular constants.

56. CO$_2$(g). $\Delta_f H^\circ$ was calculated from combustion measurements on graphite by Prosen and Rossini (1944), $-(393.58 \pm 0.06)$ kJ·mol^{-1}, Hawtin et al. (1966), $-(393.47 \pm 0.06)$ kJ·mol^{-1}, Dewey and Harper (1938), $-(393.51 \pm 0.10)$ kJ·mol^{-1}, Jessup (1938), $-(393.40 \pm 0.06)$ kJ·mol^{-1}, and Lewis et al. (1965), $-(393.47 \pm 0.04)$ kJ·mol^{-1}. Acheson spectroscopic graphite is taken as the reference state for carbon (see Note 53). The estimated uncertainty takes into account the effect of the previous histories of the graphite samples that have been studied.

The entropy and the enthalpy increment were calculated using molecular constants of the $^{12}C^{16}O_2$ molecule in the anharmonic-oscillator, non-rigid-rotator approximation. The assigned uncertainties arise mainly from the use of averaged molecular constants.

57. CO$_2$(aq, undissoc.). The enthalpy of formation and the entropy were calculated from the values given by Berg and Vanderzee (1978a, 1978b) for the reaction $CO_2(g) = CO_2(aq, undissoc.)$, $\Delta_r H^\circ = -(19.75 \pm 0.17)$ kJ·mol^{-1}; $\Delta_r G^\circ = (8.371 \pm 0.015)$ kJ·mol^{-1}. A comprehensive review of earlier data is given in the above references.

58. CO$_3^{2-}$(aq). The enthalpy of formation and the entropy were obtained from the values given by Berg and Vanderzee (1978a, 1978b) for the reaction $HCO_3^-(aq) = H^+(aq) + CO_3^{2-}(aq)$, $\Delta_r H^\circ = (14.70 \pm 0.12)$ kJ·mol^{-1}; $\Delta_r G^\circ = (58.96 \pm 0.40)$

$kJ \cdot mol^{-1}$. This value of $\Delta_r G^o$ is in excellent agreement with that given by Peiper and Pitzer (1982). The above data yield $S^o(CO_3^{2-}, aq) = -(50.0 \pm 1.0)$ $J \cdot K^{-1} \cdot mol^{-1}$.

A value for S^o was also computed from data for $Na_2CO_3(cr)$ and $Na_2CO_3 \cdot 10H_2O(cr)$. The $\Delta_{sol}H^o$ of $Na_2CO_3(cr)$, $-(26.65 \pm 0.10)$ $kJ \cdot mol^{-1}$, was calculated from the measurements by Waterfield et al. (1968) and Berg and Vanderzee (1978a). $\Delta_{sol}G^o$ for $Na_2CO_3 \cdot 10H_2O(cr)$, (4.75 ± 0.10) $kJ \cdot mol^{-1}$, was calculated from the solubility and activity data obtained by Kobe and Sheehy (1948), Khvorostin et al. (1975), Robinson and Macaskill (1979), and Saegusa (1950). $\Delta_r G^o$ for the decomposition of the decahydrate to $Na_2CO_3(cr)$ was taken from Vanderzee (1982), (11.294 ± 0.100) $kJ \cdot mol^{-1}$ (corrected to liquid H_2O). S^o for Na_2CO_3, cr, (135.0 ± 0.8) $J \cdot K^{-1} \cdot mol^{-1}$ was taken from Waterfield et al. (1968). From these data and the value for $Na^+(aq)$, (see Note 142), $S^o(CO_3^{2-}, aq) = -(49.5 \pm 1.0)$ $J \cdot K^{-1} \cdot mol^{-1}$.

$\Delta_{sol}H^o$ for $Li_2CO_3(cr)$ was calculated from the measurements by Brown and Latimer (1936), $-(18.0 \pm 1.0)$ $kJ \cdot mol^{-1}$, using a dilution correction from 0.029 M, $-(3.8 \, kJ \cdot mol^{-1})$, based on the data for $Na_2CO_3(aq)$ taken from Berg and Vanderzee (1978b). The solubility (0.170 M) was taken from the data of Saegusa (1948), Jäger (1958), and Lagarde (1945). Extrapolation of the data of Lagarde on $Li_2CO_3 + Na_2CO_3 + H_2O$ and $Li_2CO_3 + LiCl + H_2O$ and the data of Jäger on $Li_2CO_3 + LiCl + H_2O$ yielded $\ln K_{SP} = -6.863, -7.473$, and -7.070, respectively. We selected $\ln K_{SP} = -7.00$, corresponding to $\gamma_\pm = 0.36$ and $\Delta_{sol}G^o = (17.3 \pm 0.5)$ $kJ \cdot mol^{-1}$. $S^o (Li_2CO_3, cr, 298.15 \, K) = (90.30 \pm 0.30)$ $J \cdot K^{-1} \cdot mol^{-1}$ was calculated from the heat-capacity measurements of Brown and Latimer (1936), (16-300 K). From these data and the value for $Li^+(aq)$, (see Note 139), the value $S^o = -(52.6 \pm 2.0)$ $J \cdot mol^{-1} \cdot K^{-1}$ was obtained.

59. HCO_3^-(aq). $\Delta_f H^o$ and S^o were calculated from the data given by Berg and Vanderzee (1978a, 1978b) for the reaction $H_2O(l) + CO_2(g) = H^+(aq) + HCO_3^-$ (aq), $\Delta_r H^o = -(10.59 \pm 0.15)$ $kJ \cdot mol^{-1}$, $\Delta_r G^o = -(44.622 \pm 0.020)$ $kJ \cdot mol^{-1}$. See Berg and Vanderzee (1978b) for a review of earlier measurements.

60. Si(cr).   Defined reference state.

The entropy and the enthalpy increment were calculated from low-temperature heat-capacity data of Flubacher et al. (1959), (7-300 K), Kalishevich et al. (1965), (60-300 K), and Keesom and Seidel (1959), (1.2 - 4.2 K).

61. Si(g). $\Delta_f H^o$ was calculated from sublimation measurements on Si(cr) and decomposition measurements on silicon carbide by Davis et al. (1961), Grieveson and Alcock (1960), Gulbransen et al. (1966), Batdorf and Smits (1959), Drowart et al. (1958), Drowart and De Maria (1960), and Zmbov et al. (1973). See discussion by Gurvich et al. (1979), part 1, pages 206-207.

The entropy and the enthalpy increment were calculated taking the energies of the $^3P_1$ and $^3P_0$ sublevels and of the $^1D_2$ state equivalent to 77.115, 223.157, and 6298.85 cm^{-1}, respectively, from Moore (1970).

62. SiO_2(cr, α-quartz). $\Delta_f H^o$ was obtained from the following sets of measurements. Wise et al. (1963) measured the enthalpies of reaction of Si(cr) and SiO_2(cr, α-quartz) with $F_2(g)$. From the difference in the enthalpies, $\Delta_f H^o(SiO_2, cr, \alpha$-quartz) $= -(911.1 \pm 1.4)$ $kJ \cdot mol^{-1}$. Good et al. (1964) measured the enthalpy of combustion of Si(cr) in a bomb, followed by dissolution of the products in aqueous

HF. The result was combined with the measurements of Kilday and Prosen (1973) on $\Delta_{sol}H^{\circ}$ of SiO_2(cr, α-quartz) in aqueous HF to obtain $\Delta_f H^{\circ}(SiO_2$, cr, α-quartz) = -(910.4 ± 1.4) kJ·mol^{-1}.

The entropy and the enthalpy increment were calculated from the heat-capacity measurements by Westrum (1956), (6 - 300 K), Jones and Hallett (1960), (2 - 4 K), and Gurevich and Khlyustov (1979), (9 - 300 K). See also Anderson (1936), (53 - 296 K).

63. SiF$_4$(g). $\Delta_f H^{\circ}$ was derived from the measurements by Wise et al. (1963) on the reaction of Si(cr) with F$_2$(g).

The entropy and the enthalpy increment were calculated in the harmonic-oscillator, rigid-rotator approximation using the frequencies obtained by Clark and Rippon (1972) from the Raman spectrum of gaseous SiF$_4$, together with the molecular dimensions obtained by Beagley et al. (1973) and Hagen and Hedberg (1973) from electron-diffraction measurements.

64. Ge(cr). Defined reference state.

The entropy and the enthalpy increment were calculated from the heat-capacity measurements of Flubacher et al. (1959), (2 - 300 K) and Piesbergen (1963), (12 - 273 K).

65. Ge(g). Vapor-pressure measurements on Ge(l) by Searcy (1952), (1510-1882 K), Searcy and Freeman (1955), (1608-1885 K), Schultz and Searcy (1961), (1808-2070 K), and Tseplyaeva et al. (1982), (1263-1647 K) yielded for Ge(g), $\Delta_f H^{\circ}$ = (372 ± 2), (378 ± 4), (374 ± 1), and (368 ± 3) kJ·mol^{-1}, respectively.

D_o (298 K) for GeS(g) was calculated from the spectroscopic data of Drummond and Barrow (1952); $\Delta_f H^{\circ}$ for GeS(cr), -(74.8 ± 3.0) kJ·mol^{-1}, was obtained from the GeS(cr) + H$_2$(g) equilibrium measurements of Ono and Sudo (1955), -(74.8 ± 4.0) kJ·mol^{-1}, and the fluorine combustion measurements by Adams et al. (1970a), -(74.7 ± 4.0) kJ· mol^{-1}; the enthalpy of sublimation of GeS(cr), (168.7 ± 3.0) kJ·mol^{-1}, was taken from the review by Gurvich et al. (1979), part 1, page 289. Combination of these values gave $\Delta_f H^{\circ}$ = (369.2 ± 4.0) kJ·mol^{-1}.

The entropy and the enthalpy increment were calculated using the atomic energy levels $^3P_1$, $^3P_2$, and $^1D_2$ equivalent to 557.134, 1409.961, and 7125.299 cm^{-1}, respectively, as recommended by Moore (1970).

66. GeO$_2$(cr, tetrag.). $\Delta_f H^{\circ}$ was calculated from the results of fluorine combustion of Ge(cr) and GeO$_2$(cr, tetrag.) by Gross et al. (1966), -(580.2 ± 1.2) kJ·mol^{-1}, equilibrium measurements by Faktor and Carasso (1965), -(579.1 ± 2.0) kJ·mol^{-1}, and high-temperature cell measurements by Jacob and Alcock (1974), -(579.5 ± 2.0) kJ·mol^{-1}, Katayama et al. (1975), -(579.2 ± 2.0) kJ·mol^{-1}, Katayama et al. (1979), -(583.8 ± 2.0) kJ·mol^{-1}, and Sreedharan et al. (1979), -(577.3 ± 2.0) kJ·mol^{-1}. The oxygen combustion measurements by Mah and Adami (1962) and Becker and Roth (1932) did not yield suitable crystalline combustion products.

The entropy and the enthalpy increment were calculated from the heat-capacity measurements by Counsell and Martin (1967), (16-320 K). The entropy contribution below 16 K is 0.02 J·K^{-1}·mol^{-1}.

67. GeF$_4$(g). $\Delta_f H^{\circ}$ was calculated from the measurement of the enthalpy of the reaction of Ge(cr) with F$_2$(g) by Gross et al. (1966), -(1189.80 ± 0.63) kJ·mol^{-1},

O'Hare et al. (1969), -(1190.64 ± 0.88) kJ·mol⁻¹, and Adams et al. (1970a), -(1190.0 ± 1.6) kJ·mol⁻¹.

The entropy and the enthalpy increment were calculated in the harmonic-oscillator, rigid-rotator approximation using the frequencies obtained by Armstrong and Clark (1976) and Clark and Rippon (1972) from the Raman spectrum and by Königer et al. (1977) from the infrared spectrum of gaseous GeF_4. The molecular dimensions were obtained by Caunt et al. (1951) from electron-diffraction measurements.

68. Sn(cr, white). Defined reference state.

Although Sn(cr, white) is thermodynamically unstable below 287 K, its heat capacity has been measured down to 1.7 K by Naumov et al. (1979), (1.7-311 K); the entropy and the enthalpy increment were calculated from these measurements. References to earlier data are given in Hultgren et al. (1973) and Gurvich et al. (1979), part 1, page 295.

69. Sn(g). $\Delta_f H^o$ was calculated from vapor-pressure measurements by Searcy and Freeman (1954), (303 ± 1.5) kJ·mol⁻¹, Alcock et al. (1969), 301 kJ·mol⁻¹, Onillon and Olette (1968), (300.5 ± 0.5) kJ·mol⁻¹, Munir and Searcy (1965), (302 ± 2) kJ·mol⁻¹, Hart and Searcy (1966), (300 ± 1) kJ·mol⁻¹, Mar and Searcy (1967), 301 kJ·mol⁻¹, Skinner and Searcy (1968), 301 kJ·mol⁻¹, and Hansen et al. (1969), (302.9 ± 0.8) kJ·mol⁻¹. The uncertainty in the selected value comes mainly from the thermal functions for Sn(l). See discussion by Hultgren et al. (1973) and Gurvich et al. (1979), part 1, pages 295-296. The entropy and the enthalpy increment were calculated taking the energy levels $^3P_1$, $^3P_2$, and $^1D_2$ equivalent to 1691.8, 3427.7, and 8613.0 cm⁻¹ respectively, as recommended by Moore (1970).

70. Sn^{2+}(aq). The measurements by Vasil'ev et al. (1973a, 1973c, 1976) on $\Delta_{sol}H^o$ of Sn(cr), $SnCl_2$(cr), and $SnCl_2 \cdot 2H_2O$(cr) in aqueous HCl + H_2O_2 solutions and on $\Delta_{sol}H^o$ of the dihalides in aqueous $HClO_4$ solutions yielded $\Delta_f H^o(Sn^{2+}, aq) = $ -(8.9 ± 0.8) kJ·mol⁻¹. E.m.f. measurements on Sn^{2+} in aqueous $HClO_4$ solutions by Vasil'ev and Glavina (1973) yielded for the reaction Sn(cr) + 2 H^+(aq) = Sn^{2+}(aq) + H_2(g), $\Delta_r G^o = $ -(27.88 ± 0.08) kJ·mol⁻¹, $\Delta_r S^o = $ (51.1 ± 4.0) J·K⁻¹·mol⁻¹. The value $\Delta_f H^o = $ -(12.6 ± 1.5) kJ·mol⁻¹ was calculated from these data. The value $\Delta_f H^o = $ -(8.9 ± 1.0) kJ·mol⁻¹ was selected.

$\Delta_f G^o$ for Sn^{2+}(aq) was calculated from the cell data of Vasil'ev and Glavina (1973), (see above), and of Haring and White (1938), -(27.13 ± 0.20) kJ·mol⁻¹, and from the equilibrium measurements with Pb^{2+}(aq) of Noyes and Toabe (1917), -(26.80 ± 0.30) kJ·mol⁻¹. Combination of the equilibrium measurements of Vanderzee and Rhodes (1952) and Garrett and Heiks (1941) yielded $\Delta_f G^o = $ -(27.7 ± 1.0) kJ·mol⁻¹. The value $\Delta_f G^o = $ -(27.60 ± 0.40) kJ·mol⁻¹ was selected, then combined with the selected value of $\Delta_f H^o$ (see above) to compute S^o for the ion.

71. SnO(cr, tetrag.). $\Delta_f H^o$ was calculated from the measurements by Lavut et al. (1981) on the combustion of Sn(cr) and SnO(cr, tetrag.) in O_2(g), -(280.71 ± 0.21) kJ·mol⁻¹. Earlier combustion measurements by Humphrey and O'Brien (1953) yielded $\Delta_f H^o = $ -(285.94 ± 0.70) kJ·mol⁻¹; the cell measurements by Maier (1929), using a Hg(l) + HgO(cr, red) electrode, yielded $\Delta_f H^o = $ -(285.8 ± 1.0) kJ·mol⁻¹.

The entropy and the enthalpy increment were calculated from the heat capacity measurements of Kostryukov et al. (1977), (5 - 311 K). Values were also reported by Millar (1929), (70-293 K).

72. SnO_2(cr, tetrag.). $\Delta_f H^o$ was calculated from the measurements by Lavut et al. (1981) on the combustion of Sn(cr) in O_2(g). Earlier measurements by Humphrey and O'Brien (1953) yielded $\Delta_f H^o$ = -(580.76 $\pm$ 0.40) kJ·mol^{-1}. See Gurvich et al. (1979), part 1, page 305, for a review of the many equilibrium measurements on SnO_2(cr), which yield $\Delta_f H^o$ values from -(575 $\pm$ 3) kJ·mol^{-1} to -(583 $\pm$ 1.5) kJ·mol^{-1}.

The entropy and the enthalpy increment were calculated from the heat capacity measurements by Zhogin et al. (1980), (6-300 K) and Bachmann et al. (1981), (4-12 K).

73. Pb(cr).   Defined reference state.

The entropy and the enthalpy increment were calculated for the range 1-15 K from the heat-capacity measurements listed by Hultgren et al. (1973), for the range 15-170 K from data of Meads et al. (1941), (14-300 K), and for the range 170-298.15 K from data of Meads et al. (1941), Bronson and Wilson (1936), (203-393 K), Leadbetter (1968), (304-587 K) and Douglas and Dever (1954), (enthalpy increment above 273 K).

74. Pb(g). $\Delta_f H^o$ was calculated from vapor-pressure data reported by Hawkins (1966), (195.1 $\pm$ 0.4) kJ·mol^{-1}, Kim and Cosgarea (1966), (195.0 $\pm$ 0.4) kJ·mol^{-1}, Schins et al. (1971), (195.7 $\pm$ 0.4) kJ·mol^{-1}, Andon et al. (1971), (195.5 $\pm$ 0.8) kJ·mol^{-1}, Bohdansky and Schins (1967), (194.9 $\pm$ 1.0) kJ·mol^{-1}, and Shiu and Munir (1971b), (195.25 $\pm$ 0.40) kJ·mol^{-1}. There is considerable uncertainty in the thermal functions for Pb(l) above 1200 K. The entropy and the enthalpy increment were calculated taking the atomic energy levels $^3P_1$ and $^3P_2$ equivalent to 7819.2626 and 10650.3271 cm^{-1}, respectively, as recommended by Moore (1970).

75. Pb^{2+}(aq). $\Delta_f H^o$ was calculated from the measurements of $\Delta_{sol} H^o$ of Pb(cr) and PbO(cr, yellow) in aqueous ($HClO_4$ + H_2O_2) mixtures by Vasil'ev et al. (1971), (0.92 $\pm$ 0.25) kJ·mol^{-1}. Cell measurements by Vasil'ev and Glavina (1971) on the Pb/Pb^{2+}(aq) electrode in aqueous $HClO_4$ yielded for the reaction Pb(cr) + 2 H$^+$ (aq) = Pb^{2+}(aq) + H_2(g), $\Delta_f G^o$ = -(23.87 $\pm$ 0.15) kJ·mol^{-1}, $\Delta_f S^o$ = (84 $\pm$ 5) J·K^{-1}·mol^{-1}. From these data the value $\Delta_f H^o$ = (1.2 $\pm$ 1.5) kJ·mol^{-1} was derived. The value $\Delta_f H^o$ = (0.92 $\pm$ 0.25) kJ·mol^{-1} was selected.

Measurements were made on the cell Pb(Hg)|$PbCl_2$(aq)|AgCl(cr)|Ag(cr) by Garrels and Gucker (1949), Hannan (1936), and Carmody (1929). Taking $\Delta_f G^o$ for Pb(Hg) = -1.130 kJ·mol^{-1} from Carmody (1929) and $\Delta_f G^o$ for (AgCl, cr) = -(109.785 $\pm$ 0.020) kJ·mol^{-1}, (see Note 111), we calculated for Pb^{2+}(aq), $\Delta_f G^o$ = -(24.39 $\pm$ 0.20), -(24.48 $\pm$ 0.15), and -(24.53 $\pm$ 0.20) kJ·mol^{-1}, respectively. The data of Vasil'ev and Glavina (1971) on the Pb electrode in aqueous $HClO_4$ yielded $\Delta_f G^o$ = -(23.87 $\pm$ 0.15) kJ·mol^{-1}. The thermodynamic solubility product for $PbSO_4$(cr), (1.60 x 10^{-8} mol^2·kg^{-2}) has been calculated from the measurements by Cowperthwaite and LaMer (1931) and Singh (1955), from which we calculated $\Delta_{sol} G^o$ = (44.50 $\pm$ 0.30) kJ·mol^{-1}. Taking $\Delta_f G^o$ for $PbSO_4$(cr) = -(813.10 $\pm$ 0.20) kJ·mol^{-1}, (see Note 76), we calculated $\Delta_f G^o$(Pb^{2+}, aq) = -(24.49 $\pm$ 0.40) kJ·mol^{-1}.

We have selected $\Delta_f G^o = -(24.20 \pm 0.30)$ kJ·mol^{-1}. S^o was calculated from the selected values of $\Delta_f H^o$ and $\Delta_f G^o$.

76. PbSO$_4$(cr). $\Delta_r G^o$ for the reaction Pb(Hg) + Hg$_2$SO$_4$(cr) = PbSO$_4$(cr) + 2 Hg(l) was measured by Henderson and Stegeman (1918), -(186.206 $\pm$ 0.050) kJ·mol^{-1}, Mellon and Henderson (1920), -(186.212 $\pm$ 0.050) kJ·mcl^{-1}, and Harned and Hamer (1935), -(186.206 $\pm$ 0.050) kJ·mol^{-1}; see Hamer (1972). Taking $\Delta_f G^o$ for Pb(Hg) = -1.130 kJ·mol^{-1} from Carmody (1929) and $\Delta_f G^o$ for Hg$_2$SO$_4$(cr) = -(625.85 $\pm$ 0.15) kJ·mol^{-1}, (see Note 103), we calculated for PbSO$_4$(cr), $\Delta_f G^o$ = -(813.186 $\pm$ 0.200), -(813.192 $\pm$ 0.200), and -(813.186 $\pm$ 0.200) kJ·mol^{-1}, respectively. Henderson and Stegeman (1918) also measured the reaction Pb(cr) + Hg$_2$SO$_4$(cr) = PbSO$_4$(cr) + 2 Hg(l), for which $\Delta_r G^o$ = -(187.190 $\pm$ 0.040) kJ·mol^{-1} and $\Delta_f G^o$ (PbSO$_4$,cr) = -(813.040 $\pm$ 0.15) kJ·mol^{-1}. $\Delta_r G^o$ for the reaction Pb(Hg) + H$_2$SO$_4$(aq) = PbSO$_4$(cr) + H$_2$(g) was measured by Shrawder and Cowperthwaite (1934) (re-extrapolated by Pitzer et al. (1977)), -(68.080 $\pm$ 0.040) kJ·mol^{-1}) and by Lilley and Briggs (1976), -(68.040 $\pm$ 0.060) kJ·mol^{-1}, from which $\Delta_f G^o$ = -(813.32 $\pm$ 0.15) and -(813.28 $\pm$ 0.15) kJ·mol^{-1}, respectively. The measurements by Vosburgh and Craig (1929) on the reaction Pb(cr) + PbO$_2$(cr) + 2 H$_2$SO$_4$(aq) = 2 PbSO$_4$(cr) + 2 H$_2$O(l), $\Delta_r G^o$ = -(395.108 $\pm$ 0.050) kJ·mol^{-1}, were combined with the data of Covington et al. (1965a) on the reaction H$_2$(g) + H$_2$SO$_4$(aq) + PbO$_2$(cr) = PbSO$_4$(cr) + 2 H$_2$O(l), $\Delta_r G^o$ = -(326.243 $\pm$ 0.050) kJ·mol^{-1}, to obtain $\Delta_f G^o$ PbSO$_4$(cr) = -(812.98 $\pm$ 0.10) kJ·mol^{-1}. The value $\Delta_f G^o$ = -(813.10 $\pm$ 0.20) kJ·mol^{-1} was selected.

The entropy and the enthalpy increment were calculated from the heat capacity data from Gallagher et al. (1960), (12-324 K). The entropy contribution below 12 K =1.50 J·K^{-1}·mol^{-1}. $\Delta_f H^o$ was calculated from the selected values of $\Delta_f G^o$ and S^o.

77. B(cr). Defined reference state.
The entropy and the enthalpy increment were calculated from the data of Bogdanov et al. (1970), (16-280 K) below 50 K, and from 50 K to 298.15 K from the data of Bogdanov et al. and Johnston et al. (1951), (17-308 K).

78. B(g). $\Delta_f H^o$ was calculated from the vapor-pressure data obtained by Akishin et al. (1959), (573 $\pm$ 4) kJ·mol^{-1}, Paule and Margrave (1963), (570 $\pm$ 3) kJ·mol^{-1}, Robson and Gilles (1964), (562 $\pm$ 3) kJ·mol^{-1}, Hildenbrand and Hall (1964), (564 $\pm$ 2) kJ·mol^{-1}, Mar and Bedford (1976), (561 $\pm$ 3) kJ·mol^{-1}, and Storms and Mueller (1977), (575 $\pm$ 3) kJ·mol^{-1}. These measurements were selected from the many others as being the most self-consistent. See the bibliography and discussion in Gurvich et al. (1981), part 1, page 10. The uncertainty takes into account uncertainties in the thermal functions of B(cr) and B(l).

The entropy and the enthalpy increment were calculated taking the energy of sublevel $^2P_{\frac{3}{2}}$ equivalent to 15.25 cm^{-1}, as recommended by Edlen et al. (1970). The uncertainty in S^o(298.15 K) takes into account the variability of the isotopic composition of boron.

79. B$_2$O$_3$(cr). $\Delta_f H^o$ of B$_2$O$_3$(cr), H$_3$BO$_3$(cr), H$_3$BO$_3$(aq, undissoc.), and BF$_3$(g) were obtained from a simultaneous evaluation by least squares of the following reactions.

- B(cr) + 1.5 F$_2$(g) = BF$_3$(g), measured by Gross et al. (1967), -(1134.7 ± 2.0) kJ·mol^{-1}, Domalski and Armstrong (1967), -(1133.9 ± 2.1) kJ·mol^{-1}, Gross et al. (1969), -(1136.2 ± 0.8) kJ·mol^{-1}, and Johnson et al. (1966), -(1136.58 ± 0.90) kJ·mol^{-1}.

- B$_2$O$_3$(cr) + 3 F$_2$(g) = 2 BF$_3$(g) + 1.5 O$_2$(g), measured by Johnson and Hubbard (1969), -(999.22 ± 1.60) kJ·mol^{-1}.

- B(cr) + 0.75 O$_2$(g) + 1.5 H$_2$O(l) = H$_3$BO$_3$(cr), from the measurements made by Good and Månsson (1966), -(665.24 ± 0.80) kJ·mol^{-1}.

- B(am) + 1.5 Cl$_2$(g) = BCl$_3$(l), measured by Johnson et al. (1959), -(429.7 ± 1.5) kJ·mol^{-1}, and Gal'chenko et al. (1960a), -(430.2 ± 2.0) kJ·mol^{-1}.

- B$_2$H$_6$(g) = 2 B(am) + 3 H$_2$(g), from Prosen et al. (1958), -(28.2 ± 2.0) kJ·mol^{-1}.

- B$_2$H$_6$(g) + 6 H$_2$O(l) = 2 H$_3$BO$_3$(aq, undissoc.) + 6 H$_2$(g), from Prosen et al. (1959), -(466.3 ± 2.3) kJ·mol^{-1}, Gunn and Green (1960), (corrected in Gunn (1965)), -(468.36 ± 0.60) kJ·mol^{-1}), and Skinner et al. (1961), -(466.5 ± 1.8) kJ·mol^{-1}.

- BCl$_3$(l) + 3 H$_2$O(l) = H$_3$BO$_3$(aq, undissoc.) + 3 HCl(aq), from Gunn and Green (1960), (corrected in Gunn (1965)), -(286.60 ± 0.40) kJ·mol^{-1}, Skinner and Smith (1953), -(287.0 ± 1.0) kJ·mol^{-1}, and Fasolino (1965), -(288.0 ± 2.0) kJ·mol^{-1}.

- B$_2$O$_3$(cr) + 3 H$_2$O(l) = 2 H$_3$BO$_3$(aq, undissoc.), from Johnson and Hubbard (1969), -(14.71 ± 0.10) kJ·mol^{-1}, Fasolino (1965), -(14.24 ± 0.20) kJ·mol^{-1}, and Southard (1941), -(14.49 ± 0.20) kJ·mol^{-1}.

- H$_3$BO$_3$(cr) = H$_3$BO$_3$(aq, undissoc.), from Devina et al. (1982a), (22.00 ± 0.02) kJ·mol^{-1}, Smisko and Mason (1950), (22.08 ± 0.05) kJ·mol^{-1}, Kilday and Prosen (1964), (22.03 ± 0.18) kJ·mol^{-1}, Fasolino (1965), (22.83 ± 0.05) kJ·mol^{-1}, and Van Artsdalen and Anderson (1951), (21.61 ± 0.15) kJ·mol^{-1}.

- B(am) + 0.5 N$_2$(g) = BN(cr), measured by Hildenbrand and Hall (1963), -(252.09 ± 0.40) kJ·mol^{-1}, and Gal'chenko et al. (1960b), -(254.0 ± 3.0) kJ·mol^{-1}.

- BN(cr) + 1.5 F$_2$(g) = BF$_3$(g) + 0.5 N$_2$(g), measured by Wise et al. (1966), -(885.6 ± 1.2) kJ·mol^{-1}.

- Dilution data for H$_3$BO$_3$(aq) were taken from Lange and Miederer (1957a) and Ward and Millero (1973).

- The entropy and the enthalpy increment of B$_2$O$_3$(cr) were calculated from the heat-capacity measurements reported by Kerr et al. (1950), (18-297 K), and Kelley (1941), (52-296 K). The contribution to the entropy below 18 K was taken as 0.13 J·K^{-1}·mol^{-1}.

80. H$_3$BO$_3$(cr).   For $\Delta_f H°$ see Note 79.

The entropy and the enthalpy increment were calculated from the heat capacity measurements reported by Oguni et al. (1977), (12-374 K) and Johnston and Kerr (1950), (17-301 K). The entropy at 12 K (0.91 J·K^{-1}·mol^{-1}) includes a zero-point contribution of (0.56 ± 0.30) J·K^{-1}·mol^{-1}. See Oguni et al. (1977).

81. H$_3$BO$_3$(aq, undissoc.).   For $\Delta_f H°$ see Note 79.

$\Delta_{sol} G°$ for H$_3$BO$_3$(cr) = (0.392 ± 0.020) kJ·mol^{-1} was calculated from the solubility data reviewed by Linke (1958), (0.928 M), (see also Nies and Hulbert (1967), Platford (1969), and Devina et al. (1982a)), and the activity coefficient, (0.92), from Platford (1969). See Note 79 for $\Delta_{sol} H°$(H$_3$BO$_3$, cr), (22.00 ± 0.05) kJ·mol^{-1}, and see Note 80 for S^o of H_3BO_3 (cr). All these values were then combined to calculate $S^o(H_3BO_3$, aq, undissoc.).

82. $BF_3(g)$. For $\Delta_f H^o$ see Note 79.

The entropy and the enthalpy increment were calculated using the rigid-rotator, harmonic-oscillator approximation, with molecular constants obtained from the infrared and Raman spectral data of Ginn et al. (1968a, 1968b, 1970), Yost et al. (1938), Brown and Overend (1969), Konaka et al. (1966), and Kuchitsu and Konaka (1966) for $^{10}BF_3$ and $^{11}BF_3$. A compatible value, S^o(298.15 K) = (254.4 ± 2.0) J·K^{-1}·mol^{-1}, was obtained from the heat-capacity and phase-transition data of Eucken and Schröder (1938). See Spencer (1946).

83. $Al(cr)$. Defined reference state.

The entropy and the enthalpy increment were calculated from the heat-capacity measurements made by Downie and Martin (1980), (7-294 K), Giauque and Meads (1941), (15-300 K), and Berg (1968), (3-20 K).

84. $Al(g)$. $\Delta_f H^o$ was calculated from vapor-pressure measurements by Brewer and Searcy (1951), (329.2 ± 1.7) kJ·mol^{-1}, Johnson et al. (1956), (327 ± 2) kJ·mol^{-1}, Potter et al. (1970), (332 ± 2) kJ·mol^{-1}, and Rao and Motzfeldt (1970), 333 kJ·mol^{-1}.

The entropy and the enthalpy increment were calculated taking the energy of sublevel $^2P_{\frac{3}{2}}$ equivalent to 112.061 cm^{-1}, as recommended by Moore (1970).

85. $Al^{3+}(aq)$. $\Delta_f H^o$ for $AlCl_3(cr)$ was calculated from the measurements by Coughlin (1958) on the enthalpy of solution of $Al(cr)$ and $AlCl_3(cr)$ in aqueous HCl, $\Delta_f H^o$ = -(705.0 ± 0.6) kJ·mol^{-1}, and by Gross and Hayman (1970) on the reaction of $Al(cr)$ with $Cl_2(g)$, $\Delta_f H^o$ = -(706.3 ± 0.7) kJ·mol^{-1}. The value $\Delta_f H^o$ = -(705.6 ± 1.0) kJ·mol^{-1} was selected.

Measurements by Coughlin (1958) on $\Delta_{sol} H^o$ of $AlCl_3(cr)$ and $AlCl_3·6H_2O(cr)$ in aqueous HCl yielded for $AlCl_3·6H_2O(cr)$, $\Delta_f H^o$ = -(2693.1 ± 1.5) kJ·mol^{-1}. Smith and Bass (1963) measured the enthalpy of solution of $AlCl_3·6H_2O(cr)$ in various concentrations of aqueous HCl. Extrapolation to pure $H_2O(l)$ yielded for the process $AlCl_3·6H_2O(cr)$ + 224 $H_2O(l)$ = $AlCl_3$(aq, in 230H_2O), $\Delta_r H$ = -(55.10 ± 0.25) kJ·mol^{-1}. Dilution data were taken from Lange and Miederer (1957b). Correction for the hydrolysis reaction $Al^{3+}(aq)$ + $H_2O(l)$ = $AlOH^{2+}(aq)$ + $H^+(aq)$ was made using $\Delta_r G^o$ = (30.0 ± 0.20) kJ·mol^{-1}, (Frink and Peech (1962, 1963)), and $\Delta_r H^o$ = (44.7 ± 2.5) kJ·mol^{-1}, calculated from the data of Lange and Miederer (1957b). Thus for the reaction $AlCl_3·6H_2O(cr)$ = $Al^{3+}(aq)$ + 3 $Cl^-(aq)$ + 6$H_2O(l)$, $\Delta_r H^o$ = -(61.4 ± 0.9) kJ·mol^{-1}, and $\Delta_f H^o$ (Al^{3+}, aq) = -(538.3 ± 2.0) kJ·mol^{-1}.

Measurements by Krivtsov et al. (1971) on the enthalpy of solution of $AlCl_3(cr)$ in water were corrected as described above and yielded for the reaction $AlCl_3(cr)$ = $Al^{3+}(aq)$ + 3 $Cl^-(aq)$, $\Delta_r H^o$ = -(334.1 ± 0.7) kJ·mol^{-1}, and $\Delta_f H^o$ (Al^{3+}, aq) = -(538.5 ± 1.3) kJ·mol^{-1}.

$\Delta_{sol} G^o$ of $AlCl_3·6H_2O(cr)$, -(44.8 ± 1.0) kJ·mol^{-1}, was calculated from the solubility (3.382 M), (Brown et al. (1979)), the activity coefficient of the solute ($\gamma_\pm$ = 46.4 ± 5.0) and the activity of H_2O (a_w = 0.404 ± 0.010); see Mason (1938, 1941), and Fricke and Havestadt (1927). Combination with the selected value of $\Delta_{sol} H^o$

(see above) and the value of $S^o(\mathrm{AlCl_3 \cdot 6H_2O}$, cr, 298.15 K) measured by Stull et al. (1970), (318.1 $\pm$ 2.5) $\mathrm{J \cdot K^{-1} \cdot mol^{-1}}$, yielded $S^o(\mathrm{Al^{3+}}$, aq) = -(327 $\pm$ 6) $\mathrm{J \cdot K^{-1} \cdot mol^{-1}}$.

The measurements by Hemingway and Robie (1977) on the enthalpy of solution of Al(cr) and $\mathrm{Al(OH)_3}$(cr, gibbsite) in aqueous HF yielded $\Delta_f H^o(\mathrm{Al(OH)_3}$, cr, gibbsite) = -(1293.1 $\pm$ 1.2) kJ. $\mathrm{mol^{-1}}$. Measurements by Gross et al. (1970) on $\Delta_{sol} H^o$ of $\mathrm{AlCl_3}$(cr) and $\mathrm{Al(OH)_3}$(cr, gibbsite) in aqueous HCl yielded $\Delta_f H^o(\mathrm{Al(OH)_3}$, cr, gibbsite) = -(1293.5 $\pm$ 1.5) $\mathrm{kJ \cdot mol^{-1}}$. The value $\Delta_f H^o$ = -(1293.3 $\pm$ 1.0) $\mathrm{kJ \cdot mol^{-1}}$ was selected.

$\Delta_{sol} G^o$ of gibbsite, (194.1 $\pm$ 2.5) $\mathrm{kJ \cdot mol^{-1}}$, was calculated from the solubility data reviewed by Parks (1972). $\Delta_{sol} H^o$ of gibbsite, (64.9 $\pm$ 3.0) $\mathrm{kJ \cdot mol^{-1}}$, calculated from $\Delta_f H^o$ selected above and other appropriate data, was combined with $\Delta_{sol} G^o$ obtained above and with $S^o(\mathrm{Al(OH)_3}$, cr, gibbsite, 298.15 K), measured by Hemingway et al. (1977), (68.44 $\pm$ 0.17) $\mathrm{J \cdot K^{-1} \cdot mol^{-1}}$, to yield $S^o(\mathrm{Al^{3+}}$, aq) = -(332 $\pm$ 7) $\mathrm{J \cdot K^{-1} \cdot mol^{-1}}$.

$S^o(\mathrm{Al^{3+}}$, aq) was also calculated from measurements by Latimer and Greensfelder (1928) of S^o for $\mathrm{CsAl(SO_4)_2 \cdot 12H_2O}$(cr), (686.1 $\pm$ 2.5) $\mathrm{J \cdot K^{-1} \cdot mol^{-1}}$, and values for $\Delta_{sol} H^o$(56.74 $\pm$ 0.13) $\mathrm{kJ \cdot mol^{-1}}$, the solubility (0.01403 M), and the activity coefficient at saturation (0.217 $\pm$ 0.010) of $\mathrm{CsAl(SO_4)_2 \cdot 12H_2O}$, from which $S^o(\mathrm{Al^{3+}}$, aq) = -(313 $\pm$ 6) $\mathrm{J \cdot K^{-1} \cdot mol^{-1}}$.

86. $\mathrm{Al_2O_3}$(cr, corundum). $\Delta_f H^o$ was computed from the enthalpy of combustion measurements of Al(cr) in $\mathrm{O_2}$(g) by Mah (1957), -(1675.7 $\pm$ 1.0) $\mathrm{kJ \cdot mol^{-1}}$, and by Holley and Huber (1951), -(1675.7 $\pm$ 1.3) kJ. $\mathrm{mol^{-1}}$. Measurements have also been reported by Schneider and Gattow (1954), Snyder and Seltz (1945), and Roth et al. (1940). The selected value is confirmed by dehydration and solution studies on gibbsite, $\mathrm{Al(OH)_3}$(cr), and diaspore, $\mathrm{AlO(OH)}$(cr); see review by Khodakovsky et al. (1980).

The entropy and the enthalpy increment were calculated from the heat-capacity measurements by Furukawa et al. (1956), (13-300 K) and Fugate and Swenson (1969), (2-25 K).

87. $\mathrm{AlF_3}$(cr). $\Delta_f H^o$ was derived from the enthalpy of fluorination of Al(cr), as measured by Rudzitis et al. (1967), -(1510.4 $\pm$ 1.3) $\mathrm{kJ \cdot mol^{-1}}$. The measurements by Domalski and Armstrong (1965) yielded $\Delta_f H^o$ = -(1507.9 $\pm$ 6.7) $\mathrm{kJ \cdot mol^{-1}}$. See Gurvich et al. (1981), part 1, page 126, for a review of other measurements.

The entropy and the enthalpy increment were calculated from the heat-capacity measurements by King (1957), (54-296 K). The entropy increment for 0-54 K was taken as (2.7 $\pm$ 0.3) $\mathrm{J \cdot K^{-1} \cdot mol^{-1}}$.

88. Zn(cr). Defined reference state.

The entropy and the enthalpy increment were calculated from measurements of the heat capacity by Cetas et al. (1969), (1-25 K), Griffiths and Griffiths (1914b), (120-390 K), Clusius and Harteck (1928), (12-202 K), Bronson and Wilson (1936), (193-393 K), and Eichenauer and Schulze (1959), (12-273 K).

89. Zn(g). $\Delta_f H^o$ was calculated from the vapor-pressure data of Rodebush and Dixon (1925a), (130.54 $\pm$ 0.30) $\mathrm{kJ \cdot mol^{-1}}$, Jenkins (1926), (130.34 $\pm$ 0.60) $\mathrm{kJ \cdot mol^{-1}}$, Barrow et al. (1955), (130.43 $\pm$ 0.50) $\mathrm{kJ \cdot mol^{-1}}$, Tsvetkov and Edel'stein (1963), (130.30 $\pm$ 0.10) $\mathrm{kJ \cdot mol^{-1}}$, Aldred and Platt (1963), (129.80 $\pm$ 0.30 $\mathrm{kJ \cdot mol^{-1}}$), Cordes and Cammenga (1965), (129.81 ± 0.50 kJ·mol^{-1}), Chernyaev and Ershova (1964), (130.20 ± 0.10) kJ·mol^{-1}, Baker (1966), (130.02 ± 0.40) kJ· mol^{-1}, Piacente and De Maria (1969), (130.30 ± 0.20) kJ·mol^{-1}, Mar and Searcy (1970), (130.93 ± 0.50) kJ·mol^{-1}), Shiu and Munir (1971b), (130.83 ± 0.25) kJ·mol^{-1}, and McCreary and Thorn (1969), (130.42 ± 0.80) kJ·mol^{-1}.

The entropy and the enthalpy increment were calculated taking the electronic partition function equal to 1.

90. Zn^{2+}(aq). $\Delta_f H^o$ was calculated from measurements of the enthalpy of solution of ZnO(cr), (see Note 91), in aqueous HClO$_4$ by Berg and Vanderzee (1975). See also Davies and Staveley (1972).

$\Delta_f G^o$ was calculated from e.m.f. measurements made by Bates (1938), -(147.27 ± 0.05) kJ·mol^{-1}, Corsaro and Stephans (1957), -(146.94 ± 0.12) kJ·mol^{-1}, Scatchard and Tefft (1930), -(147.20 ± 0.05) kJ·mol^{-1}, Stokes and Stokes (1945), -(147.24 ± 0.03) kJ·mol^{-1}, and Lutfullah et al. (1976), -(147.03 ± 0.05) kJ·mol^{-1}. The value $\Delta_f G^o$ = -(147.20 ± 0.05) kJ·mol^{-1} was selected. S^o was then calculated from the selected values of $\Delta_f H^o$ and $\Delta_f G^o$. The measurements by Barieau and Giauque (1950) and Giauque et al. (1950) on ZnSO$_4$.7 H$_2$O(cr) yield S^o = -(110.0 ± 1.5) J·K^{-1}·mol^{-1}; the cell data of Stokes and Stokes (1945) yield S^o = -(109.6 ± 1.0) J·K^{-1}·mol^{-1}.

91. ZnO(cr). $\Delta_f H^o$ was calculated from the measurements by Vorob'ev and Broier (1971a) of the enthalpy of solution of Zn(cr) in aqueous HCl combined with the measurements by Berg and Vanderzee (1975) on the solution of ZnO(cr) in aqueous HCl of the same concentration. The measurements by Peppler and Newman (1951) yield $\Delta_f H^o$ = -(350.70 ± 0.40) kJ·mol^{-1}. The combustion measurements of Zn(cr) in O$_2$(g) by Becker and Roth (1933) yield $\Delta_f H^o$ = -(347.8 ± 4.0) kJ·mol^{-1}. The thermal decomposition equilibrium of ZnO(cr) was studied by Hirschwald et al. (1964) and Anthrop and Searcy (1964), yielding $\Delta_f H^o$ = -(350.2 ± 1.0) and -(348.6 ± 1.5) kJ·mol^{-1}, respectively. See also Maier (1930), Shchukarev et al. (1953), Maier et al. (1926), and Kitchener and Ignatowicz (1951).

The entropy and the enthalpy increment were calculated from the heat-capacity measurements made by Millar (1928), (89-298 K), Maier et al. (1926), (88-295 K), and Clusius and Harteck (1928), (30-200 K). The entropy contribution (0 - 30 K) = 0.70 J·K^{-1}·mol^{-1}.

92. Cd(cr).   Defined reference state.

The entropy and the enthalpy increment were calculated from the heat-capacity measurements reported by Cetas et al. (1969), (1-30 K), Craig et al. (1954), (12 - 320 K), Bronson and Wilson (1936), (193-393 K), and Griffiths and Griffiths (1914a), (108-371 K).

93. Cd(g). $\Delta_f H^o$ was calculated from the vapor-pressure data reported by Piacente and De Maria (1969), (111.84 ± 0.30) kJ·mol^{-1}, McCreary and Thorn (1969), (111.29 ± 0.20) kJ·mol^{-1}, Bonderman et al. (1970), (112.13 ± 0.40) kJ·mol^{-1}, Malaspina et al. (1971), (111.59 ± 0.10) kJ·mol^{-1}, Paule and Mandel (1971), (111.55 ± 0.60) kJ·mol^{-1}, Muradova and Muradov (1972), (111.62 ± 0.10) kJ·mol^{-1}, and Schuffenecker et al. (1970), (111.88 ± 0.08) kJ·mol^{-1}. Malespina et al. also reported calorimetric measurements of the enthalpy of sublimation at temperatures between 498 K and 590 K; correction to 298.15 K yielded $\Delta_f H^o = (111.990 \pm 0.050)$ kJ·mol^{-1}.

The entropy and the enthalpy increment were calculated taking the electronic partition function equal to 1.

94. Cd^{2+}(aq). $\Delta_f G^o$ was calculated from the e.m.f.s of the following types of cells containing Cd/Cd^{2+} electrodes.

- Measurements versus the Ag|AgCl|Cl$^-$(aq) electrode by Harned and Fitzgerald (1936), Treumann and Ferris (1958), and Quintin (1936) yielded $\Delta_f G^o = $ -(77.58 $\pm$ 0.05), -(77.67 $\pm$ 0.10) and -(77.72 $\pm$ 0.15) kJ·mol^{-1}, respectively. Measurements versus Hg/Hg(I) acetate (aq) by Choudhary and Prasad (1975) gave $\Delta_f G^o = $ -(77.59 $\pm$ 0.02) kJ·mol^{-1}.
- Measurements by Priepke and Vosburgh (1930) and Taylor and Perrott (1921) yielded for CdCl$_2\cdot\frac{5}{2}$H$_2$O(cr), $\Delta_f G^o = $ -(944.10 $\pm$ 0.05) kJ·mol^{-1}. From the solubility given by Linke (1958), (6.574 M) and activity data from Robinson (1940) and Ishikawa et al. (1932), ($\gamma_{\pm} = 0.0257$; $a_w = 0.827$), we obtained for Cd^{2+}(aq), $\Delta_f G^o = $ -(77.65 $\pm$ 0.10) kJ·mol^{-1}.
- Measurements against the Ag|AgBr|Br$^-$(aq)| electrode by Bates (1939) yielded $\Delta_f G^o = $ -(77.60 $\pm$ 0.10) kJ·mol^{-1}.
- Measurements by Ishikawa and Ueda (1933) and Ishikawa and Takai (1937), using the Hg|Hg$_2$Br$_2$|Br$^-$(aq) electrode, yielded for CdBr$_2\cdot$4H$_2$O(cr), $\Delta_f G^o = $ -(1248.10 $\pm$ 0.40) kJ·mol^{-1}. From their solubility data, (4.125 M), and activity data from Robinson (1940), ($\gamma_{\pm} = 0.0276$; $a_w = 0.890$), we calculated $\Delta_f G^o = $ -(77.51 $\pm$ 0.40) kJ·mol^{-1}.
- Measurements by LaMer and Parks (1931) , using a Pb(Hg) + PbSO$_4$ electrode, gave $\Delta_f G^o = $ -(78.01 $\pm$ 0.30) kJ·mol^{-1}.
- From data obtained by Hamer (1972), using a Hg|Hg$_2$SO$_4$|SO$_4^{2-}$(aq) electrode, a value for $\Delta_f G^o$(CdSO$_4\cdot\frac{8}{3}$H$_2$O, cr) equal to -(1465.38 $\pm$ 0.50) kJ·mol^{-1} was calculated. Using solubility data from Brickwedde (1946), (3.679 M), and activity data from Pitzer and Mayorga (1974) and Ishikawa and Murooka (1933), ($\gamma_{\pm} = 0.0364$; $a_w = 0.891$), we calculated $\Delta_{sol} G^o = $ (10.73 $\pm$ 0.15) kJ·mol^{-1}, and $\Delta_f G^o$ (Cd^{2+}, aq) = -(78.03 $\pm$ 0.60) kJ·mol^{-1}.
- Similar measurements by Vinal and Brickwedde (1941) yielded $\Delta_f G^o$ (CdSO$_4\cdot$H$_2$O, cr) = -(1069.02 $\pm$ 0.50) kJ·mol^{-1}. Using their values for the solubility, (4.03 M), and data from Pitzer and Mayorga (1974), ($\gamma_{\pm} = 0.0395$; $a_w = 0.811$), we calculated $\Delta_{sol} G^o$ as (9.630 $\pm$ 0.100) kJ·mol^{-1}, giving $\Delta_f G^o$(Cd^{2+}, aq) = -(78.09 $\pm$ 0.60) kJ·mol^{-1}.

The final selected value of $\Delta_f G^o$ for Cd^{2+}(aq) is -(77.75 $\pm$ 0.15) kJ·mol^{-1}.

The entropy of the ion was calculated from the following data sets.

- S^o(298.15 K) of CdBr$_2$(cr), (138.82 $\pm$ 0.30) J·K^{-1}·mol^{-1}, was measured by Itskevich and Strelkov (1960) and $\Delta_{sol} H^o$, -(2.81 $\pm$ 0.20) kJ·mol^{-1} by Ishikawa and Ueda (1930). $\Delta_{sol} G^o$ (10.76 $\pm$ 0.10) kJ·mol^{-1} was calculated from the data of Ishikawa and Takai (1937), (see above), and the decomposition pressure measurements on the tetrahydrate by Ishikawa and Ueda (1930). From these values we derived $S^o = $ -(71.8 $\pm$ 1.0) J·K^{-1}·mol^{-1}.
- Measurements of S^o(298.15 K) of CdSO$_4\cdot\frac{8}{3}$H$_2$O(cr), (229.65 $\pm$ 0.40) J·K^{-1}·mol^{-1} (see Note 96), and of $\Delta_{sol} H^o$, -(18.35 $\pm$ 0.10) kJ·mol^{-1}, by Papadopoulos and Giauque (1955), and of $\Delta_{sol} G^o$, (10.73 $\pm$ 0.15) kJ·mol^{-1}, (see above), yielded $S^o = $ -(72.92 $\pm$ 0.50) J·K^{-1}·mol^{-1}. See also Larson et al. (1968).

- Measurements of S^o(298.15 K) of $CdSO_4 \cdot H_2O$(cr), (154.03 ± 0.40) $J \cdot K^{-1} \cdot mol^{-1}$ and $\Delta_{sol} H^o$ -(31.73 ± 0.10) $kJ \cdot mol^{-1}$ by Papadopoulos and Giauque (1955), and of $\Delta_{sol} G^o$ (9.630 ± 0.100) $kJ \cdot mol^{-1}$, (see above), yielded S^o = -(73.14 ± 0.60) $J \cdot K^{-1} \cdot mol^{-1}$.

- Combination of $\Delta_{sol} G^o$ for $CdSO_4 \cdot H_2O$(cr), (see above), with $\Delta_r G^o$ for decomposition of the monohydrate to $CdSO_4$(cr), (9.038 ± 0.100) $kJ \cdot mol^{-1}$, from Ishikawa and Murooka (1930) yielded for $CdSO_4$(cr), $\Delta_{sol} G^o$ = 0.59 $kJ \cdot mol^{-1}$. Calculation of $\Delta_r G^o$ for the decomposition of the monohydrate from $\Delta_r H^o$, (20.42 ± 0.15) $kJ \cdot mol^{-1}$, and the appropriate entropy values, (see above), yielded $\Delta_r G^o$ = (8.81 ± 0.20) $kJ \cdot mol^{-1}$ and $\Delta_{sol} G^o$ = 0.83 $kJ \cdot mol^{-1}$. We have selected $\Delta_{sol} G^o$ = (0.73 ± 0.30) $kJ \cdot mol^{-1}$. Papadopoulos and Giauque (1955) have measured S^o for $CdSO_4$(cr), (123.04 ± 0.40) $J \cdot K^{-1} \cdot mol^{-1}$, and $\Delta_{sol} H^o$, -(52.15 ± 0.15) $kJ \cdot mol^{-1}$. From these we calculated S^o for Cd^{+2}(aq) = -(72.8 ± 1.0) $J \cdot K^{-1} \cdot mol^{-1}$.

The selected value of S^o, -(72.8 ± 1.5) $J \cdot K^{-1} \cdot mol^{-1}$, was combined with the selected value of $\Delta_f G^o$ to calculate $\Delta_f H^o$.

95. CdO(cr). $\Delta_f H^o$ was calculated from the measurements by Adami and King (1965) of the enthalpy of solution of CdO(cr) and $CdSO_4$(cr) in aqueous H_2SO_4, yielding $\Delta_f H^o$ = -(258.34 ± 0.40) $kJ \cdot mol^{-1}$, and from the measurements by Provost and Wulff (1970) of the enthalpy of solution of CdO(cr) in aqueous $HClO_4$, yielding $\Delta_f H^o$ = -(258.36 ± 0.60) $kJ \cdot mol^{-1}$.

The entropy and the enthalpy increment were calculated from the heat- capacity measurements by Millar (1928), (71-291 K). The entropy contribution below 71 K was taken as 9.0 $J \cdot K^{-1} \cdot mol^{-1}$.

96. $CdSO_4 \cdot \frac{8}{3}H_2O$(cr). A value of $\Delta_f H^o$ equal to -(1729.50 ± 0.50) $kJ \cdot mol^{-1}$ was calculated from the value of $\Delta_f G^o$ obtained by Hamer (1972), (see Note 94), and the appropriate entropy data. A value of $\Delta_f H^o$ equal to -(1729.12 ± 0.80) $kJ \cdot mol^{-1}$ was obtained from the values of $\Delta_{sol} H^o$, (see Note 94), and of $\Delta_f H^o$ of the ions and H_2O(l).

The entropy and the enthalpy increment were calculated from the heat-capacity measurements by Papadopoulos and Giauque (1955), (15-299 K). The contribution to the entropy below 15 K = 1.09 $J \cdot K^{-1} mol^{-1}$.

97. Hg(l). Defined reference state.

The entropy and the enthalpy increment were calculated from the heat- capacity data of Busey and Giauque (1953), (15-330 K), Smith and Wolcott (1956), (1-20 K), and Amitin et al. (1979), (5-300 K).

98. Hg(g). $\Delta_f H^o$ was calculated from the vapor-pressure measurements by Rodebush and Dixon (1925b), (61.40 ± 0.08) $kJ \cdot mol^{-1}$, Mayer (1931), (61.44 ± 0.20) $kJ \cdot mol^{-1}$, Filosofo et al. (1950), (61.30 ± 0.20) $kJ \cdot mol^{-1}$, Sugawara et al. (1962), (see Hubbard and Ross (1982)), (61.360 ± 0.080) $kJ \cdot mol^{-1}$, and Ambrose and Sprake (1972), (61.400 ± 0.040) $kJ \cdot mol^{-1}$. The data treated here were for pressures sufficiently low that the vapor could be treated as ideal; for data at higher pressures, where the vapor is no longer ideal, see the review by Hultgren et al. (1973).

The entropy and the enthalpy increment were calculated taking the electronic partition function equal to 1.

99. $Hg^{2+}(aq)$. $\Delta_f H^o$ was calculated from the measured enthalpy of solution of $HgO(cr)$, (see Note 101), in $HClO_4(aq)$ reported by Vanderzee et al. (1974), $\Delta_r H^o$ = -(24.83 ± 0.16) $kJ \cdot mol^{-1}$. $\Delta_f G^o$ was calculated from $\Delta_r G^o$ = -(11.10 ± 0.10) $kJ \cdot mol^{-1}$ for the reaction $Hg(l) + Hg^{2+}(aq) = Hg_2^{2+}(aq)$, as reported by Hietanen and Sillén (1956), (see Note 100). S^o was calculated from the selected values of $\Delta_f H^o$ and $\Delta_f G^o$.

100. $Hg_2^{2+}(aq)$. $\Delta_f H^o$ was calculated from the data reported by Vanderzee and Swanson (1974) for the reaction $Hg_2^{2+}(aq) + 2\,Cl^-(aq) = Hg_2Cl_2(cr)$, $\Delta_r H^o$ = -(98.08 ± 0.18) $kJ \cdot mol^{-1}$, $\Delta_r G^o$ = -(101.86 ± 0.10) $kJ \cdot mol^{-1}$. Taking $\Delta_f H^o$ and $\Delta_f G^o$ for $Hg_2Cl_2(cr)$ = -(265.37 ± 0.40) $kJ \cdot mol^{-1}$ and -(210.760 ± 0.020) $kJ \cdot mol^{-1}$, respectively, (see Note 102), we obtained for $Hg_2^{2+}(aq)$, $\Delta_f H^o$ = (166.87 ± 0.50) $kJ \cdot mol^{-1}$ and $\Delta_f G^o$ = (153.60 ± 0.20) $kJ \cdot mol^{-1}$. From the measurements by Gerke and Geddes (1927) and Gupta et al. (1963b) on the $Hg_2Br_2 +$ Hg electrode we obtained for $Hg_2Br_2(cr)$, $\Delta_f G^o$ = -(180.88 ± 0.30) $kJ \cdot mol^{-1}$. From the solubility product data reported by Brodsky (1929) we have calculated for the reaction $Hg_2Br_2(cr) = Hg_2^{2+}(aq) + 2\,Br^-(aq)$, $\Delta_r G^o$ = (127.06 ± 0.50) $kJ \cdot mol^{-1}$, whence for $Hg_2^{2+}(aq)$, $\Delta_f G^o$ = (153.93 ± 0.70) $kJ \cdot mol^{-1}$. We have selected the value $\Delta_f G^o$ = (153.60 ± 0.20) $kJ \cdot mol^{-1}$. The value of S^o was calculated from the selected values of $\Delta_f H^o$ and $\Delta_f G^o$.

See Vanderzee and Swanson (1974) for a review of earlier measurements.

101. $HgO(cr, red)$. $\Delta_f G^o$ was derived from cell measurements on $HgO(cr, red)$ reported by Ishikawa and Kimura (1927), -(58.508 ± 0.080) $kJ \cdot mol^{-1}$, Shibata et al. (1931a), -(58.555 ± 0.080) $kJ \cdot mol^{-1}$, and Shibata and Murata (1931a), -(58.545 ± 0.080) $kJ \cdot mol^{-1}$. Values of $\Delta_f G^o$ were also derived from the cell measurements on $HgO(cr, yellow)$ reported by Chow (1920) and Fried (1926). These values were converted to those for $HgO(cr, red)$, yielding $\Delta_f G^o$ = -(58.489 ± 0.090) and -(58.570 ± 0.090) $kJ \cdot mol^{-1}$, respectively. For the transformation $HgO(cr, yellow) = HgO(cr, red)$, the value $\Delta_r G$ = -(0.125 ± 0.020) $kJ \cdot mol^{-1}$ was calculated from the data of Garrett and Hirschler (1938), Aurivillius and von Heidenstam (1961), and Ishikawa and Kimura (1927). The value $\Delta_f G^o$ = -(58.540 ± 0.050) $kJ \cdot mol^{-1}$ was selected.

The entropy and the enthalpy increment were calculated from the heat-capacity measurements by Bauer and Johnston (1953), (15-298 K). Values of S^o may also be calculated from the temperature coefficients of the e.m.f.'s measured by Shibata et al. (1931a), (70.7 ± 1.0) $J \cdot K^{-1} \cdot mol^{-1}$, and Shibata and Murata (1931a), (69.5 ± 1.0) $J \cdot K^{-1} \cdot mol^{-1}$. The value of $\Delta_f H^o$ was derived from $\Delta_f G^o$ and appropriate entropy values.

102. $Hg_2Cl_2(cr)$. $\Delta_f G^o$ was calculated from e.m.f. measurements on the $Hg(l) | Hg_2Cl_2 | Cl^-(aq)$ electrode by Ives and Prasad (1970), -(210.754 ± 0.020) $kJ \cdot mol^{-1}$, Gupta et al. (1963a), -(210.76 ± 0.02) $kJ \cdot mol^{-1}$, Grzybowski (1958a), -(210.76 ± 0.05) $kJ \cdot mol^{-1}$, Sharma et al. (1968), -(210.72 ± 0.05) $kJ \cdot mol^{-1}$, Covington et al. (1967a), -(210.760 ± 0.020) $kJ \cdot mol^{-1}$, Gerke (1922), -(210.78 ± 0.08) $kJ \cdot mol^{-1}$, and Lietzke and Vaughen (1955), -(210.64 ± 0.15) $kJ \cdot mol^{-1}$. Gerke (1922) obtained $\Delta_f G^o$ = -(210.46 ± 0.20) $kJ \cdot mol^{-1}$ from e.m.f. measurements on the direct combination of $Hg(l)$ and $Cl_2(g)$ in an electrochemical cell. See also Ahluwahlia and Cobble (1964). The value $\Delta_f G^o$ = -(210.760 ± 0.020) $kJ \cdot mol^{-1}$ was selected.

The entropy was obtained from measurements of the temperature coefficient of the $Hg(l)|Hg_2Cl_2(cr)|Cl^-(aq)$ electrode by Ives and Prasad (1970), (191.8 $\pm$ 0.5) $J \cdot K^{-1} \cdot mol^{-1}$, Gupta et al. (1963a), (191.8 $\pm$ 0.5) $J \cdot K^{-1} \cdot mol^{-1}$, Ahluwahlia and Cobble (1964), (191.4 $\pm$ 0.4) $J \cdot K^{-1} \cdot mol^{-1}$, Lietzke and Vaughen (1955), (191.7 $\pm$ 0.8) $J \cdot K^{-1} \cdot mol^{-1}$, Das and Ives (1962), (191.7 $\pm$ 0.8) $J \cdot K^{-1} \cdot mol^{-1}$, and Covington et al. (1967b), (191.4 $\pm$ 0.5) $J \cdot K^{-1} \cdot mol^{-1}$. See also Gerke (1922). The value S^o (298.15 K) = (191.6 $\pm$ 0.8) $J \cdot K^{-1} \cdot mol^{-1}$ was selected. By use of this selected value, the heat-capacity data of Pollitzer (1911, 1913), (22-198 K), were extrapolated to 298.15 K. The enthalpy increment was obtained by integration of the extrapolated curve.

103. $Hg_2SO_4(cr)$. $\Delta_f G^o$ for $Hg_2SO_4(cr)$ was obtained as -(625.85 $\pm$ 0.15) $kJ \cdot mol^{-1}$ from the e.m.f. measurements on the reaction $H_2(g) + Hg_2SO_4(cr) = 2 Hg(l) + H_2SO_4(aq)$ by Covington et al. (1965a), -(625.910 $\pm$ 0.050) $kJ \cdot mol^{-1}$, and by Sharma and Prasad (1970b), (using a quinhydrone electrode), -(625.80 $\pm$ 0.15) $kJ \cdot mol^{-1}$. The data of Hamer (1972) for the same reaction yielded $\Delta_f G^o$ = -(625.36 $\pm$ 0.20) $kJ \cdot mol^{-1}$. Values of $\Delta_r G^o$ for the reaction $Hg_2SO_4(cr) = Hg_2^{2+}(aq) + SO_4^{2-}(aq)$ were obtained from the data of Sharma and Prasad (1970a), (34.777 $\pm$ 0.080) $kJ \cdot mol^{-1}$, and Brown and Land (1957), (35.20 $\pm$ 0.20) $kJ \cdot mol^{-1}$, yielding for $Hg_2SO_4(cr)$, $\Delta_f G^o$ = -(625.28 $\pm$ 0.60) and -(625.71 $\pm$ 0.70) $kJ \cdot mol^{-1}$, respectively. See also Anan'eva and Maloshuk (1975) and Gardner et al. (1969b).

The entropy and the enthalpy increment were calculated from the heat-capacity measurements by Brackett et al. (1960), (5-21 K), and Papadopoulos and Giauque (1962), (16-298 K). Values of S^o in good agreement were obtained from the temperature coefficients of e.m.f.'s reported by Sharma and Prasad (1970b), (198.6 $\pm$ 3.0) $J \cdot K^{-1} \cdot mol^{-1}$, and Anan'eva and Maloshuk (1975), (199.8 $\pm$ 1.0) and (199.1 $\pm$ 1.0) $J \cdot K^{-1} \cdot mol^{-1}$.

104. $Cu(cr)$. Defined reference state.

The entropy and the enthalpy increment were calculated from the data given in the review by Furukawa et al. (1968) and Robie et al. (1976).

105. $Cu(g)$. $\Delta_f H^o$ was calculated from vapor-pressure measurements reported by Hersh (1953), (340.54 $\pm$ 0.60) $kJ \cdot mol^{-1}$, Edwards et al. (1953), (337.27 $\pm$ 0.80) $kJ \cdot mol^{-1}$, Morris and Zellars (1956), (337.00 $\pm$ 0.60) $kJ \cdot mol^{-1}$, Krupkowski and Golonka (1964), (335.50 $\pm$ 0.80) $kJ \cdot mol^{-1}$, McCormack et al. (1965), (338.9 $\pm$ 1.2) $kJ \cdot mol^{-1}$, Ponslet and Bariaux (1966), (336.20 $\pm$ 0.60) $kJ \cdot mol^{-1}$, and Myles and Darby (1968), (336.50 $\pm$ 0.20) $kJ \cdot mol^{-1}$.

The entropy and the enthalpy increment were calculated taking the electronic partition function equal to 2.

106. $Cu^{2+}(aq)$. $\Delta_f G^o$ of $Cu^{2+}(aq)$ was calculated from cell measurements made against the $Hg(l)|Hg_2SO_4(cr)|SO_4^{2-}$ electrode by Wetmore and Gordon (1937), (65.115 $\pm$ 0.060) $kJ \cdot mol^{-1}$, Müller and Reuther (1942), (65.096 $\pm$ 0.080) $kJ \cdot mol^{-1}$, and Tomassi and Wroblowa (1956), (65.077 $\pm$ 0.080) $kJ \cdot mol^{-1}$. The value $\Delta_f G^o$ = (65.100 $\pm$ 0.050) $kJ \cdot mol^{-1}$ was selected.

Values of S^o for the ion were calculated from the following sets of data.

- Combination of S^o(298.15 K) for $CuSO_4 \cdot 5H_2O(cr)$, measured by Stout (unpublished measurements; see Larson et al. (1968)), 301.2 $J \cdot K^{-1} \cdot mol^{-1}$, with $\Delta_{sol} H^o$

measured by Larson et al. (1968), (5.98 ± 0.20) kJ·mol^{-1}, and with $\Delta_{sol} G^{o}$ (15.14 ± 0.050) kJ·mol^{-1}, as calculated from solubility and activity data reported by Wetmore and Gordon (1937), Robinson and Jones (1936), Miles and Menzies (1937), and Ishikawa and Murooka (1930) and from the cell data of Müller and Reuther (1942), (solubility = 1.39 M; $\gamma_{\pm}$ = 0.0365; a_{w} = 0.971), gave S^{o} (Cu^{2+}, aq) = -(97.8 ± 1.5) J·K^{-1}· mol^{-1}.

- S^{o}(298.15 K) for CuSO$_4$(cr), (see Note 107), was combined with $\Delta_{sol} H^{o}$ measured by Larson et al. (1968) and Gedansky et al. (1970), -(73.00 ± 0.30) kJ·mol^{-1}, and with $\Delta_{sol} G^{o}$, -(16.0 ± 2.0) kJ·mol^{-1}. The latter value was calculated from that for the pentahydrate (see above), and from ΔG^{o} for the decomposition of the pentahydrate to the anhydrous salt (31.1 ± 1.0) kJ·mol^{-1}, obtained from the decomposition-pressure data of Sano (1936), Collins and Menzies (1936), Wilson (1921), Bell (1940), Kohler and Zäske (1964), and Menzies and Hitchcock (1931). From all these data the value S^{o}(Cu^{2+},aq) = -(100.5 ± 2.0) J·K^{-1}·mol^{-1} was derived.

- S^{o}(298.15 K) for CuCl$_2$(cr), calculated from the heat-capacity measurements by Stout and Chisholm (1962), (108.1 ± 0.2) J·K^{-1}· mol^{-1}, was combined with $\Delta_{sol} H^{o}$ measured by Gedansky et al. (1970), -(51.38 ± 0.40) kJ·mol^{-1}, and with $\Delta_{sol} G^{o}$, -(25.00 ± 1.00) kJ·mol^{-1}. The latter value was calculated from that for the dihydrate obtained from solubility and activity data reported by Stokes (1948), Chrétien and Weil (1935), and Blidin and Gordienko (1954), (solubility = 5.75 M; $\gamma_{\pm}$ = 0.669; a_{w} = 0.678), and from decomposition-pressure data reported by Perret (1966), Derby and Yngve (1916), and Bell (1940), $\Delta_{r} G^{o}$ = (13.47 ± 0.80) kJ·mol^{-1}. From these data the value S^{o}(Cu^{2+}, aq) = -(93.6 ± 2.0) J·K^{-1}·mol^{-1} was obtained.

- The value S^{o}(Cu^{2+}, aq) = -(99.3 ± 1.5) J·K^{-1}· mol^{-1} was calculated from the temperature coefficient of the e.m.f. of the cell studied by Wetmore and Gordon (1937), (see above).

The average of the above four values was selected.

$\Delta_f H^{o}$ of the ion was calculated from the selected values of $\Delta_f G^{o}$ and S^{o}. A value in good agreement was calculated from the measurements by Ishikawa and Murooka (1930) on a cell containing a solution saturated with respect to CuSO$_4$·5H$_2$O(cr), from which $\Delta_f H^{o}$ of the pentahydrate was calculated to be -(2279.66 ± 0.80) kJ·mol^{-1}. This value was combined with $\Delta_{sol} H^{o}$ from Larson et al. (1968), (5.98 ± 0.20) kJ·mol^{-1}, to yield $\Delta_f H^{o}$ = (65.1 ± 0.7) kJ·mol^{-1}. The measurements by Vasil'ev and Kunin (1972b) yield $\Delta_f H^{o}$ = (67.0 ± 1.0) kJ·mol^{-1}.

107. CuSO$_4$(cr). $\Delta_f H^{o}$ was calculated from $\Delta_{sol} H^{o}$, -(73.00 ± 0.30) kJ·mol^{-1}, and the values for the appropriate ions, (see Note 106).

The entropy and the enthalpy increment were calculated from the heat-capacity measurements made by Stout (1941), (15-50 K), and Weller (1965), (51-298 K). The entropy contribution below 15 K was taken as 0.37 J·K^{-1}·mol^{-1}.

108. Ag(cr). Defined reference state.

Values of the entropy and the enthalpy increment were taken from the critical literature review by Furukawa et al. (1968).

109. Ag(g). $\Delta_f H^{o}$ was calculated from the vapor-pressure measurements reported by Panish (1961), (284.5 ± 2.0) kJ·mol^{-1}, Boyer and Meadowcroft (1965), (285.10

$\pm$ 0.40) kJ·mol^{-1}, Bohdansky and Schins (1965, 1967), (284.4 $\pm$ 1.0) kJ·mol^{-1}, Krupkowski and Golonka (1964), (284.47 $\pm$ 0.80) kJ·mol^{-1}, Woolf et al. (1960), (284.30 $\pm$ 0.40) kJ·mol^{-1}, Zavitsanos (1964), (285.2 $\pm$ 1.2) kJ·mol^{-1}, and McCabe and Birchenall (1953), (285.1 $\pm$ 0.30) kJ·mol^{-1}. See also Paule and Mandel (1971).

The entropy and the enthalpy increment were calculated taking the electronic partition function equal to 2.

110. Ag$^+$(aq). $\Delta_f G^o$ was derived from the e.m.f. measurements by Owen and Brinkley (1938), 77.103 kJ·mol^{-1}, Bates and Bower (1954), 77.103 kJ·mol^{-1}, Harned et al. (1936), 77.069 kJ·mol^{-1}, Hetzer et al. (1962), 77.067 kJ·mol^{-1}, Hetzer et al. (1964), 77.124 kJ·mol^{-1}, and Kortüm and Häussermann (1965), 77.161 kJ·mol^{-1}. See also Lietzke and Vaughen (1955), and Topol (1968). The value $\Delta_f G^o = (77.112 \pm 0.020)$ kJ·mol^{-1} was selected.

S^o of the ion was calculated from measurements of the temperature coefficients of the above cells plus measurements of $\Delta_{sol} H^o$, $\Delta_{sol} G^o$ and S^o(298.15 K) of AgCl(cr), AgBr(cr), AgI(cr), Ag$_2$SO$_4$(cr), Ag$_2$O(cr) and AgNO$_3$(cr). See Annex II for individual selected values, and see Wagman and Kilday (1973) for a review of the literature.

$\Delta_f H^o$ was calculated from the selected value of $\Delta_f G^o$ and appropriate S^o values.

111. AgCl(cr). $\Delta_f G^o$ was obtained from cell measurements made by Bates and Bower (1954), -(109.789 $\pm$ 0.040) kJ·mol^{-1}, and Faita et al. (1967), -(109.60 $\pm$ 0.20) kJ·mol^{-1}. Also, the value $\Delta_f G^o$ = -(109.77 $\pm$ 0.40) kJ·mol^{-1} was obtained from the solubility data reviewed by Wagman and Kilday (1973). The value $\Delta_f G^o$ = -(109.785 $\pm$ 0.020) kJ·mol^{-1} was selected.

Entropy and enthalpy increment values were calculated from the heat-capacity measurements by Eastman and Milner, (1933), (15-293 K), Clusius and Harteck (1928), (10-126 K), and Berg (1976), (2-20 K), yielding S^o(298.15 K) = (96.25 $\pm$ 0.20) J·K^{-1}·mol^{-1}. Values of S^o were also calculated from the temperature dependence of e.m.f., as measured by Bates and Bower (1954), (96.18 $\pm$ 0.40) J·K^{-1}·mol^{-1} and by Faita et al. (1967), (96.33 $\pm$ 0.10) J·K^{-1}·mol^{-1}.

$\Delta_f H^o$ was calculated from the selected value of $\Delta_f G^o$ and appropriate S^o values. $\Delta_f H^o$ may also be derived from the measurements of the enthalpy of precipitation of AgCl(cr), as reported by Wagman and Kilday (1973), -(127.05 $\pm$ 0.15) kJ·mol^{-1}.

112. Ti(cr). Defined reference state.

The entropy and the enthalpy increment were calculated from heat-capacity measurements by Kothen and Johnston (1953), (15-306 K), Clusius and Franzosini (1958), (14-272 K), Stalinski and Bieganski (1961), (27-360 K), Hake and Cape (1964), (1.2-4.5 K), Agarwal and Betterton (1974b), (1.2-4.5 K), Aven et al. (1956), (4-16 K), Wolcott (1957), (1.2-21 K) and Dummer (1965), (0.9-12 K).

113. Ti(g). $\Delta_f H^o$ was calculated from measurements of the vapor pressure of Ti(cr) by Wu and Wahlbeck (1971), (473.6 $\pm$ 2.0) kJ·mol^{-1}, and by Edwards et al. (1953), (473.2 $\pm$ 1.0) kJ·mol^{-1}. Vapor-pressure data on Ti(l) have been measured by Bernstein and Kaufman (1966), (471 $\pm$ 8) kJ·mol^{-1}, Koch et al. (1969), (469 $\pm$ 4) kJ·mol^{-1}, and Strassmair and Stark (1967), (467 $\pm$ 4) kJ·mol^{-1}. The uncertainty in the calculated values depends principally on the uncertainties in the thermal functions of the condensed phases of Ti.

The entropy and the enthalpy increment were calculated using the atomic energy levels recommended by Corliss and Sugar (1979).

114. TiO$_2$(cr, rutile). $\Delta_f H^o$ was calculated as a weighted average of the enthalpy of combustion measurements by Mah et al. (1957), -(944.2 ± 0.4) kJ. mol^{-1}, and by Humphrey (1951), -(943.3 ± 1.0) kJ·mol^{-1}. Measurements in good agreement have been reported by Ariya et al. (1957) and Neumann et al. (1934).

The entropy and the enthalpy increment were calculated from the heat-capacity measurements by Shomate (1947), (52-298 K), Dugdale et al. (1954), (12-270 K), Mitsuhashi and Takahashi (1980), (80-1100 K), and Sandin and Keesom (1969), (0.3-20 K).

115. TiCl$_4$(g). $\Delta_f H^o$(g) was calculated from measurements of the enthalpy of reaction of Ti(cr) with Cl$_2$(g) by Johnson et al. (1959b). Values for TiCl$_4$(l) were obtained for the same reaction by Gross et al. (1957), -(801.0 ± 1.5) kJ·mol^{-1}, Skinner and Ruehrwein (1955), -(796.2 ± 2.0) kJ·mol^{-1}, and Krieve et al. (1956), -(795.0 ± 2.0) kJ·mol^{-1}. $\Delta_{vap} H^o$ (40.60 ± 0.10) kJ·mol^{-1} was calculated from the vapor-pressure data of Weed (1957) and Schäfer and Zeppernick (1953). However the values of $\Delta_f H^o$ for TiCl$_4$(l) obtained by direct chlorination involve a large and uncertain correction for the enthalpy of solution of Cl$_2$(g) in TiCl$_4$(l). Application of a Third Law analysis to the equilibrium measurements by Farber and Darnell (1955) shows a significant trend in the value of $\Delta_r H^o$(298.15 K) with the temperature of measurement. See Gross et al. (1957).

The entropy and the enthalpy increment were calculated using the harmonic-oscillator, rigid-rotator approximation, with vibrational frequencies obtained from the Raman spectrum by Clark et al. (1972) and from the infrared spectrum by Königer et al. (1976). The molecular dimensions were obtained by Morino and Uehara (1966) from electron diffraction measurements. The uncertainties come principally from the neglect both of the anharmonicity and of the rotation-vibration interaction.

116. U(cr). Defined reference state.

The entropy and the enthalpy increment were calculated from the heat-capacity measurements by Flotow and Osborne (1966), (1.7-23 K) and Flotow and Lohr (1960), (5.7-348 K). Measurements by Jones et al. (1952), (15-300 K) and Nakamura et al. (1980), (80-1000 K) were in satisfactory agreement.

117. U(g). $\Delta_f H^o$ was calculated from vapor-pressure measurements by De Maria et al. (1960), (537 ± 10) kJ·mol^{-1}, Pattoret et al. (1969), (542 ± 5) kJ·mol^{-1}, Ackermann and Rauh (1969), (532 ± 5) kJ·mol^{-1}, and Das et al. (1985), (528 ± 4) kJ· mol^{-1}. For a detailed discussion and bibliography on the sublimation of U(cr) see Storms (1966, 1967), Pattoret et al. (1969), and Oetting et al. (1976).

The entropy and the enthalpy increment were calculated using the energy levels given by Blaise and Radziemski (1976).

118. UO$_2$(cr). $\Delta_f H^o$ was calculated from the combustion measurements by Huber and Holley (1969), -(1085.0 ± 3.0) kJ·mol^{-1}, and Johnson and Steele (1981), -(1085.1 ± 2.5) kJ·mol^{-1}.

The entropy and the enthalpy increment were calculated from the heat-capacity measurements made by Huntzicker and Westrum (1971), (5-350 K). Measurements by Jones et al. (1952), (15-300 K) are in good agreement, although their sample was significantly less pure than that of Huntzicker and Westrum.

119. UO_2^{2+}(aq). $\Delta_f H^\circ(UO_2(NO_3)_2 \cdot 6H_2O$, cr), -(3167.5 ± 1.5) $kJ \cdot mol^{-1}$, was calculated from the measurements by Cordfunke (1964) on $\Delta_{sol} H^\circ$ of UO_3(cr, gamma), (for $\Delta_f H^\circ$ see Note 120), and $UO_2(NO_3)_2 \cdot 6H_2O$(cr) in aqueous HNO_3. $\Delta_{sol} H^\circ$ of the hexahydrate in H_2O(l), (19.35 ± 0.20) $kJ \cdot mol^{-1}$, is based on the value reported by Fuger and Oetting (1976), (19.58 ± 0.25) $kJ \cdot mol^{-1}$, and the newer measurements by Devina et al. (1982b), (19.18 ± 0.25) $kJ. mol^{-1}$. From these data we obtained for UO_2^{2+}(aq), $\Delta_f H^\circ$ = -(1019.5 ± 1.5) $kJ \cdot mol^{-1}$. A value of $\Delta_f H^\circ$ for UO_2Cl_2(cr), -(1243.12 ± 0.90) $kJ \cdot mol^{-1}$, was obtained by Cordfunke et al. (1976) from the dissolution of UO_3(cr, gamma) and UO_2Cl_2(cr) in aqueous HCl. They also obtained a value, -(1246.4 ± 3.5) $kJ \cdot mol^{-1}$, from measurements of the enthalpy of solution of UCl_4(cr) and UO_2Cl_2(cr) in aqueous $FeCl_3$. We have selected for UO_2Cl_2(cr), $\Delta_f H^\circ$ = -(1243.6 ± 1.5) $kJ \cdot mol^{-1}$. $\Delta_{sol} H^\circ$ of UO_2Cl_2(cr), -(109.20 ± 0.20) $kJ \cdot mol^{-1}$, was calculated from unpublished measurements made by Cordfunke; see Fuger et al. (1983). From these values we obtained for UO_2^{2+}(aq), $\Delta_f H^\circ$ = -(1018.6 ± 1.6) $kJ \cdot mol^{-1}$.

See Fuger and Oetting (1976), pages 20-26, and Fuger et al. (1983), pages 106-110, for a detailed review and analysis of the available data for UO_2Cl_2(cr) and UO_2^{2+}(aq).

$\Delta_{sol} G^\circ$ of $UO_2(NO_3)_2 \cdot 6H_2O$(cr), -(13.24 ± 0.10) $kJ \cdot mol^{-1}$, was calculated from values of solubility (3.24 M), $\gamma_\pm$ (2.149) and a_w (0.7325) reported by Goldberg (1979). S°(298.15 K) for the hexahydrate, (505.6 ± 2.0) $J \cdot K^{-1} \cdot mol^{-1}$, was measured by Coulter et al. (1940), (13-299 K). Combining these values with the enthalpy of solution, (19.35 ± 0.20) $kJ \cdot mol^{-1}$ (see above), we obtain $S^\circ(UO_2^{2+}$, aq) = -(98.2 ± 3.0) $J \cdot K^{-1} \cdot mol^{-1}$.

120. UO_3(cr, gamma). $\Delta_f H^\circ$ was calculated from the measurements by Fitzgibbon et al. (1967) on the solution enthalpy of uranium oxides in ceric solutions, corrected to stoichiometric UO_3(cr), -(1223.8 ± 2.0) $kJ \cdot mol^{-1}$, and from the measurements by Johnson and O'Hare (1978) on the enthalpy of solution of UO_3(cr, gamma) and UF_6(cr) in aqueous HF, -(1223.9 ± 4.3) $kJ \cdot mol^{-1}$. Equilibrium measurements by Cordfunke and Aling (1965) on the 3 UO_3(cr) = U_3O_8(cr) + $\frac{1}{2}$ O_2(g) system yielded $\Delta_f H^\circ$ = -(1223.7 ± 3.0) $kJ \cdot mol^{-1}$.

The entropy and the enthalpy increment were calculated from the heat-capacity measurements made by Westrum (1966), (5-350 K).

121. U_3O_8(cr). $\Delta_f H^\circ$ was derived from the combustion measurements made by Huber and Holley (1969). The measurements by Popov and Ivanov (1957) yielded $\Delta_f H^\circ$ = -(3584 ± 13) $kJ \cdot mol^{-1}$.

The entropy and the enthalpy increment were calculated from the heat capacity measurements reported by Westrum and Grønvold (1959), (5-350 K).

122. Th(cr). Defined reference state.

The entropy and the enthalpy increment were calculated using the heat-capacity data obtained by Smith and Wolcott (1955), (1.4 - 20 K), Schmidt and Wolf (1975),

(2-11 K), Griffel and Skochdopole (1953), (18-300 K), and Nakamura et al. (1980), (80-1000 K). There is considerable variation in the values of heat capacity reported above 80 K.

123. Th(g). $\Delta_f H^\circ$ was calculated from vapor-pressure effusion measurements by Ackermann and Rauh (1972b), (602 ± 5) kJ·mol^{-1}. The uncertainty in this value depends mainly on the inaccuracy of the thermal functions of Th(cr). The entropy and the enthalpy increment were calculated using the atomic energy levels given by Zalubas (1976). See the review by Oetting et al. (1976).

124. ThO$_2$(cr). $\Delta_f H^\circ$ was derived from the combustion measurements by Huber et al. (1952). Roth and Becker (1932) obtained $\Delta_f H^\circ = -(1226 \pm 5)$ kJ·mol^{-1}.

The entropy and the enthalpy increment were calculated from the heat-capacity measurements made by Osborne and Westrum (1953), (10-305 K). The entropy contribution below 10 K = 0.05 J·K^{-1}·mol^{-1}

125. Be(cr). Defined reference state.

The entropy and the enthalpy increment were calculated from the heat-capacity measurements made by Hill and Smith (1953), (4-300 K) and Ahlers (1966), (1.4 - 30 K).

126. Be(g). $\Delta_f H^\circ$ was calculated by the Third Law method from the vapor- pressure data obtained by Holden et al. (1948), (325.8 ± 1.5) kJ·mol^{-1}, Gulbransen and Andrew (1950), (327.3 ± 1.2) kJ·mol^{-1}, Kovtun et al. (1964), (319.2 ± 1.5) kJ·mol^{-1}, and Hildenbrand and Murad (1966), (323.8 ± 2.0) kJ·mol^{-1}.

The entropy and the enthalpy increment were calculated taking the electronic partition function equal to 1.

127. BeO(cr). $\Delta_f H^\circ$ was calculated from measurements of the enthalpy of solution of Be(cr) in aqueous HF by Bear and Turnbull (1965) and Armstrong and Coyle (1965), and on the enthalpy of solution of BeO(cr) in aqueous HF by Fricke and Wüllhorst (1932), Kolesov et al. (1959), and Kilday et al. (1973), which yielded $\Delta_f H^\circ$(BeO, cr) = -(609.2 ± 2.5) kJ·mol^{-1}. Measurements of $\Delta_{sol} H^\circ$ of Be(cr) and BeO(cr) in aqueous HCl by Thompson et al. (1962), Blachnik et al. (1968) and Kilday et al. (1973) yielded $\Delta_f H^\circ = -(609.6 \pm 2.5)$ kJ·mol^{-1}. Combustion measurements on Be(cr) in an oxygen bomb (Cosgrove and Snyder (1953)) yielded results of low reliability, because of lack of analysis of the combustion products. See Parker (1973) for a complete review and analysis of these data.

The entropy and the enthalpy increment were calculated from the heat-capacity measurements by Furukawa and Reilly (1968), (15-300 K). See also Gmelin (1966), (5-75 K).

128. Mg(cr). Defined reference state.

The entropy and the enthalpy increment were calculated from the heat-capacity measurements by Craig et al. (1954), (12-320 K), Männchen and Bornkessel (1959), (12-300 K), and Clusius and Vaughen (1930), (11-228 K). The entropy contribution below 11 K amounts to 0.015 J·K^{-1}·mol^{-1}.

129. Mg(g). $\Delta_f H^\circ$ was calculated from the results of the Third Law treatment of the vapor-pressure data obtained by Coleman and Egerton (1935), (147.6 ± 0.8)

$kJ \cdot mol^{-1}$, Priselkov (1954), (144.6 ± 0.3) $kJ \cdot mol^{-1}$, Schmahl and Sieben (1960), (146.4 ± 0.4) $kJ \cdot mol^{-1}$, Greenbank and Argent (1965), (147.4 ± 1.0) $kJ \cdot mol^{-1}$, Gilbreath (1965), (147.9 ± 1.0) $kJ \cdot mol^{-1}$, McCreary and Thorn (1971), (147.11 ± 0.40) $kJ \cdot mol^{-1}$, Mashovets and Puchkov (1965), (148.3 ± 0.6) $kJ \cdot mol^{-1}$, and Prasad et al. (1978), (148.1 ± 0.4) $kJ \cdot mol^{-1}$.

The entropy and the enthalpy increment were calculated taking the electronic partition function equal to 1.

130. Mg^{2+}(aq). Data for Mg^{2+}(aq) were obtained by the following routes.

a. A value of $\Delta_f H^o$ was calculated from measurements of the enthalpy of solution of (i) Mg(cr) in HCl (1.027 M) by Shomate and Huffman (1943), -(465.48 ± 0.17) $kJ \cdot mol^{-1}$, and (ii) of $MgCl_2$(cr) in the same acid concentration by Shin and Criss (1979), -(149.98 ± 0.18) $kJ \cdot mol^{-1}$, yielding $\Delta_f H^o$($MgCl_2$, cr) = -(644.28 ± 0.69) $kJ \cdot mol^{-1}$. (Shomate and Huffman's data for $MgCl_2$(cr) were rejected; see Shin and Criss (1979)). $\Delta_{sol} H^o$ of $MgCl_2$(cr) in H_2O(l), -(156.40 ± 0.80) $kJ \cdot mol^{-1}$, was selected from the measurements by Greyson and Snell (1969), -(156.03 ± 0.80) $kJ \cdot mol^{-1}$, Vorob'ev et al. (1974), -(157.50 ± 0.60) $kJ \cdot mol^{-1}$, and Shin and Criss (1979), -(155.84 ± 0.40) $kJ \cdot mol^{-1}$. From these data we calculated $\Delta_f H^o$ for Mg^{2+}(aq) as -(466.5 ± 1.0) $kJ \cdot mol^{-1}$.

b. Measurements of $\Delta_{sol} H$ of Mg(cr) in dilute HCl and dilute $HClO_4$ were made by Monaenkova et al. (1971). Extrapolation to the standard state yielded for Mg^{2+}(aq), $\Delta_f H^o$ = -(465.3 ± 1.0) and -(468.86 ± 0.80) $kJ \cdot mol^{-1}$, respectively. Similar measurements by Vasil'ev et al. (1981) with the two reagents yielded -(467.56 ± 0.80) and -(467.39 ± 0.90) $kJ \cdot mol^{-1}$, respectively. Coffy and Olofsson (1979) measured $\Delta_{sol} H$ in aqueous $HClO_4$; extrapolation to the standard state yielded -(466.8 ± 1.2) $kJ \cdot mol^{-1}$. Because of possible complexing in HCl solutions, only the values for the $HClO_4$ solutions were used here.

c. A value of S^o for Mg^{2+}(aq) was calculated from data for $MgSO_4 \cdot 6H_2O$(cr), for which Cox et al. (1955) measured S^o(298.15 K) as (348.11 ± 0.80) $J \cdot K^{-1} \cdot mol^{-1}$. $\Delta_{sol} G^o$ of the hexahydrate (9.85 ± 0.20) $kJ \cdot mol^{-1}$ was calculated from the solubility (3.17 M, taken from Linke (1965)) and from the values $\gamma_\pm$ = 0.0604, a_w = 0.895 from Wu et al. (1969), Yokoyama and Yamatera (1975), and Robinson and Jones (1936). $\Delta_{sol} H^o$ of the hexahydrate, -(4.1 ± 1.0) $kJ \cdot mol^{-1}$, was calculated from the value for the heptahydrate, (11.84 ± 0.40) $kJ \cdot mol^{-1}$, based on data by Cappellina and Napolitano (1966) and Kaganovich and Mishchenko (1951) and the enthalpy of dehydration to the hexahydrate, (15.9 ± 2.0) $kJ \cdot mol^{-1}$, taken from the measurements by Kohler and Zäske (1964) and Thomsen (1882). These values yielded S^o(Mg^{2+}, aq) = -(136.9 ± 5.0) $J \cdot K^{-1} \cdot mol^{-1}$.

d. A value of S^o was calculated from data on $Mg(OH)_2$(cr) as follows. S^o(298.15 K) = (63.18 ± 0.40) $J \cdot K^{-1} \cdot mol^{-1}$ was measured by Giauque and Archibald (1937). A value for $\Delta_{sol} G^o$, (63.64 ± 0.10) $kJ \cdot mol^{-1}$, was calculated from the data of Hostetler (1963). $\Delta_{sol} H^o$ = -(2.58 ± 0.70) $kJ \cdot mol^{-1}$ was calculated (i) from the value of $\Delta_f H^o$ for MgO(cr), (see Note 131), (ii) the difference in the enthalpies of reaction of MgO(cr) and $Mg(OH)_2$(cr) with acids, as measured by Torgeson and Sahama (1948), Giauque and Archibald (1937), and Taylor and Wells (1938), and (iii) the values of $\Delta_f H^o$ for Mg^{2+}(aq), (see above), and for OH^-(aq), (see Note 6). From these data S^o(Mg^{2+}, aq) = -(137.1 ± 4.0) $J \cdot K^{-1} \cdot mol^{-1}$.

e. S^o(298.15 K) of $MgCl_2 \cdot 6H_2O$(cr) was measured by Kelley and Moore (1943), (366.1 ± 4.0) $J \cdot K^{-1} \cdot mol^{-1}$, whilst the value (365.3 ± 5.0) $J \cdot K^{-1} \cdot mol^{-1}$ may be calculated from $S^o(MgCl_2.4 \ H_2O, cr) = (264.0 \pm 3.0)$ J·K^{-1}·mol^{-1}, measured by Kelley and Moore (1943), and measurements on the decomposition of the hexahydrate to the tetrahydrate by Derby and Yngve (1916), Kondirev and Berezovskii (1935), and the solution calorimetry of Shomate and Huffman (1943) on these hydrates. We have selected the value $S^o = (366.0 \pm 4.0)$ J·K^{-1}·mol^{-1}. $\Delta_{sol}H^o$ of the hexahydrate -(16.10 ± 0.40) kJ·mol^{-1} was calculated from the data of Tsvetkov and Rabinovich (1969), -(16.32 ± 0.20) kJ·mol^{-1}, and Mishchenko and Yakovlev (1959), -(15.63 ± 0.40) kJ·mol^{-1}. $\Delta_{sol}G^o$, -(25.75 ± 0.20) kJ·mol^{-1}, was calculated from the solubility (5.84 M), $\gamma_\pm$ (31.6) and a_w (0.330) obtained by Stokes (1945, 1948). These values yielded $S^o(Mg^{2+}, aq) = $ -(134.5 ± 6.0) J·K^{-1}mol^{-1}.

131. MgO(cr). The species is the macrocrystalline cubic (periclase) form. $\Delta_f H^o$ was calculated from measurements of the enthalpy of solution of Mg(cr) and MgO(cr) in aqueous acids by Shomate and Huffman (1943), -(601.58 ± 0.25) kJ·mol^{-1}, and by Vasil'ev et al. (1981), -(601.66 ± 0.60) kJ·mol^{-1}. The oxygen-combustion measurements by Holley and Huber (1951), -(600.9 ± 1.5) kJ·mol^{-1}, and Vorob'ev and Skuratov (1958), -(601.8 ± 1.0) kJ·mol^{-1}, are in good agreement, but these authors did not consider the influence of crystal size on their results. See Giauque and Archibald (1937) and also Shchukarev et al. (1960) and von Wartenberg (1909).

The entropy and the enthalpy increment were calculated from the heat-capacity measurements made by Barron et al. (1959), (3-270 K). See also Gmelin (1969).

132. MgF₂(cr). $\Delta_f H^o$ was calculated from the enthalpy of reaction of Mg(cr) with $F_2(g)$ as measured by Rudzitis et al. (1964), -(1124.2 ± 1.2) kJ·mol^{-1}. Measurements by Rezukhina et al. (1973) on cells containing Mg(cr) and AlF$_3$(cr) yielded $\Delta_f H^o = $ -(1122.3 ± 2.4) kJ·mol^{-1}. Measurements by Torgeson and Sahama (1948) on the enthalpy of solution of MgO(cr) in aqueous HF yielded $\Delta_f H^o = $ -(1123.8 ± 2.0) kJ·mol^{-1}. See also Gross et al. (1954) and Domange (1937).

133. Ca(cr). Defined reference state.

The entropy and the enthalpy increment were calculated from the heat capacities measured by Griffel et al. (1957), (2-4 K), Agarwal and Betterton (1974a), (1-4 K), Roberts (1957), (2-20 K), and Clusius and Vaughen (1930), (10-201 K). Values above 200 K were obtained by extrapolation of the heat-capacity curve to fit smoothly with the high temperature data of Jauch (1946) and Eastman et al. (1924).*

134. Ca(g). $\Delta_f H^o$ was calculated by a Third Law treatment of the vapor-pressure data of Priselkov and Nesmeyanov (1954), (176.9 ± 1.5) kJ·mol^{-1}, Douglas (1954), (177.1 ± 1.5) kJ·mol^{-1}, Muradov and Gel'd (1965), (178.0 ± 1.7) and (177.9 ± 1.7) kJ·mol^{-1}, Smith (1962), (179.5 ± 1.5) kJ·mol^{-1}, Bohdansky and Schins (1967), (178.2 ± 3.0) kJ·mol^{-1}, Bogoslovskii et al. (1969), (174.4 ± 4.0) kJ·mol^{-1}, Schins et al. (1971), (178.2 ± 4.0) kJ·mol^{-1}, and De Maria and Piacenti (1974), (178.9 ± 6.0) kJ·mol^{-1}. The uncertainties in the given values are due mainly to the inaccuracy of the thermal functions for Ca(cr or l), which increases rapidly at high temperatures.

---

* Note added in proof: New experimental measurements by R. Robie of the U.S. Geological Survey (private communication. 1985) indicate that S^o and H^o(298.15 K) - H^o(0) selected here may be as much as 1.3 J·K^{-1}·mol^{-1} and 0.08 kJ·mol^{-1} low.

The entropy and the enthalpy increment were calculated taking the electronic partition function equal to 1.

135. Ca^{2+}(aq). $\Delta_f H^o$ was calculated from the following sets of data:

a. $\Delta_{sol} H^o$ of $Ca(OH)_2$(cr) was measured by Hopkins and Wulff (1965b), -(18.24 $\pm$ 0.20) $kJ \cdot mol^{-1}$. The value -(65.25 $\pm$ 0.30) $kJ \cdot mol^{-1}$ for the enthalpy of the reaction CaO(cr) + H_2O(l) = $Ca(OH)_2$(cr) was calculated from data reported by Thorvaldson et al. (1930a, 1930b), -(65.65 $\pm$ 0.80), -(65.02 $\pm$ 0.60) $kJ \cdot mol^{-1}$, Wells and Taylor (1937), -(64.98 $\pm$ 0.40) $kJ \cdot mol^{-1}$, Taylor and Wells (1938), -(65.19 $\pm$ 0.40), -(64.56 $\pm$ 0.40) $kJ \cdot mol^{-1}$, and Schwiete and Pranschke (1935), -(65.31 $\pm$ 0.40) $kJ \cdot mol^{-1}$. From these data, and taking $\Delta_f H^o$ (CaO, cr) = -(634.92 $\pm$ 0.90) $kJ \cdot mol^{-1}$, (see Note 136), we obtained $\Delta_f H^o$ (Ca^{2+}, aq) = -(544.2 $\pm$ 1.2) $kJ \cdot mol^{-1}$.

b. $\Delta_r H^o$ for the reaction CaO(cr) + 2 H^+(aq) = Ca^{2+}(aq) + H_2O(l) was calculated from the measurements with aqueous $HClO_4$ reported by Monaenkova and Vorob'ev (1972), -(194.40 $\pm$ 0.30) $kJ \cdot mol^{-1}$. Measurements with dilute HCl by Taylor and Wells (1938) yielded -(193.55 $\pm$ 0.40) $kJ \cdot mol^{-1}$. From these values, $\Delta_f H^o$(Ca^{2+}, aq) = -(543.5 $\pm$ 1.0) and (542.6 $\pm$ 1.0) $kJ \cdot mol^{-1}$, respectively.

c.   Ehrlich et al. (1963) measured the enthalpy of solution of Ca(cr) in dilute HCl. Correction to standard states (with no allowance for possible complexing) yielded $\Delta_f H^o$(Ca^{2+}, aq) = -(546.0 $\pm$ 2.0) $kJ \cdot mol^{-1}$.

d.   Young (1944) determined $\Delta_f H^o$ for $Ca(NO_3)_2$(cr), -(937.20 $\pm$ 0.80) $kJ \cdot mol^{-1}$, by solution of CaO(cr) and $Ca(NO_3)_2$(cr) in aqueous acid. $\Delta_{sol} H^o$ of $Ca(NO_3)_2$(cr) in H_2O(l), -(18.60 $\pm$ 0.20) $kJ \cdot mol^{-1}$, was taken from the measurements by Partington and Soper (1929), -(18.32 $\pm$ 0.40) $kJ \cdot mol^{-1}$, Boerio and O'Hare (1976), -(18.66 $\pm$ 0.08) $kJ \cdot mol^{-1}$, and Muldrow and Hepler (1958), -(18.45 $\pm$ 0.20) $kJ \cdot mol^{-1}$. See also Ewing et al. (1932), Scholle and Brunclikova (1968), and Krestov and Egorova (1967b). These values yield $\Delta_f H^o$(Ca^{2+}, aq) = -(542.10 $\pm$ 0.90) $kJ \cdot mol^{-1}$.

e. $\Delta_{sol} H^o$ of $CaCO_3$(cr, calcite), -(10.25 $\pm$ 0.60) $kJ \cdot mol^{-1}$, in H_2O(l) was calculated from the work of Plummer and Busenberg (1982), -(9.61 $\pm$ 1.20) $kJ \cdot mol^{-1}$, Jacobson and Langmuir (1974), -(10.82 $\pm$ 1.20) $kJ \cdot mol^{-1}$, Bäckström (1925), -(9.41 $\pm$ 1.00) $kJ \cdot mol^{-1}$, and Hull and Turnbull (1973), -(10.35 $\pm$ 0.60) $kJ \cdot mol^{-1}$. The value -(10.94 $\pm$ 0.50) $kJ \cdot mol^{-1}$ was calculated from $\Delta_r H^o$ of CaO(cr) with H^+(aq), -(194.00 $\pm$ 0.50) $kJ \cdot mol^{-1}$ (see above), and from the decomposition data for $CaCO_3$(cr, calcite), as reviewed by Ko et al. (1982). $\Delta_f H^o$ for $CaCO_3$(cr) was derived from the review by Ko et al. (1982). From all these data $\Delta_f H^o$ (Ca^{2+}, aq) = -(542.4 $\pm$ 1.5) $kJ \cdot mol^{-1}$.

The entropy of the ion was calculated from the following sets of data.

f.   Hopkins and Wulff (1965b) reported values for the standard enthalpy, -(18.24 $\pm$ 0.20) $kJ \cdot mol^{-1}$, and Gibbs energy, (29.71 $\pm$ 0.20) $kJ \cdot mol^{-1}$, for the solution of $Ca(OH)_2$(cr) in water. S^o($Ca(OH)_2$, cr, 298.15 K) = (83.39 $\pm$ 0.40) $J \cdot K^{-1} \cdot mol^{-1}$ was calculated from the heat-capacity measurements by Hatton et al. (1959), (19-330 K). From these data S^o(Ca^{2+}, aq) = -(55.6 $\pm$ 1.0) $J \cdot K^{-1} \cdot mol^{-1}$.

g. $\Delta_{sol} H^o$ for $CaCO_3$(cr, calcite) was taken as -(10.25 $\pm$ 0.60) $kJ \cdot mol^{-1}$ (see above); the value (48.39 $\pm$ 0.10) $kJ \cdot mol^{-1}$ for $\Delta_{sol} G^o$ was taken from Jacobson and Langmuir (1974), (48.37 $\pm$ 0.30) $kJ \cdot mol^{-1}$, and Plummer and Busenberg (1982), (48.40 $\pm$ 0.10) $kJ \cdot mol^{-1}$. S^o($CaCO_3$, cr, 298.15 K) = (91.71 $\pm$ 0.30) $J \cdot K^{-1} \cdot mol^{-1}$ was obtained from the heat-capacity data reported by Staveley and Linford

(1969), (11-303 K). From these data $S^o(Ca^{2+}, aq) = -(55.0 \pm 2.0)$ J·K^{-1}·mol^{-1}.

h. $\Delta_{sol}H^o$ for $Ca(NO_3)_2(cr)$ was taken as $-(18.60 \pm 0.20)$ kJ·mol^{-1} (see above). A value for $\Delta_{sol}G^o$, $-(31.27 \pm 0.40)$ kJ·mol^{-1}, was calculated from the solubility (20.75 M) given by Linke (1958) and the value of $\gamma_\pm$ (2.035) calculated from the measurements by Yakimov et al. (1969), Pearce (1936), and Stokes and Robinson (1948), (see Guggenheim and Stokes (1958)). A value, $-(31.61 \pm 0.60)$ kJ·mol^{-1}, was calculated from data on $Ca(NO_3)_2 \cdot 4H_2O(cr)$ (solubility = 8.41 M from Pearce (1936) and Linke (1958); $\gamma_\pm = 0.86$ and $a_w = 0.49$ (see above)), and decomposition-pressure measurements on the hydrates reported by Ewing (1927). The selected value for $\Delta_{sol}G^o$ was $-(31.40 \pm 0.40)$ kJ·mol^{-1}. S^o $(Ca(NO_3)_2, cr, 298.15 K), = (193.2 \pm 2.0)$ J·K^{-1}·mol^{-1}, was calculated from the heat-capacity data reported by Shomate and Kelley (1944), (53-312 K). These values yield $S^o(Ca^{2+}, aq) = -(57.3 \pm 3.0)$ J·K^{-1}·mol^{-1}.

i. The value $-(1.13 \pm 0.08)$ kJ·mol^{-1} for $\Delta_{sol}H^o$ of $CaSO_4 \cdot 2H_2O(cr, gypsum)$ was calculated from the measurements by Wallace (1946), $-(1.13 \pm 0.10)$ kJ·mol^{-1}, and Lange and Monheim (1930), $-(1.14 \pm 0.20)$ kJ·mol^{-1}. $\Delta_{sol}G^o$, (26.14 ± 0.10) kJ·mol^{-1}, was calculated from the solubility (0.0152 M) from Power et al. (1964), (0.0156 M), Zdanovskii and Spiridonov (1967), (0.0153 M), Bock (1961), (0.0152 M), Madgin and Swales (1956), (0.0151 M), and Marshall and Slusher (1966), (0.0151 M), and values for $\gamma_\pm$ (0.338) and a_w (0.9995) taken from Lilley and Briggs (1976). See also Tanji (1969) and Nakayama and Rasnick (1967). $S^o(CaSO_4 \cdot 2 H_2O, cr, 298.15 K) = (193.9 \pm 1.5)$ J·K^{-1}·mol^{-1} was calculated from data reported by Kelley et al. (1941), (53-298 K) and Latimer et al. (1933a), (18-303 K). From these data $S^o(Ca^{2+}, aq) = -(56.0 \pm 2.5)$ J·K^{-1}·mol^{-1}.

j. The value $-(17.99 \pm 0.12)$ kJ·mol^{-1} for $\Delta_{sol}H^o$ of $CaSO_4(cr, anhydrite)$ was calculated by combining the enthalpy of solution measurements by Wallace (1946) on $CaSO_4 \cdot 2H_2O(cr)$, (see above) with the dehydration data of Kelley et al. (1941). The value (25.223 ± 0.080) kJ·mol^{-1} was calculated for $\Delta_{sol}G^o$ from the solubility (0.0200 M), –from Madgin and Swales (1956), (0.0200 M), Power et al. (1964), (0.0191 M), Bock (1961), (0.0202 M), and Zdanovskii and Spiridonov (1967), (0.0205 M)–, and the value 0.3087 for $\gamma_\pm$, extrapolated from the data of Lilley and Briggs (1976) and Yokoyama and Yamatera (1975). $S^o(CaSO_4, cr, 298.15 K) = (106.5 \pm 2.0)$ J·K^{-1}·mol^{-1} was calculated from the heat-capacity data reported by Kelley et al. (1941), (50-296 K). From these data $S^o(Ca^{2+}, aq) = -(56.9 \pm 2.5)$ J·K^{-1}·mol^{-1}.

136. CaO(cr). $\Delta_f H^o$ was derived from measurements of the enthalpy of combustion of Ca(cr) in $O_2(g)$ by Huber and Holley (1956). Measurement of $\Delta_{sol}H^o$ of Ca(cr) and CaO(cr) in aqueous acids leads to values of lower accuracy. See Gurvich et al. (1981), part 1, page 308.

The entropy and the enthalpy increment were calculated from the heat- capacity measurements made by Gmelin (1969), (1-300 K), adjusted for the fact that similar measurements by Gmelin (1966, 1969) on BeO(cr) and MgO(cr) were considered by the Task Group to be slightly too high.

137. Li(cr). Defined reference state.

The entropy and the enthalpy increment were calculated from heat-capacity measurements by Martin (1960b, 1962), (20-300 K) and Filby and Martin (1963), (3-30 K), taking into account the polymorphic transition at 80 K.

138. Li(g). $\Delta_f H^\circ$ was calculated using a Third Law treatment of the vapor-pressure data obtained by Bohdansky and Schins (1965, 1967), (159.5 ± 1.5) kJ·mol^{-1}, Hartmann and Schneider (1929), (159.0 ± 0.4) kJ·mol^{-1}, Schins et al. (1971), (159.2 ± 1.0) kJ·mol^{-1}, Anisimov and Volyak (1969), (159.6 ± 0.7) kJ·mol^{-1}, Shpil'rain and Belova (1968), (159.1 ± 1.8) kJ·mol^{-1}, and Rigney et al. (1965), (159.3 ± 1.1) kJ·mol^{-1}. The calculations were done for the ideal gas mixture of monatomic and diatomic lithium using $D_o(0)(Li_2) = (101.0 ± 1.0)$ kJ·mol^{-1} (Hessel and Vidal (1979), Stwalley (1970)). The uncertainty in the evaluated data comes principally from uncertainty in the thermal functions of Li(l). Equivalent results may be obtained by treating the vapor as an imperfect gas. See Hultgren et al. (1973).

The entropy and the enthalpy increment were calculated taking the electronic partition function equal to 2.

139. Li$^+$(aq).   Calculated from the data given in Annex II.

140. Na(cr).   Defined reference state.

The entropy and the enthalpy increment were calculated from heat-capacity measurements by Martin (1960a), (20-300 K), Martin (1961), (0.4-1.5 K), and Filby and Martin (1963), (3-30 K), taking into account the transition at 51 K. See Gurvich et al. (1982), part 1, page 309.

141. Na(g). $\Delta_f H^\circ$ was calculated by a Third Law treatment of the vapor-pressure data obtained by Heycock and Lamplough (1912), (107.5 ± 0.6) kJ·mol^{-1}, Rodebush and DeVries (1925), (107.4 ± 0.5) kJ·mol^{-1}, Kistiakowsky (1941), (107.4 ± 0.7) kJ·mol^{-1}, Makansi et al. (1955), (107.2 ± 0.7) kJ·mol^{-1}, Shpil'rain and Zvereva (1963), (107.3 ± 0.5) kJ·mol^{-1}, Stone et al. (1966), (107.7 ± 0.7) kJ·mol^{-1}, Achener and Jouthas (1966), (107.8 ± 0.6) kJ·mol^{-1}, Vinogradov and Volyak (1966), (107.6 ± 0.5) kJ·mol^{-1}, Bohdansky and Schins (1967), (107.5 ± 0.7) kJ·mol^{-1}, and Schins et al. (1971), (107.4 ± 0.7) kJ·mol^{-1}. The computations were done for a gas mixture of Na and Na$_2$ taking $D_o(0)(Na_2) = (70.60 ± 0.25)$ kJ·mol^{-1} (Kusch and Hessel (1978)); at higher temperatures the correction for non-ideality of the gaseous mixture was taken into account. The uncertainty in the evaluated data depends mainly on uncertainty in the thermal functions of Na(l) at temperatures above 800 K. Equivalent results may be obtained by treating the vapor as an imperfect gas. See Hultgren et al. (1973) and Gurvich et al. (1982) part 1, pages 521-522.

The entropy and the enthalpy increment were calculated taking the electronic partition function equal to 2.

142. Na$^+$(aq).   Calculated from the data given in Annex II.

143. K(cr).   Defined reference state.

The entropy and the enthalpy increment were calculated from heat-capacity measurements by Krier et al. (1957), (12-320 K), Filby and Martin (1965), (0.4 - 26 K), and Dauphinee et al. (1955), (30-330 K).

144. K(g). $\Delta_f H^\circ$ was calculated from a Third Law treatment of vapor-pressure data obtained by Mayer (1931), (88.58 ± 0.50) kJ·mol^{-1}, Edmondson and Egerton (1927), (89.71 ± 0.25) kJ·mol^{-1}, Neumann and Völker (1932), (89.56 ± 0.25) kJ·mol^{-1}, Roeder and Morawietz (1956), (89.59 ± 0.40) kJ·mol^{-1}, Bonilla (1961),

(88.45 ± 0.80) kJ·mol^{-1}, Achener (1965), (89.07 ± 0.70) kJ·mol^{-1}, Buck and Pauly (1965), (89.72 ± 0.40) kJ·mol^{-1}, Vinogradov and Volyak (1966), (89.20 ± 0.40) kJ·mol^{-1}, Shpil'rain et al. (1968), (88.94 ± 0.50) kJ·mol^{-1}, Schins et al. (1971), (89.04 ± 0.50) kJ·mol^{-1}, and Belova et al. (1980), (89.9 ± 1.0) kJ·mol^{-1}. Computations were done for a gas mixture of K and K$_2$ using the value $D_o(0)(K_2) = (52 \pm 3)$ kJ·mol^{-1} based on the spectroscopic data of Loomis and Nusbaum (1932), (49.8 ± 4.0) kJ·mol^{-1}, the calculations of Müller and Meyer (1984), 51.8 kJ·mol^{-1}, and a value, (53.8 ± 0.8) kJ·mol^{-1}, calculated from vapor-pressure data by Volyak et al. (1975). A value of $\Delta_f H^\circ$, (89.24 ± 0.20) kJ·mol^{-1}, was obtained by Hultgren et al. (1973) from a virial equation treatment of measurements at high temperature ($>$ 800 K).

The entropy and the enthalpy increment were calculated taking the electronic partition function equal to 2.

145. K$^+$(aq).   Calculated from the data given in Annex II.

146. Rb(cr).   Defined reference state.
The entropy and the enthalpy increment were calculated from heat-capacity measurements by Martin (1970), (0.4-3 K) and Filby and Martin (1965), (0.4-320 K).

147. Rb(g). $\Delta_f H^\circ$ was calculated from a Third Law treatment of the vapor-pressure data obtained by Bonilla et al. (1962), (80.9 ± 0.8) kJ·mol^{-1}, Tepper et al. (1963), (81.0 ± 0.8) kJ·mol^{-1}, Achener (1964), (81.0 ± 0.7) kJ·mol^{-1}, Volyak et al. (1968), (80.9 ± 0.3) kJ·mol^{-1}, Schins et al. (1971), (80.8 ± 0.6) kJ·mol^{-1}, and Shpil'rain and Nikanorov (1971), (80.8 ± 0.8) kJ·mol^{-1}. The computations were done for a gas mixture of Rb and Rb$_2$; the partial pressures of Rb were obtained using $D_o(0)(Rb_2) = (46.77 \pm 0.12)$ kJ·mol^{-1}, (Breford and Engelke (1980)). At higher temperatures the correction for non-ideality was taken into account. The uncertainties in the listed values depend mainly on uncertainties in the thermal functions of Rb(l). Equivalent results may be obtained by treating the vapor as an imperfect gas. See Hultgren et al. (1973) and Gurvich et al. (1982), part 1, pages 521-522.

The entropy and the enthalpy increment were calculated taking the electronic partition function equal to 2.

148. Rb$^+$(aq).   Calculated from the data given in Annex II.

149. Cs(cr).   Defined reference state.
The entropy and the enthalpy increment were calculated from heat-capacity measurements by Filby and Martin (1965), (0.4-301 K) and Martin (1970), (0.4-3 K).

150. Cs(g). $\Delta_f H^\circ$ was calculated by a Third Law treatment of the vapor-pressure data obtained by Taylor and Langmuir (1937), (76.7 ± 0.2) kJ·mol^{-1}, Bonilla et al. (1962), (76.2 ± 0.8) kJ·mol^{-1}, Achener (1964), (76.6 ± 0.7) kJ·mol^{-1}, Buck and Pauly (1965), (76.7 ± 0.3) kJ·mol^{-1}, Ewing et al. (1966), (76.4 ± 1.2) kJ·mol^{-1}, Stone et al. (1966), (76.5 ± 1.2) kJ·mol^{-1}, Shpil'rain and Belova (1967), (76.6 ± 0.8) kJ·mol^{-1}, Volyak et al. (1968), (76.5 ± 0.6) kJ·mol^{-1}, Novikov and Roshchupkin (1969), (76.6 ± 0.9) kJ·mol^{-1}, Shpil'rain et al. (1970a), (76.5 ± 0.8) kJ·mol^{-1}, Schins et al. (1971), (76.3 ± 0.7) kJ·mol^{-1}, Shpil'rain and Nikanorov (1972), $(76.5$

± 0.9) kJ·mol⁻¹, and Gushchin et al. (1975), (76.8 ± 0.4) kJ·mol⁻¹. The calculations were done for a gas mixture of Cs and Cs_2, using $D^o(0)(Cs_2) = (42.23 \pm 0.24)$ kJ·mol⁻¹ (Honig et al. (1979)); at higher temperatures the correction for non-ideality was taken into account. The uncertainties in the listed values arise mainly from uncertainties in the thermal functions of Cs(l), which increase rapidly with temperature. Equivalent results may be obtained by treating the vapor as an imperfect gas. See Hultgren et al. (1973) and Gurvich et al. (1982), part 1, pp. 521-522.

The entropy and the enthalpy increment were calculated taking the electronic partition function equal to 2.

151. Cs^+(aq).   Calculated from the data given in Annex II.

Chapter 3

BIBLIOGRAPHY

Abel, E.; Schmid, H.; Stein, M. *Z. Elektrochem.* **1930,** 36, 692.

Abramowitz, S. Private communication. **1977.**

Abrosimov, V.K.; Krestov, G.A. *Zhur. Fiz. Khim.* **1967,** 41, 3150.

Achener, P.Y. *U.S. Aerojet General Corp. Rept. AGN-8090.* Vol I. San Ramon. **1964.**

Achener, P.Y. *U.S. A.E.C. Rept. AGN-8141* **1965.**

Achener, P.Y.; Fisher, D.L. *U.S. A.E.C. Rept. AGN-8191* Vol. 2. **1967.**

Achener, P.Y.; Jouthas, J.T. *U.S. A.E.C. Rept. AGN-8191* **1966.**

Ackermann, R.J.; Rauh, E.G. *J. Phys. Chem.* **1969,** 73, 769.

Ackermann, R.J.; Rauh, E.G. *High Temp. Sci.* **1972a,** 4, 496.

Ackermann, R.J.; Rauh, E.G. *J. Chem. Thermodynamics* **1972b,** 4, 521.

Adami, L.H.; King, E.G. *U.S. Bur. Mines Rept. Inv. 6617* **1965.**

Adams, G.P.; Charlu, T.V.; Margrave, J.L. *J. Chem. Eng. Data* **1970a,** 15, 42.

Adams, G.P.; Margrave, J.L.; Wilson, P.W. *J. Chem. Thermodynamics* **1970b,** 2, 591.

Affortit, C. *Commis. Energ. Atom. Rept. CEA-R-4266* **1967.**

Agarwal, K.L.; Betterton, J.O., Jr. *J. Low Temp. Phys.* **1974a,** 17, 509.

Agarwal, K.L.; Betterton, J.O., Jr. *J. Low Temp. Phys.* **1974b,** 17, 515.

Ahlers, G. *Phys. Rev.* **1966,** 145, 419.

Ahluwalia, J.C.; Cobble, J.W. *J. Am. Chem. Soc.* **1964,** 86, 5381.

Ahrland, S. *Helv. Chim. Acta* **1967,** 50, 306.

Ahrland, S.; Kullberg, L. *Acta Chem. Scand.* **1971a,** 25, 3471.

Ahrland, S.; Kullberg, L. *Acta Chem. Scand.* **1971b,** 25, 3677.

Åkerlöf, G.; Turck, H.E. *J. Am. Chem. Soc.* **1935,** 57, 1746.

Akishin, P.A.; Nikitin, O.T.; Gorokhov, L.N. *Dokl. Akad. Nauk SSSR* **1959,** 129, 1075.

Alcock, C.B.; Grieveson, P. *J. Inst. Metals* **1961,** 90, 304.

Alcock, C.B.; Sridhar, R.; Svedberg, R.C. *Acta Met.* **1969,** 17, 839.

Aldred, A.T.; Pratt, J.N. *J. Chem. Eng. Data* **1963,** 8, 429.

Ambrose, D.; Sprake, C.H.S. *J. Chem. Thermodynamics* **1972,** 4, 603.

Amitin, E.E.; Kovalevskaya, Y.A.; Lebedeva, E.G.; Paukov, I.E. *Zhur. Fiz. Khim.* **1974,** 48, 1880.

Amitin E.E.; Lebedeva, E.G.; Paukov, I.E. *Zhur. Fiz. Khim.* **1979,** 53, 2666.

Anan'eva, E.A.; Maloshuk, V.V. *Thermodynamics and Structure of Solutions. Part 2.* Ivanova. **1975,** p.148.

Anderson, C.T. *J. Am. Chem. Soc.* **1933,** 55, 3621.

Anderson, C.T. *J. Am. Chem. Soc.* **1936,** 58, 568.

Anderson, K.P.; Butler, E.A.; Woolley, E.M. *J. Phys. Chem.* **1971,** 75, 93.

Andon, R.J.L.; Counsell, J.F.; McKerrell, H.; Martin, J.F. *Trans. Faraday Soc.* **1963,** 59, 2702.

Andon, R.J.L.; Martin, J.F.; Mills, K.C. *J. Chem. Soc. A* **1971,** 1788.

Andon, R.J.L.; Mills, K.C. *J. Chem. Thermodynamics* **1971,** 3, 583.

Anisimov, V.M.; Volyak, L.D. *Teplofiz. Vys. Temp.* **1969,** 7, 317.

Anthrop, D.F.; Searcy, A.W. *J. Phys. Chem.* **1964,** 68, 2385.

Applebey, M.P.; Crawford, F.H.; Gordon, K. *J. Chem. Soc.* **1934,** 1665.

Applebey, M.P.; Reid, R.D. *J. Chem. Soc.* **1922,** 121, 2129.

Arii, K. *Bull. Inst. Phys. Chem. Research (Tokyo)* **1929,** 8, 714; *Sci. Repts. Tohoku Imp. Univ.* **1933,** 22, 182.

Ariya, S.M.; Morosova, M.P.; Vol'f, E. *Zhur. Neorg. Khim.* **1957,** 2, 13.

Armbruster, M.H.; Crenshaw, J.L. *J. Am. Chem. Soc.* **1934,** 56, 2525.

Armstrong, G.T.; Coyle, C.F. *U.S. A.E.C. AED-Conf.65-358-1* **1965,** 35.

Armstrong, R.S.; Clark, R.J.H. *J. Chem. Soc. Faraday Trans. II* **1976,** 72, 11.

Arnett, E.M.; McKelvey, D.R. *J. Am. Chem. Soc.* **1966,** 88, 2598.

Aston, J.G.; Gittler, F.L. *J. Am. Chem. Soc.* **1955,** 77, 3173.

Aurivillius, K.; von Heidenstam, O. *Acta Chem. Scand.* **1961,** 15, 1993.

Austin, J.M.; Mair, A.D. *J. Phys. Chem.* **1962,** 66, 519.

Aven, M.H.; Craig, R.S; Waite, T.R.; Wallace, W.E. *Phys. Rev.* **1956,** 102, 1263.

Awbery, J.H.; Griffiths, E. *Proc. Roy. Soc., (London)* **1933,** A 141, 1.

Ayerst, R.P.; Phillips, M.I. *J. Chem. Eng. Data* **1966,** 11, 494.

Bach, R.O.; Boardman, W.W.; Forsyth, M.W. *Chimia* **1964,** 18, 110.

Bachmann, K.J.; Buehler, E. *J. Electrochem. Soc.* **1974,** 121, 835.

Bachmann, K.J.; Hsu, F.S.L.; Remeika, J.P. *Phys. Status Solidi* **1981,** (A) 67, K39.

Bäckström, H.L.J. *J. Am. Chem. Soc.* **1925,** 47, 2443.

Bailey, A.R.; Larson, J.W. *J. Phys. Chem.* **1971,** 75, 2368.

Baker, E.H. *J. Appl. Chem.* **1966,** 16, 321.

Barany, R.; Adami, L.H. *U.S. Bur. Mines Rept. Inv. 6873* **1966.**

Barbero, J.A.; McCurdy, K.G.; Tremaine, P.R. *Can. J. Chem.* **1982,** 60, 1872.

Barieau, R.E.; Giauque, W.F. *J. Am. Chem. Soc.* **1950,** 72, 5676.

Barron, T.H.K.; Berg, W.T.; Morrison, J.A. *Proc. Roy. Soc. (London)* **1959,** A 250, 70.

Barron, T.H.K.; Leadbetter, A.J.; Morrison, J.A. *Proc. Roy. Soc. (London)* **1964,** A 279, 62.

Barrow, R.F.; Clark, T.C.; Coxon, J.A.; Yee, K.K. *J. Mol. Spectr.* **1974,** 51, 428.

Barrow, R.F.; Dodsworth, P.G.; Downie, A.R.; Jefferies, E.A.N.S.; Pugh, A.C.P.; Smith, F.J.; Swinstead, J.M. *Trans. Faraday Soc.* **1955,** 51, 1354.

Barrow, R.F.; du Parcq, R.P. in *Elemental Sulfur,* Meyer, B.: editor. Interscience: New York. **1965.**

Barrow, R.F.; du Parcq, R.P.; Ricks, J.M. *J. Phys. B.* **1969,** 2, 413.

Barrow, R.F.; Stamper, J.G. *Proc. Roy. Soc. (London)* **1961,** A 263, 277.

Barrow, R.F.; Yee, K.K. *J. Chem. Soc. Faraday Trans. II* **1973,** 69, 684.

Bartel, J.J.; Callanan, J.E.; Westrum, E.F., Jr. *J. Chem. Thermodynamics* **1980,** 12, 753.

Batdorf, R.L.; Smits, F.M. *J. Appl. Phys.* **1959,** 30, 259.

Bates, R.G. *J. Am. Chem. Soc.* **1938,** 60, 2983.

Bates, R.G. *J. Am. Chem. Soc.* **1939,** 61, 308.

Bates, R.G.; Acree, S.F. *J. Res. Nat. Bur. Stand.* **1943,** 30, 129.

Bates, R.G.; Bower, V.E. *J. Res. Nat. Bur. Stand.* **1954,** 53, 283.

Bates, R.G.; Pinching, G.D. *J. Am. Chem. Soc.* **1950,** 72, 1393.

Bates, R.G.; Pinching, G.D. *J. Res. Nat. Bur. Stand.* **1949,** 42, 419.

Bates, R.G.; Vosburgh, W.C. *J. Am. Chem. Soc.* **1937,** 59, 1188.

Bates, S.J.; Kirschman, H.D. *J. Am. Chem. Soc.* **1919,** 41, 1991.

Bauer, T.W.; Johnston, H.L. *J. Am. Chem. Soc.* **1953,** 75, 2217.

Bauer, T.W.; Johnston, H.L.; Kerr, E.C. *J. Am. Chem. Soc.* **1950,** 72, 5174.

Baumann, E.W. *J. Inorg. Nucl. Chem.* **1969,** 31, 3155.

Baxter, G.P.; Grose, M.R. *J. Am. Chem. Soc.* **1915,** 37, 1061.

Baxter, G.P.; Hickey, C.H.; Holmes, W.C. *J. Am. Chem. Soc.* **1907,** 29, 127.

Baxter, G.P.; Lansing, J.E. *J. Am. Chem. Soc.* **1920,** 42,419.

Baxter, G.P.; Starkweather, H.W. *J. Am. Chem. Soc.* **1916,** 38, 2038.

Bazlova, I.V.; Stakhanova, M.S.; Karapet'yants, M.Kh.; Vlasenko, K.K. *Zhur. Fiz. Khim.* **1965,** 39, 1245.

Beagley, B.; Brown, D.P.; Freeman, J.M. *J. Molec. Struct.* **1973,** 18, 337.

Bear, I.J.; Turnbull, A.G. *J. Phys. Chem.* **1965,** 69, 2828.

Beattie, I.R.; Ozin, G.A.; Perry, R.O. *J. Chem. Soc. A* **1970,** 2071.

Beck, W.H.; Dobson, J.V.; Wynne-Jones, W.F.K. *Trans. Faraday Soc.* **1960,** 56, 1172.

Becker, G.; Roth, W.A. *Z. Phys. Chem.* **1932,** A 161, 69.

Becker, G.; Roth, W.A. *Z. Phys. Chem.* **1933,** A 167, 1.

Becker, G.; Roth, W.A. *Z. Elektrochem.* **1934,** 40, 836.

Behrens, R.G.; Woodrow, H.O.; Aronson, S. *J. Chem. Thermodynamics* **1977,** 9, 1035.

Bell, J. *J. Chem. Soc.* **1937,** 459.

Bell, J. *J. Chem. Soc.* **1940,** 72.

Belova, A.M.; Shpil'rain, E.E.; Shkermontov, V.I.; Mozgovai, A.G. *High Temp. (USSR)* **1980,** 18, 233.

Belyaev, I.N.; Le Tyuk *Zhur. Neorg. Khim.* **1966,** 11, 1919.

Benson, G.C.; Benson, G.W. *Rev. Sci. Instruments* **1955,** 26, 477.

Bent, H.E.; Forbes, G.S.; Forziati, A.F. *J. Am. Chem. Soc.* **1939,** 61, 709.

Bent, H.E.; Swift, E., Jr. *J. Am. Chem. Soc.* **1936,** 58, 2216.

Berezovskii, G.A.; Paukov, I.E. *Zhur. Fiz. Khim.* **1978,** 52, 2677.

Berezovskii, G.A.; Paukov, I.E. *Zhur. Fiz. Khim.* **1982,** 56, 1783.

Berg, R.L.; Vanderzee, C.E. *J. Chem. Thermodynamics* **1975,** 7, 229.

Berg, R.L.; Vanderzee, C.E. *J. Chem. Thermodynamics* **1978a,** 10, 1113.

Berg, R.L.; Vanderzee, C.E. *J. Chem. Thermodynamics* **1978b,** 10, 1049.

Berg, W.T. *Phys. Rev.* **1968,** 167, 583.

Berg, W.T. *Phys. Rev.* **1976,** B13, 2641.

Berg, W.T.; Morrison, J.A. *Proc. Roy. Soc. (London)* **1957,** A 242, 467.

Berkeley, Earl of *Phil. Trans. Roy. Soc.* **1904,** A 203, 189.

Berkowitz, J.; Chupka, W.A. *J. Chem. Phys.* **1969,** 50, 4245.

Berkowitz, J.; Chupka, W.A.; Guyon, P.M.; Holloway, J.H., Spohr, R. *J. Chem. Phys.* **1971,** 54, 5165.

Bernstein, H.; Kaufman, L. *Rept. AFML-TR66-193* **1966.**

Bezboruah, C.P.; Camoes, M.F.; Covington, A.K.; Dobson, J.V. *J. Chem. Soc., Faraday Trans. I* **1973,** 69, 949.

Bhise, V.S.; Bonilla, C.F. *Proc. Intern. Conf. Liquid Metal Technol. Energy Prod., U.S. D.o.E. Rept. COO-3027-21* **1977,** 2, 657.

Biktimirov, R.S.; Kuzovkina, L.A. *Zhur. Neorg. Khim.* **1970,** 15, 240.

Bills, J.L.; Cotton, F.A. *J. Phys. Chem.* **1964,** 68, 802.

Birch, J.A.; Collins, J.G.; White, G.K. *Austral. J. Phys.* **1979,** 32, 463.

Birky, M.M.; Hepler, L.G. *J. Phys. Chem.* **1960,** 64, 686.

Birley, G.I.; Skinner, H.A. *Trans. Faraday Soc.* **1968,** 64, 3232.

Blachnik, R.O.G.; Gross, P.; Hayman, C. *Fulmer Res. Inst. Sci. Rept. No. 5, Contract AF 61 (052)-863* **1968.**

Blaise, J.; Radziemski, L.J., Jr. *J. Opt. Soc. Am.* **1976,** 66, 644.

Blidin, V.P.; Gordienko, V.I. *Dokl. Akad. Nauk SSSR* **1954,** 94, 1081.

Bock, E. *Can. J. Chem.* **1961,** 39, 1746.

Bodenstein, M. *Z. Phys. Chem.* **1894,** 13, 56.

Boerio, J.; O'Hare, P.A.G. *J. Chem. Thermodynamics* **1976,** 8, 725.

Bogdanov, V.I.; Vekilov, Yu.K.; Tsagarishvili, G.V.; Zhgent, N.M. *Fiz. Tverdogo Tela* **1970,** 12, 3333.

Bogoslovskii, S.S.; Kulifeev, V.K.; Stikhin, A.N.; Ukhlinov, G.A.; Krestovnikov, A.N. *Izvest. Vys. Ucheb. Zaveden., Tsvet. Met.* **1969,** 12, 6, 52.

Bogros, A. *Ann. Phys.,* **1929,** 10, 17; *C.R. Acad. Sci. (Paris)* **1930,** 191, 322.

Bohdansky, J.; Schins, H.E.J. *J. Appl. Phys.* **1965,** 36, 3683.

Bohdansky, J.; Schins, H.E.J. *J. Phys. Chem.* **1967,** 71, 215.

Bonderman, D.; Cater, E.D.; Bennett, W.E. *J. Chem. Eng. Data* **1970,** 15, 396.

Bonilla, C.F. *U.S. A.E.C. Rept. ORNL-3605* **1961,** 1, 286.

Bonilla, C.F. *U.S. D.o.E. Rept. COO-3027-30* **1977.**

Bonilla, C.F.; Sawhney, D.L.; Makansi, M.M. *Trans. Am. Soc. Metals* **1962,** 55, 877.

Bosworth, Y.M.; Clark, R.J.H.; Rippon, D.M. *J. Mol. Spectr.* **1973,** 18, 337.

Bowles, K.J. *N.A.S.A. Tech. Note D-4535.* **1968.**

Bowles, K.J.; Rosenblum, L. *J. Chem. Eng. Data* **1965,** 10, 321; *N.A.S.A., Tech. Note D-2849.* **1965.**

Boyd, D.R.J.; Thompson, H.W. *Spectrochim. Acta* **1952,** 5, 308.

Boyer, A.J.; Meadowcroft, T.R. *Trans. Met. Soc., A.I.M.E.* **1965,** 233; 388.

Brackett, T.E.; Hornung, E.W.; Hopkins, T.E. *J. Am. Chem. Soc.* **1960,** 82, 4155.

Brandner, J.D.; Junk, N.M.; Lawrence, J.W.; Robins, J. *J. Chem. Eng. Data* **1962,** 7, 227.

Bräuer, K.; Strehlow, H. *Z. Phys. Chem. (Frankfurt)* **1958,** 17, 346.

Braune, H.; Peter, S.; Neveling, V. *Z. Naturforsch.* **1951,** 6a, 32.

Bredig, M.A.; Johnson, J.W.; Bronstein, H.R.; Smith, W.T. *U.S. A.E.C. Rept. ORNL-1674* **1953,** 27.

Breford, E.J.; Engelke, F. *Chem. Phys. Letters* **1980,** 75, 132.

Brewer, L.; Searcy, A.W. *J. Am. Chem. Soc.* **1951,** 73, 5308.

Brickwedde, L.H. *J. Res. Nat. Bur. Stand.* **1946,** 36, 377.

Briggs, T.R.; Conrad, C.C.; Gregg, C.C.; Reed, W.H. *J. Phys. Chem.* **1941,** 45, 614.

Briggs, T.R.; Greenawald, J.A.; Leonard, J.W. *J. Phys. Chem.* **1930,** 34, 1951.

Bright, N.F.H.; Hagerty, R.P. *Trans. Faraday Soc.* **1947,** 43, 697.

Brix, P.; Herzberg, G. *Can. J. Phys.* **1954,** 32, 110.

Broadbank, R.W.C.; Dhabanandana, S.; Morcum, K.W. *J. Chem. Soc. A* **1968a,** 213.

Broadbank, R.W.C.; Dhabanandana, S.; Morcum, K.W.; Muju, B.L. *Trans. Faraday Soc.* **1968b,** 64, 3311.

Broadbank, R.W.C.; Hayes, R.W; Morcum, K.W. *J. Chem. Thermodynamics* **1977,** 9, 269.

Brodale, G.; Giauque, W.F. *J. Phys. Chem.* **1972,** 76, 737.

Brodale, G.; Giauque, W.F. *J. Am. Chem. Soc.* **1958,** 80, 2042.

Brodsky, A.E. *Z. Phys. Chem.* **1926,** 121, 1.

Brodsky, A.E. *Z. Elektrochem.* **1929,** 35, 833.

Broene, H.H.; De Vries, T. *J. Am. Chem. Soc.* **1947,** 69, 1644.

Bronson, H.L.; Wilson, A.J.C. *Can. J. Res.* **1936,** A 14, 181.

Brown, C.W.; Overend, J. *Spectrochim. Acta* **1969,** A 25, 1535.

Brown, O.L.I.; Latimer, W.M. *J. Am. Chem. Soc.* **1936,** 58, 2228.

Brown, R.R.; Daut, G.E.; Mrazek, R.V.; Gokcen, N.A. *U.S. Bur. Mines Rept. Inv.* *8379* **1979**.

Brown, S.A.; Land, J.E. *J. Am. Chem. Soc.* **1957**, 79, 3015.

Brown, W.G. *Phys. Rev.* **1931**, 38, 709.

Buck, U.; Pauly, H. *Z. Phys. Chem. (Frankfurt)* **1965**, 44, 345.

Budininkas, P.; Edwards, R.K.; Wahlbeck, P.G. *J. Chem. Phys.* **1968**, 48, 2859.

Burger, J.D.; Liebhafsky, H.A. *Anal. Chem.* **1973**, 45, 600.

Burnett, J.L.; Zirin, M.H. *J. Inorg. Nucl. Chem.* **1966**, 28, 902.

Busey, R.H.; Giauque, W.F. *J. Am. Chem. Soc.* **1953**, 75, 806.

Büttenbender, G.; Herzberg, G. *Ann. Phys.* **1935**, 21, 577.

Callear, A.B.; Pilling, M.J. *Trans. Faraday Soc.* **1970**, 66, 1618.

Calvet, E. *J. Chim. Phys.* **1933**, 30, 140.

Calvo, C.; Simons, E.L. *J. Am. Chem. Soc.* **1952**, 74, 1202.

Campbell, A.N.; Griffiths, J.E. *Can. J. Chem.* **1956**, 34, 1647.

Caneiro, A.; Fouletier, J.; Kleitz, M. *J. Chem. Thermodynamics* **1981**, 13, 823.

Cappelina, F.; Napolitano, G. *Ann. Chim. (Rome)* **1966**, 56, 30.

Capwell, K.J.; Rosenblatt, G.M. *J. Mol. Spectr.* **1970**, 33, 525.

Carmody, W.R. *J. Am. Chem. Soc.* **1929**, 51, 2905.

Carpenter, C.D.; Jette, E.R. *J. Am. Chem. Soc.* **1923**, 45, 578.

Caunt, A.D.; Mackle, H.; Sutton, L.E. *Trans. Faraday Soc.* **1951**, 47, 943.

Centnerszwer, M. *Z. Phys. Chem.* **1913**, 85, 99.

Cerquetti, A.; Longhi, P.; Mussini, T. *J. Chem. Eng. Data* **1968**, 13, 458.

Cerutti, P.J.; Ko, H.C.; McCurdy, K.G.; Hepler, L.G. *Can. J. Chem.* **1978**, 56, 3084.

Cetas, T.C.; Holste, J.C.; Swenson, C.A. *Phys. Rev.* **1969**, 182, 679.

Cezairliyan, A.; Müller, A.B. *High Temp.-High Press.* **1977**, 9, 319.

Chadwick, J.R.; Mansfield, J.D.; Scott, T.J. *J. Inorg. Nucl. Chem.* **1966**, 28, 2449.

Chambers, F.S., Jr.; Sherwood, T.K. *J. Am. Chem. Soc.* **1937**, 59, 316.

Chang, Y.A.; Wilhelm, G.C.; Lathrop, M.; Gyuk, I. *Acta Met.* **1971**, 19, 795.

Charette, G.G.; Flengas, S.N. *J. Electrochem. Soc.* **1968**, 115, 796.

Chernyaev, V.S.; Ershova, S. *Izvest. Vys. Ucheb. Zaved., Tsvet. Met.* **1964**, 76.

Chernyaeva, L.I.; Proskurin, V.N. *Teplofiz.Vys. Temp.* **1972**, 10, 674.

Chiang, T.L.; Hsieh, Y.Y. *J. Chin. Chem. Soc.* **1949**a, 16, 65.

Chiang, T.L.; Hsieh, Y.Y. *J. Chin. Chem. Soc.* **1949**b, 16, 10.

Chihara, H.; Nakamura, M. *Bull. Chem. Soc. Japan* **1972**, 45, 133.

Childs, C.W.; Downes, C.J.; Platford, R.F. *Austral. J. Chem.* **1973**, 26, 863.

Choudhary, B.K.; Prasad, B. *J. Indian Chem. Soc.* **1973**, 50, 153.

Choudhary, B.K.; Prasad, B. *J. Indian Chem. Soc.* **1975**, 52, 679.

Chow, M. *J. Am. Chem. Soc.* **1920**, 42, 497.

Chrétien, A.; Weil, R. *Bull. Soc. Chim. France* **1935**, [5], 2, 1577.

Christ, C.L.; Hostetler, P.B.; Silbert, R.M. *J. Res. U.S. Geol. Survey* **1974**, 2, 175.

Christensen, J.J.; Izatt, R.M. *J. Phys. Chem.* **1962**, 66, 1030.

Christensen, J.J.; Izatt, R.M.; Hansen, L.D.; Partridge, J.H. *J. Phys. Chem.* **1966**, 70, 2003.

Ciavatta, L.; Palombari, R. *Gazz. Chim. Ital.* **1972**, 102, 1098.

Clark, R.J.H.; Hunter, B.K.; Rippon, D.M. *Inorg. Chem.* **1972**, 11, 56.

Clark, R.J.H.; Rippon, D.M. *Chem. Commun.* **1971**, 1295.

Clark, R.J.H.; Rippon, D.M. *J. Mol. Spectr.* **1972**, 44, 479.

Clarke, E.C.W.; Glew, D.N. *Can. J. Chem.* **1971**, 49, 691.

Clusius, K.; Eichenauer, W. *Z. Naturforsch.* **1956**, 11a, 715.

Clusius, K.; Franzosini, P. *Z. Phys. Chem. (Frankfurt)* **1958**, 16, 194.

Clusius, K.; Goldman, J.; Perlick, A. *Z. Naturforsch.* **1949**, 4a, 424.

Clusius, K.; Harteck, P. *Z. Phys. Chem.* **1928**, 134, 243.

Clusius, K.; Vaughen, J.V. *J. Am. Chem. Soc.* **1930**, 52, 4686.

Clynne, M.A.; Potter, R.W. *J. Chem. Eng. Data* **1979**, 24, 338.

CODATA Task Group on Key Values for Thermodynamics. *CODATA Bulletin No.5* **1971**.

CODATA Task Group on Key Values for Thermodynamics. *J. Chem. Thermodynamics,* **1972**, 4, 331; *Teplofiz. Vys. Temp.* **1971**, 9, 657.

CODATA Task Group on Key Values for Thermodynamics. *CODATA Bulletin No. 10* **1973**; *J. Chem. Thermodynamics* **1975**, 7, 1.

CODATA Task Group on Key Values for Thermodynamics. *CODATA Bulletin No. 17* **1976**.

CODATA Task Group on Key Values for Thermodynamics. *CODATA Bulletin No. 22* **1977**.

CODATA Task Group on Key Values for Thermodynamics. Recalculation from referenced data, **1984.**

CODATA Thermodynamic Tables. Selections for Some Compounds of Cacium and Related Mixtures. Garvinj, D.; Parker, V.B.; White, H.J., Jr. editors. Hemisphere Publ. Co., New York, 1987.

Coffy, G.; Olofsson, G. *J. Chem. Thermodynamics* **1979**, 11, 141.

Cogley, D.R.; Butler, J.N. *J. Phys. Chem.* **1968**, 72, 1017.

Colbourn, E.A.; Dagenais, M.; Douglas, A.E.; Raymonda, J.W. *Can. J. Phys.* **1976**, 54, 1343.

Coleman, F.F.; Egerton, A.C. *Phil. Trans. Roc. Soc. (London)* **1935**, A 234, 177.

Collins, E.M.; Menzies, A.W.C. *J. Phys. Chem.* **1936**, 40, 379.

Colomina, M.; Nicolas, J. *An. Soc. Españ. Fis. Quim., B* **1949**, 45, 137.

Cook, R.O.; Davies, A.; Staveley, L.A.K. *J. Chem. Soc. Faraday Trans. I* **1972**, 68, 1384.

Cordes, H.; Commenga, H. *Z. Physik. Chem. [N.F]* **1965**, 45, 196.

Cordfunke, E.H.P. *J. Phys. Chem.* **1964**, 68, 3353.

Cordfunke, E.H.P.; Aling, P. *Trans. Faraday Soc.* **1965**, 61, 50.

Cordfunke, E.H.P.; Ouweltjes, W.; Prins, G. *J. Chem. Thermodynamics* **1976**, 8, 241.

Corliss, C.; Sugar, J. *J. Phys. Chem. Ref. Data* **1979**, 8, 1.

Corsaro, G.; Stephans, H.L. *J. Electrochem. Soc.* **1957**, 104, 512.

Cosgrove, L.A.; Snyder, P.E. *J. Am. Chem. Soc.* **1953**, 75, 3102.

Coughlin, J.P. *J. Am. Chem. Soc.* **1955**, 77, 868.

Coughlin, J.P. *J. Phys. Chem.* **1958**, 62, 419.

Coulter, L.V.; Pitzer, K.S.; Latimer, W.M. *J. Am. Chem. Soc.* **1940**, 62, 2845.

Counsell, J.F.; Martin, J.F. *J. Chem. Soc. A* **1967**, 560.

Covington, A.K.; Dobson, J.V.; Srinivasan, K.V. *J. Chem. Soc., Faraday Trans. I* **1973**, 69, 94.

Covington, A.K.; Dobson, J.V.; Wynne-Jones, W.F.K. *Trans. Faraday Soc.* **1965**a, 61, 2050.

Covington, A.K.; Dobson, J.V.; Wynne-Jones, W.F.K. *Trans. Faraday Soc.* **1965**b, 61, 2057.

Covington, A.K.; Dobson, J.V.; Wynne-Jones, W.F.K. *Electrochim. Acta* **1967**a, 12, 513.

Covington, A.K.; Dobson, J.V.; Wynne-Jones, W.F.K. *Electrochim. Acta* **1967**b, 12, 525.

Cowan, M.; Gordy, W. *Phys. Rev.* **1956**, 104, 551.

Cowperthwaite, I.A.; La Mer, V.K. *J. Am. Chem. Soc.* **1931**, 53, 4333.

Cox, J.D.; Harrop, D. *Trans Faraday Soc.* **1965**, 61, 1328.

Cox, J.D.; Harrop, D.; Head, A.J. *J. Chem. Thermodynamics* **1979**, 11, 811.

Cox, W.P.; Hornung, E.W.; Giauque, W.F. *J. Am. Chem. Soc.* **1955**, 77, 3935.

Craig, R.S.; Krier, C.A.; Coffer, L.W.; Bates, E.A.; Wallace, W.E. *J. Am. Chem. Soc.* **1954**, 76, 238.

Crookston, R.B.; Canjar, L.N. *J. Chem. Eng. Data* **1963**, 8, 544.

Dadgar, A.; Taherian, M.R. *J. Chem. Thermodynamics* **1977**, 9, 711.

Dainton, F.S.; Kimberley, H.M. *Trans. Faraday Soc.* **1950**, 46, 912.

Dakin, T.W.; Ewing, D.T. *J. Am. Chem. Soc.* **1940**, 62, 2280.

Darnell, A.J.; McCollum, W.A.; Milne, T.A. *J. Phys. Chem.* **1960**, 64, 341.

Das, D.; Dharwadkar, S.R; Chandrasekharaiah, M.S *J. Nucl. Mater.* **1985**, 130, 217.

Das, S.N.; Ives, D.J.G. *J. Chem. Soc.* **1962**, 1619.

Dauphinee, T.M.; Martin, D.L.; Preston-Thomas, H. *Proc. Roy. Soc. (London)* **1955**, A 233, 214.

Davies, A.; Staveley, L.A.K. *J. Chem. Thermodynamics* **1972**, 4, 267.

Davies, D.H.; Benson, G.C. *Can. J. Chem.* **1965**, 43, 3100.

Davies, J.; Lacher, J.R.; Park, J.D. *Trans. Faraday Soc.* **1965**, 61, 2143.

Davies, M.; Gwynne, E. *J. Am. Chem. Soc.* **1952**, 74, 2748.

Davis, S.G.; Anthrop, D.F.; Searcy, A.W. *J. Chem. Phys.* **1961**, 34, 659.

Davis, W., Jr.; de Bruin, H.J. *J. Inorg. Nucl. Chem.* **1964**, 26, 1069.

De Maria, G.; Burns, R.P.; Drowart, J.; Inghram, M. *J. Chem. Phys.* **1960**, 32, 1373.

De Maria, G.; Piacente, V. *J. Chem. Thermodynamics* **1974**, 6, 1.

de Passille, A.; Séon, M. *C.R. Acad. Sci. (Paris)* **1934**, 199, 417.

de Passille, A. *Ann. Chim. II* **1936**, 5, 83.

Derby, I.H.; Yngve, V. *J. Am. Chem. Soc.* **1916**, 38, 1439.

DeSorbo, W. *J. Chem. Phys.* **1955**a, 23, 1970.

DeSorbo, W. *J. Am. Chem. Soc.* **1955**b, 77, 4713.

Davies, M.; Gwynne, E. *J. Am. Chem. Soc.* **1952**, 74, 2748.

DeSorbo, W.; Nichols, G.E. *J. Phys. Chem. Solids* **1958**, 6, 352.

DeSorbo, W.; Tyler, W.W. *J. Chem. Phys.* **1953**, 21, 1660.

Devina, O.A.; Efimov, M.E.; Medvedev, V.A.; Khodakovsky, I.L. *Geokhim.* **1982**a, (4), 550.

Devina, O.A.; Efimov, M.E.; Medvedev, V.A.; Khodakovsky, I.L. *Geokhim.* **1982**b, (10), 1454.

Devina, O.A.; Khodakovsky, I.L.; Efimov, M.E.; Medvedev, V.A. *Proc. Seventh All-Union Conf. Calorimetry.* Moscow. **1977**a, p. 231.

Devina, O.A.; Khodakovsky, I.L.; Efimov, M.E.; Medvedev, V.A. *Proc. Seventh All-Union Conf. Calorimetry.* Moscow. **1977**b, 231.

DeVisser, C.; Somsen, G. *J. Chem. Soc. Faraday Trans. I* **1973**, 69, 1440.

DeVries, T.; Rodebush, W.H. *J. Am. Chem. Soc.* **1927**, 49, 656.

Dewey, P.H.; Harper, D.R. *J. Res. Nat. Bur. Stand.* **1938**, 21, 457.

Dill, A.J.; Itzkowitz, L.M.; Popovych, O. *J. Phys. Chem.* **1968**, 72, 4580.

Di Lonardo, G.; Douglas, A.E. *Can. J. Phys.* **1973**, 51, 434.

Di Lonardo, G.; Douglas, A.E. *J. Chem. Phys.* **1972**, 56, 5185.

Dingemans, P. *Rec. Trav. Chim.* **1938**a, 57, 144.

Dingemans, P. *Rec. Trav. Chim.* **1938**b, 57, 703.

Dingle, T.W.; Freedman, P.A.; Gelernt, B.; Jones, W.J.; Smith, I.W.M. *Chem. Phys.* **1975**, 8, 171.

Dixit, M.N. *Proc. Indian Acad. Sci.* **1967**, A, 66, 325.

Doescher, R.N. *J. Chem. Phys.* **1952**, 20, 330.

Domalski, E.S.; Armstrong, G.T. *J. Res. Nat. Bur. Stand.* **1965**, 69 A, 137.

Domalski, E.S.; Armstrong, G.T. *J. Res. Nat. Bur. Stand.* **1967**, 71 A, 105.

Domange, L. *Ann. Chim. II* **1937**, 7, 225.

Douglas, A.E.; Hoy, R.R. *Can. J. Phys.* **1975**, 53, 1965.

Douglas, A.E.; Møller, C.K. *Can. J. Phys.* **1955**, 33, 125.

Douglas, A.E.; Møller, C.K.; Stoicheff, B.P. *Can. J. Phys.* **1963**, 41, 1174.

Douglas, A.E.; Rao, K.S. *Can. J. Phys.* **1958**, 36, 565.

Douglas, P.E. *Proc. Phys. Soc.* **1954**, 67B, 783.

Douglas, T.B.; Ditmars, D.A. *J. Res. Nat. Bur. Stand.* **1967**, 71A, 185.

Douglas, T.B.; Dever, J.L. *J. Am. Chem. Soc.* **1954**, 76, 4824.

Downie, D.B.; Martin, J.F. *J. Chem. Thermodynamics* **1980**, 12, 779.

Drakin, S.I.; Chang Yu-min *Zhur. Fiz. Khim.* **1964**, 38, 2800.

Drowart, J.; De Maria, G. *Proc. Conf. on Silicon Carbide.* Pergamon: Boston. **1960**, p.16.

Drowart, J.; De Maria, G.; Inghram, M.G. *J. Chem. Phys.* **1958**, 29, 1015.

Drowart, J.; Goldfinger, P. *J. Chim. Phys.* **1958**, 55, 721.

Drowart, J.; Goldfinger, P.; Detry, D.; Rickert, H.; Keller, H. *Advan. Mass Spectr.* **1968**, 4, 499.

Drummond, G.; Barrow, R.F. *Proc. Phys. Soc.* **1952**, 65 A, 277.

Dry, M.E.; Gledhill, J.A. *Trans. Faraday Soc.* **1955**, 51, 1119.

Duecker, H.C.; Haller, W. *J. Phys. Chem.* **1962**, 66, 225.

Dugdale, J.S.; Morrison, J.A.; Patterson, D. *Proc. Roy. Soc. (London)* **1954**, A 224, 228.

Duisman, J.A.; Stern, S.A. *J. Chem. Eng. Data* **1969**, 14, 457.

Dummer, G.Z. *Z. Phys.* **1965**, 186, 249.

Duncan, J.L. *J. Mol. Spectr.* **1967**, 22, 247.

Dunsmore, H.S.; Nancollas, G.H. *J. Phys. Chem.* **1964**, 68, 1579.

Durham, G.S.; Rock, E.J.; Frayn, J.S. *J. Am. Chem. Soc.* **1953**, 75, 5792.

Eastman, E.D.; McGavock, W.C. *J. Am. Chem. Soc.* **1937**, 59, 145.

Eastman, E.D.; Milner, R.T. *J. Chem. Phys.* **1933**, 1, 444.

Eastman, E.D.; Williams, A.M.; Young, T.F. *J. Am. Chem. Soc.* **1924**, 46, 1178.

Eckman, J.R.; Rossini, F.D. *J. Res. Nat. Bur. Stand.* **1929**, 3, 597.

Edlen, B.; Ölme, A.; Herzberg, G.; Johns, J.W.C. *J. Opt. Soc. Am.* **1970**, 60, 889.

Edmondson, W.; Egerton, A. *Proc. Roy. Soc. (London)* **1927**, A 113, 520.

Edwards, H.G.M; Long, D.O.; Love, R. *Adv. Raman Spectr.* **1973**, 1, 504.

Edwards, J.W.; Johnston, H.L.; Ditmars, W.E. *J. Am. Chem. Soc.* **1953**, 75, 2467.

Efimov, M.E.; Klevaichuk, G.N.; Medvedev, V.A.; Kilday, M.V. *J. Res. Nat. Bur. Stand.* **1979**, 84, 273.

Efimov, M.E.; Medvedev, V.A. *J. Chem. Thermodynamics* **1975**, 7, 719.

Egan, E.P., Jr.; Luff, B.B. *J. Phys. Chem.* **1961**, 65, 523.

Egan, E.P., Jr.; Luff, B.B. *J. Chem. Eng. Data* **1963**a, 8, 181.

Egan, E.P, Jr.; Luff, B.B. *Heats of Formation of Phosphorus Oxides.* Progress Report, June 1 to November 30, AD-429008, National Technical Information Services: Springfield, V.A., **1963**b.

Ehrlich, P.; Koknat, F.W.; Seifert, H.-J. *Z. Anorg. Allg. Chem.* **1965**, 341, 281.

Ehrlich, P.; Peik, K.; Koch, E. *Z. Anorg. Allg. Chem.* **1963**, 324, 113.

Eichenauer, W.; Schultze, M. *Z. Naturforsch.* **1959**, 14a, 28.

Ellis, A.J. *J. Chem. Soc.* **1963**, 4300.

Ellis, A.J.; Golding, R.M. *J. Chem. Soc.* **1959**, 127.

Ellis, A.J.; Milestone, N.B. *Geochim. Cosmochim. Acta* **1967**, 31, 615.

Elsemongy, M.M.; Fouda, A.S. *J. Chem. Thermodynamics* **1981**, 13, 725.

Elsemongy, M.M.; Fouda, A.S. *J. Chem. Thermodynamics* **1982**, 14, 1.

Emons, H.-H.; Röser, H. *Z. Anorg. Allg. Chem.* **1966**, 346, 225.

Ender, F.; Teltschik, W.; Schäfer, K. *Z. Elektrochem.* **1957**, 61, 775.

Espada, L.; Pilcher, G.; Skinner, H.A. *J. Chem. Thermodynamics* **1970**, 2, 647.

Esval, O.E.; Tyree, S.Y., Jr. *J. Phys. Chem.* **1962**, 66, 940.

Eucken, A.; Clusius, K.; Woitinek, H. *Z. Anorg. Allg. Chem.* **1931**, 203, 39.

Eucken, A.; Schröder, E. *Z. Phys. Chem.* **1938**, B 41, 307.

Evans, C.E.; Monk, C.B. *Trans. Faraday Soc.* **1971**, 67, 2652.

Everett, D.H.; Landsman, D.A. *Trans. Faraday Soc.* **1954**, 50, 1221.

Ewing, C.T.; Spann, J.R.; Stone, J.P.; Steinkuller, E.W.; Miller, R.R. *J. Chem. Eng. Data* **1970**, 15, 508.

Ewing, C.T.; Stone, J.P.; Spann, J.R.; Miller, R.R. *J. Chem. Eng. Data* **1966**, 11, 460.

Ewing, W.W. *J. Am. Chem. Soc.* **1927**, 49, 1963.

Ewing, W.W.; Rogers, A.N.; Miller, J.Z.; McGovern, E. *J. Am. Chem. Soc.* **1932**, 54, 1335.

Faita, G.; Longhi, P.; Mussini, T. *J. Electrochem. Soc.* **1967**, 114, 340.

Fajans, K.; Karagunis, G. cited in Meyer, K.H.; Dunkel, M. *Z. Physik. Chem. Bodenstein Festband* **1931**, p. 553.

Faktor, M.M.; Carasso, J.I. *J. Electrochem. Soc.* **1965**, 112, 817.

Farber, M.; Darnell, A.J. *J. Chem. Phys.* **1955**, 23, 1460.

Fasolino, L.G. *J. Chem. Eng. Data* **1965**, 10, 371.

Feakins, D.; Watson, P. *J. Chem. Soc.* **1963**, 4686.

Fedorov, A.S.; Sil'chenko, G.F. *Ukrain. Khim. Zhur.* **1937**, 12, 53.

Feick, G. *J. Am. Chem. Soc.* **1954**, 76, 5858.

Fenning, R.W.; Cotton, F.T. *Proc. Roy. Soc. (London)* **1933**, A 141, 17.

Fenwick, J.T.F.; Wilson, J.W. *J. Chem. Soc. Dalton Trans.* **1972**, 1324.

Filby, J.D.; Martin, D.L. *Proc. Roy. Soc. (London)* **1963**, A 276, 187.

Filby, J.D.; Martin, D.L. *Proc. Roy. Soc. (London)* **1965**, A 284, 83.

Filosofo, I.; Merlin, M.; Rostagni, A. *Nuovo Cim.* **1950**, 7, 69.

Fiock, E.F.; Rodebush, W.H. *J. Am. Chem. Soc.* **1926**, 48, 2522.

Fischbeck, K.; Eich, H. *Z. Anorg. Allg. Chem.* **1937**, 235, 83.

Fischer, A.K. *Rev. Sci. Instrum.* **1966**, 37, 717.

Fisher, J.R.; Barnes, H.L. *J. Phys. Chem.* **1972**, 76, 90.

Fitzgibbon, G.C.; Pavone, D.; Holley, C.E., Jr. *J. Chem. Eng. Data* **1967**, 12, 122.

Flatt, R.; Burkhardt, G. *Helv. Chim. Acta* **1944**, 27, 1605.

Fletcher, A.N. *J. Inorg. Nucl. Chem.* **1964**, 26, 955.

Flotow, H.E.; Lohr, H.R. *J. Phys. Chem.* **1960**, 64, 904.

Flotow, H.E.; O'Hare, P.A.G.; Boerio-Goates, J. *J. Chem. Thermodynamics* **1981**, 13, 477.

Flotow, H.E.; Osborne, D.W. *Phys. Rev.* **1966**, 151, 564.

Flubacher, P.; Leadbetter, A.J.; Morrison J.A. *Phil. Mag.* **1959**, 4, 273.

Flubacher, P.; Leadbetter, A.J.; Morrison, J.A. *J. Chem. Phys.* **1960,** 33, 1751.

Foote, H.W.; Saxton, B.; Dixon, J.K. *J. Am. Chem. Soc.* **1932,** 54, 563.

Forsythe, W.R.; Giauque, W.F. *J. Am. Chem. Soc.* **1942,** 64, 48.

Fortier, J.-L.; Leduc, P.A.; Desnoyers, J.E. *J. Solution Chem.* **1974,** 3, 323.

Fosbinder, R.J. *J. Am. Chem. Soc.* **1929,** 51, 1345.

Freeth, F.A. *Phil. Trans. Roy. Soc. (London)* **1922,** A 223, 35.

Freyland, W.F.; Hensel, F. *Ber. Bunsenges. Phys. Chem.* **1972,** 76, 16.

Fricke, R.; Havestadt, L. *Z. Elektrochem.* **1927,** 33, 441.

Fricke, R.; Wüllhorst, B. *Z. Anorg. Allg. Chem.* **1932,** 205, 127.

Fried, F. *Z. Phys. Chem.* **1926,** 123, 406.

Friedman, H.L.; Kahlweit, M. *J. Am. Chem. Soc.* **1956,** 78, 4243.

Friedman, H.L.; Schug, K. *J. Am. Chem. Soc.* **1956,** 78, 3881.

Frink, C.R.; Peech, M. *Soil Sci. Soc. Am. Proc.* **1962,** 26, 346.

Frink, C.R.; Peech, M. *Inorg. Chem.* **1963,** 2, 473.

Frisch, M.A.; Margrave, J.L. *J. Phys. Chem.* **1965,** 69, 3863.

Frolova, G.I.; Paukov, I.E. *Zhur. Fiz. Khim.* **1983,** 57, 2314.

Frost, G.B.; Moon, K.A.; Tompkins, E.H. *Can. J. Chem.* **1951,** 29, 604.

Füchtbauer, C.; Bartels, H. *Z. Phys.* **1921,** 4, 337.

Fugate, R.Q.; Swenson, C.A. *J. Appl. Phys.* **1969,** 40, 3034.

Fuger, J.; Oetting, F.L. *The Chemical Thermodynamics of Actinide Elements and Compounds. Part 2. The Actinide Aqueous Ions.* I.A.E.A.: Vienna. **1976.**

Fuger, J.; Parker, V.B.; Hubbard, W.N.; Oetting, F.L. *The Chemical Thermodynamics of Actinide Elements and Compounds. Part 8. The Actinide Halides.* I.A.E.A.: Vienna. **1983.**

Furukawa, G.T. Private communication. **1977.**

Furukawa, G.T.; Douglas, T.B.; McCoskey, R.E.; Ginnings, D.C. *J. Res. Nat. Bur. Stand.* **1956,** 57, 67.

Furukawa, G.T.; Reilly, M.L. *Nat. Bur. Stand. Report 9905* **1968.**

Furukawa, G.T.; Saba, W.G.; Reilly, M.L. *Critical Analysis of the Heat Capacity of the Literature and Evaluation of Thermodynamic Properties of Copper, Silver and Gold from 0 to 300 K.* NSRDS-NBS 18, **1968.**

Gal'chenko, G.L; Kornilov, A.N.; Skuratov, S.M. *Zhur. Neorg. Khim.* **1960**b, 5, 2651.

Gal'chenko, G.L.; Timofeev, B.I.; Skuratov, S.M. *Zhur. Neorg. Khim.* **1960**a, 5, 2645.

Gallagher, K.; Brodale, G.E.; Hopkins, T.E. *J. Phys. Chem.* **1960,** 64, 687.

Gamo, I. *J. Mol. Spectr.* **1969,** 30, 216.

Gamsjägar, H.; Rainer, W.; Schindler, P. *Monatsh. Chem.* **1967,** 98, 1793.

Gardner, A.W.; Glueckauf, E. *Proc. Roy. Soc. (London)* **1969,** A 313, 131.

Gardner, A.W.; Glueckauf, E. *Trans. Faraday Soc.* **1970,** 66, 1081.

Gardner, T.E.; Taylor, A.R. *U.S. Bur. Mines Rept. Inv. 6435* **1964.**

Gardner, W.L.; Jekel, E.C.; Cobble, J.W. *J. Phys. Chem.* **1969**a, 73, 2017.

Gardner, W.L.; Mitchell, R.E.; Cobble, J.W. *J. Phys. Chem.* **1969**b, 73, 2021.

Garrels, R.M.; Gucker, F.T. *Chem. Rev.* **1949,** 44, 117.

Garrett, A.B.; Heiks, R.E. *J. Am. Chem. Soc.* **1941,** 63, 562.

Garrett, A.B.; Hirschler, A.E. *J. Am. Chem. Soc.* **1938,** 60, 299.

Garrett, A.B.; Vellenga, S.; Fontana, C.M. *J. Am. Chem. Soc.* **1939,** 61, 367.

Garvin, D.; Parker, V.B.; Wagman, D.D.; Evans, W.H. *NBSIR 76-1147,* Washington D.C. **1976.**

Gaydon, A.G. *Dissociation Energies and Spectra of Diatomic Molecules,* 3rd Edition. Chapman and Hall: London. **1968.**

Gayles, J.N.; Self, J. *J. Chem. Phys.* **1964,** 40, 3530.

Gebhardt, A. *Ber. Dtsch. Chem. Ges.* **1905,** 38, 184.

Gedansky, L.M.; Pearce, P.J.; Hepler, L.G. *Can. J. Chem.* **1970,** 48, 1770.

Gel'd, P.V.; Putintsev, Yu.V. *Tr. Ural. Politekh. Inst.* **1970,** 186, 199 (CA:76-145671).

Gellner, O.H.; Skinner, H.A. *J. Chem. Soc.* **1949,** 1145.

Gerasimov, Ya. I.; Voronin, G.F.; Shiu, N.T. *J. Chem. Thermodynamics* **1969,** 1, 425.

Gerding, P.; Leden, I.; Sunner, S. *Acta Chem. Scand.* **1963,** 17, 2190.

Gerke, R.H. *J. Am. Chem. Soc.* **1922,** 44, 1684.

Gerke, R.H.; Geddes, S., *J. Phys. Chem.* **1927,** 31, 886.

Gerkin, R.E.; Pitzer, K.S. *J. Am. Chem. Soc.* **1962,** 84, 2662.

Germain, P.; Perachon, G.; Thourey, J. *J. Fluorine Chem.* **1979,** 13, 141.

Giauque, W.F.; Archibald, R.C. *J. Am. Chem. Soc.* **1937,** 59, 561.

Giauque, W.F.; Barieau, R.E.; Kunzler, J.E. *J. Am. Chem. Soc.* **1950,** 72, 5685.

Giauque, W.F.; Hornung, E.W.; Kunzler, J.E.; Rubin, T.R. *J. Am. Chem. Soc.* **1960,** 82, 62.

Giauque, W.F.; Kemp, J.D. *J. Chem. Phys.* **1938,** 6, 40.

Giauque, W.F.; Meads, P.F. *J. Am. Chem. Soc.* **1941,** 63, 1897.

Giauque, W.F.; Stout, J.W. *J. Am. Chem. Soc.* **1936,** 58, 1144.

Gibbard, H.F., Jr.; Scatchard, G. *J. Chem. Eng. Data* **1973,** 18, 293.

Giguère, P.A.; Morissette, B.G.; Olmos, A.W.; Knop, O. *Can. J. Chem.* **1955,** 33, 804.

Gilbert, I.G.F.; Kitchener, J.A. *J. Chem. Soc.* **1956,** 3919.

Gilbreath, W.P. *N.A.S.A., Tech. Note D-2723,* **1965.**

Gillespie, L.J.; Fraser, L.H.D. *J. Am. Chem. Soc.* **1936,** 58, 2260.

Gilliland, A.A.; Johnson, W.H. *J. Res. Nat. Bur. Stand.* **1961,** 65 A, 67.

Gilliland, A.A.; Wagman, D.D. *J. Res. Nat. Bur. Stand.* **1965,** 69 A, 1.

Gilliland, A.A.; Wagman, D.D. *Nat. Bur. Stand. Rept. 9389* **1966,** 92.

Gingerich, K.A. *J. Chem. Phys.* **1969,** 51, 4433.

Ginn, S.G.W.; Brown, C.W.; Kenney, J.K.; Overend, J. *J. Mol. Spectr.* **1968a,** 28, 509.

Ginn, S.G.W.; Johansen, D.; Overend, J. *J. Mol. Spectr.* **1970,** 36, 448.

Ginn, S.G.W.; Kenney, J.K.; Overend, J. *J. Chem. Phys.* **1968b,** 48, 1571.

Ginter, M.L.; Tilford, S.G. *J. Mol. Spectr.* **1971,** 37, 159.

Glasstone, S. *J. Chem. Soc.* **1921,** 119, 1914.

Gledhill, J.; Malan, G.M. *Trans. Faraday Soc.* **1954,** 50, 126.

Glemser, O.; Stocker, V. *Ber. Bunsenges. Phys. Chem.* **1963,** 67, 505.

Gmelin, E. *C.R. Acad. Sci. (Paris)* **1966,** 262, 1452.

Gmelin, E. *Z. Naturforsch.* **1969,** 24a, 1794.

Gode, B.K. *Trekhkomponentnye sistemy s bornoi kislotoi.* Riga. **1969.**

Goldberg, R.N. *J. Phys. Chem. Ref. Data* **1979,** 8, 1005.

Goldberg, R.N. *J. Phys. Chem. Ref. Data* **1981,** 10, 671.

Goldberg, R.N.; Hepler, L.G. *J. Phys. Chem.* **1968,** 72, 4654.

Goldhaber, M.B.; Kaplan, I.R. *Marine Chem.* **1975,** 3, 83.

Good, W.D.; Lacina, J.L.; DePrater, B.L.; McCullough, J.P. *J. Phys. Chem.* **1964,** 68, 579.

Good, W.D.; Lacina, J.L.; McCullough, J.P. *J. Am. Chem. Soc.* **1960,** 82, 5589.

Good, W.D.; Månsson, M. *J. Phys. Chem.* **1966,** 70, 97.

Gordievskii, A.V.; Filippov, E.L.; Shterman, V.S.; Krivoshein, A.S. *Zhur. Fiz. Khim.* **1968,** 42, 1998.

Gordon, J.E.; Montgomery, H.; Noer, R.J.; Pickett, G.R.; Tobón, R. *Phys. Rev.* **1966,** 152, 432.

Gould, R.K; Vosburgh, W.C. *J. Am. Chem. Soc.* **1940,** 62, 1817.

Grachev, N.S.; Kirillov, P.L. *Inzh. Fiz. Zhur.* **1960,** (3), 6, 62.

Graham, R.L.; Hepler, L.G. *J. Am. Chem. Soc.* **1958,** 80, 3538.

Greeley, R.S.; Smith, W.T.; Stoughton, R.W.; Lietzke, M.H. *J. Phys. Chem.* **1960,** 64, 652.

Greenbank, J.C.; Argent, B.B. *Trans. Faraday Soc.* **1965,** 61, 655.

Gregor, L.V.; Pitzer, K.S. *J. Am. Chem. Soc.* **1962,** 84, 2664.

Grenthe, I.; Ots, H.; Ginstrup, O. *Acta Chem. Scand.* **1970,** 24, 1067.

Greyson, J.; Snell, H. *J. Phys. Chem.* **1969,** 73, 3208.

Grèzes, G.; Basset, M. *C.R. Acad. Sci. (Paris)* **1965,** 260, 869.

Grieveson, P.; Alcock C.B. in *Special Ceramics,* Popper, P.: editor. Academic Press: London. **1960**.

Grieveson, P.; Hooper, G.W.; Alcock, C.B. *Metallurg. Soc. Conf.* **1961,** 7, 341.

Griffel, M.; Skochdopole, R.E. *J. Am. Chem. Soc.* **1953,** 75, 5250.

Griffel, M.; Vest, R.W.; Smith, J.F. *J. Chem. Phys.* **1957,** 27, 1267.

Griffiths, E.H.; Griffiths, E. *Phil. Trans. Roy. Soc. (London)* **1914**a, A 214, 319.

Griffiths, E.H.; Griffiths, E. *Proc. Roy. Soc. (London)* **1914**b, A 90, 558.

Gritsus, B.V.; Akhumov, E.I.; Zhilina, L.P. *Zhur. Priklad. Khim.* **1969,** 42, 208.

Grønvold, F.; Drowart, J.; Westrum, E.F., Jr. *Chemical Thermodynamics of the Actinide Elements and Compounds. Part 4. The Actinide Chalcogenides.* I.A.E.A.: Vienna. **1984**.

Gross, P.; Christie, J.; Hayman, C. *U.S. Clearing House Fed. Sci. Techn. Inform. No. 711673, AD* **1970**.

Gross, P.; Hayman, C. *Trans. Faraday Soc.* **1970,** 66, 30.

Gross, P.; Hayman, C.; Bingham, J.T. *Trans. Faraday Soc.* **1966,** 62, 2388.

Gross, P.; Hayman, C.; Levi, D.L. *Trans. Faraday Soc.* **1954,** 50, 477.

Gross, P.; Hayman, C.; Levi, D.L. *Trans. Faraday Soc.* **1955,** 51, 626.

Gross, P.; Hayman, C.; Levi, D.L. *Trans. Faraday Soc.* **1957,** 53, 1601.

Gross, P.; Hayman, C.; Stuart, M.C. *Proc. Brit. Ceram. Soc.* **1967,** 8, 39.

Gross, P.; Hayman, C.; Stuart, M.C. *Trans. Faraday Soc.* **1969,** 65, 2628.

Grzybowski, A.K. *J. Phys. Chem.* **1958**a, 62, 550.

Grzybowski, A.K. *J. Phys. Chem.* **1958**b, 62, 555.

Guenther, W.B. *J. Am. Chem. Soc.* **1969,** 91, 7619.

Guggenheim, E.A.; Stokes, R.H. *Trans. Faraday Soc.* **1958,** 54, 1646.

Gulbransen, E.A.; Andrew, K.F. *J. Electrochem. Soc.* **1950,** 97, 383.

Gulbransen, E.A.; Andrew, K.F.; Brassart, F.A. *J. Electrochem. Soc.* **1966,** 113, 834.

Gunn, S.R. *U.S. A.E.C. Rept. UCRL 7992* **1964**.

Gunn, S.R. *J. Phys. Chem.* **1965,** 69, 1010.

Gunn, S.R. *J. Phys. Chem.* **1967,** 71, 1386.

Gunn, S.R. *J. Chem. Thermodynamics* **1970,** 2, 535.

Gunn, S.R.; Green, L.G. *J. Am. Chem. Soc.* **1958,** 80, 4782.

Gunn, S.R.; Green, L.G. *J. Phys. Chem.* **1960,** 64, 61.

Gunn, S.R.; Green, L.G. *J. Chem. Eng. Data* **1963,** 8, 180.

Günther, P.; Wekua, K. *Z. Phys. Chem.* **1931,** A 154, 193.

Gupta, S.R.; Hills, G.J.; Ives, D.J.G. *Trans. Faraday Soc.* **1963**a, 59, 1874.

Gupta, S.R.; Hills, G.J.; Ives, D.J.G. *Trans. Faraday Soc.* **1963**b, 59, 1886.

Gurevich, V.M.; Khlyustov, V.G. *Geokhim.* **1979,** 829.

Gurvich, L.V.; Bergman, G.A.; Veits, I.V.; Medvedev, V.A.; Khachkuryzov, G.A.; Junman, V.S. *Thermodynamic Properties of Individual Substances,* Glushko, V.P., editor. Nauka: Moscow. Vol. I, **1978;** Vol. II, **1979;** Vol. III, **1981;** Vol IV, **1982.**

Gurvich, L.V. et al. *Thermodynamic Properties of Individual Substances,* Vol. II. Glushko, V.P. et al.: editors. Nauka: Moscow. **1979**.

Gurvich, L.V. et al. *Thermodynamic Properties of Individual Substances,* Vol. III. Glushko, V.P. et al.: editors. Nauka: Moscow. **1981**.

Gurvich, L.V. et al. *Thermodynamic Properties of Individual Substances,* Vol. IV. Glushko, V.P. et al.: editors. Nauka: Moscow. **1982**.

Gurvich, L.V.; Yungman, V.S. *I.A.E.A. Symposium on Thermodynamics. Vol. 2.* I.A.E.A. Vienna. **1966,** p. 613.

Gushchin, G.I.; Subbotin, V.A.; Khachaturov, E.Kh. *Teplofiz. Vys. Temp.* **1975,** 13, 747.

Gutowsky, H.S.; Hoffman, C.J. *J. Am. Chem. Soc.* **1950,** 72, 5751.

Haar, L. *Thermodynamic Properties of Ammonia as an Ideal Gas.* NSRDS-NBS 19, **1968**.

Haar, L.; Gallagher, J.S.; Kell, G.S. *NBS-NRC Steam Tables,* Hemisphere Press: New York. **1984**.

Haar, L.; Bradley, J.C.; Friedman, A.S. *J. Res. Nat. Bur. Stand.* **1955,** 55, 285.

Haase, R.; Naas, H.; Thumm, H. *Z. Phys. Chem. (Frankfurt)* **1963,** 37, 210.

Haber, F.; Maschke, A. *Z. Elektrochem.* **1915,** 21, 128.

Haber, F.; Tamaru, S. *Z. Electrochem.* **1915,** 21, 191.

Haber, F.; Tamaru, S.; Oeholm, W. *Z. Elektrochem.* **1915**a, 21, 206.

Haber, F.; Tamaru, S.; Ponnaz, C. *Z. Elektrochem.* **1915**b, 21, 89.

Haber, F.; Zisch, W. *Z. Phys.* **1922,** 9, 302.

Hackspill, M.L. *Ann. Chim. Phys.* **1913,** 28, 613; *C.R. Acad. Sci. (Paris)* **1912,** 154, 877.

Haeusler, C.; van Thanh, N.; Barchewitz, P. *J. Phys. Radium (Paris)* **1963,** 24, 289.

Hagen, K.; Hedberg, K. *J. Chem. Phys.* **1973,** 59, 1549.

Hake, R.R.; Cape, J.A. *Phys. Rev.* **1964,** A 135, 1151.

Hale, J.D.; Izatt, R.M.; Christensen, J.J. *J. Phys. Chem.* **1963,** 67, 2605.

Hamer, W.J. *Theoretical Mean Activity Coefficients of Strong Electrolytes in Aqueous Solutions from $0°$ to $100°C.$* NSRDS-NBS 24, **1968.**

Hamer, W.J. *J. Res. Nat. Bur. Stand.* **1972,** 76 A, 185.

Hamer, W.J.; Wu, Y.C. *J. Res. Nat. Bur. Stand.* **1970,** 74 A, 761.

Hamer, W.J.; Wu, Y.C. *J. Phys. Chem. Ref. Data* **1972,** 1, 1047.

Hannan, M.C. *Thesis, Catholic Univ. of America* **1936**.

Hansen, E.E.; Munir, Z.A.; Mitchell, M.J. *J. Am. Ceram. Soc.* **1969,** 52, 610.

Hansen, C.J. *Ber. Deutsch. Chem. Ges.* **1909,** 42, 210.

Hardesty, J.O.; Ross, W.H. *Ind. Eng. Chem.* **1937,** 29, 1283.

Haring, M.M.; White, J.C. *Trans. Electrochem. Soc.* **1938,** 73, 211.

Harned, H.S.; Bonner, F.T. *J. Am. Chem. Soc.* **1945,** 67, 1026.

Harned, H.S.; Davis, R., Jr. *J. Am. Chem. Soc.* **1943,** 65, 2030.

Harned, H.S.; Fitzgerald, M.E. *J. Am. Chem. Soc.* **1936,** 58, 2624.

Harned, H.S.; Hamer, W.J. *J. Am. Chem. Soc.* **1935,** 57, 33.

Harned, H.S.; Keston, A.S.; Donelson, J.G. *J. Am. Chem. Soc.* **1936,** 58, 989.

Harned, H.S.; Owen, B.B. *Physical Chemistry of Electrolyte Solutions.* 3rd Edition. Reinhold: New York. **1958**.

Harned, H.S.; Scholes, S.R., Jr. *J. Am. Chem. Soc.* **1941**, 63, 1706.

Harrison, J.P.; Lombardo, Y.; Peressini, P.P. *J. Phys. Chem. Solids* **1968**, 29, 557.

Hart, P.E.; Searcy, A.W. *J. Phys. Chem.* **1966**, 70, 2763.

Hartmann, H.; Schneider, R. *Z. Anorg. Allg. Chem.* **1929**, 180, 275.

Hatton, W.E.; Hildenbrand, D.L.; Sinke, G.C.; Stull, D.R. *J. Am. Chem. Soc.* **1959**, 81, 5028.

Hawkins, D.T. *M.S. Thesis, UCRL No. 16514, Univ. Calif., Berkeley,* **1966**.

Hawkins, D.T.; Hultgren, R. *Trans. A.I.M.E.* **1967**, 239, 1046.

Hawtin, P.; Lewis, J.B.; Moul, N.; Phillips, R.H. *Phil. Trans. Roy. Soc. (London)* **1966**, A 261, 67.

Head, A.J.; Lewis, G.B. *J. Chem. Thermodynamics* **1970**, 2, 701.

Hefley, J.D.; Amis, E.S. *J. Phys. Chem.* **1965**, 69, 2082.

Helgeson, H.C. *J. Phys. Chem.* **1967**, 71, 3121.

Hemingway, B.S.; Robie, R.A. *J. Res. U.S. Geol. Survey* **1977**, 5, 413.

Hemingway, B.S.; Robie, R.A.; Fischer, J.R.; Wilson, W.H. *J. Res. U.S. Geol. Survey* **1977**, 5, 797.

Henderson, W.E.; Stegeman, G. *J. Am. Chem. Soc.* **1918**, 40, 84.

Hepler, L.G.; Hopkins, H.P., Jr. *Rev. Inorg. Chem.* **1979**, 1, 303.

Hepler, L.G.; Jolly, W.L.; Latimer, W.M. *J. Am. Chem. Soc.* **1953**, 75, 2809.

Hepler, L.G.; Olofsson, G. *Chem. Rev.* **1975**, 75, 585.

Hersh, H.N. *J. Am. Chem. Soc.* **1953**, 75, 1529.

Herzberg, G. *Ann. Phys.* **1932**, 15, 677.

Herzberg, G. *Molecular Spectra and Molecular Structure. I. Diatomic Molecules.* 2nd Edition. Van Nostrand: New York. **1950**.

Herzberg, G. *J. Mol. Spectr.* **1970**, 33, 147.

Hessel, M.M.; Vidal, C.R. *J. Chem. Phys.* **1979**, 70, 4439.

Hetzer, H.B.; Robinson, R.A.; Bates, R.G. *J. Phys. Chem.* **1962**, 66, 1423.

Hetzer, H.B.; Robinson, R.A.; Bates, R.G. *J. Phys. Chem.* **1964**, 68, 1929.

Heycock, C.T.; Lamplough, F.E.E. *Proc. Chem. Soc.* **1912**, 28, 3.

Hidalgo, A.F.; Orr, C., Jr. *J. Chem. Eng. Data* **1968**, 13, 49.

Hietanen, S.; Sillén, L.G. *Arkiv Kemi* **1956**, 10, 103.

Higashigaki, Y.; Chihara, H. *Bull. Chem. Soc. Japan* **1976**, 49, 2089.

Higuchi, T. *J. Chem. Soc. Japan* **1937**, 58, 193.

Hildenbrand, D.L.; Hall, W.F. *J. Phys. Chem.* **1963**, 67, 888.

Hildenbrand, D.L.; Hall, W.F. *J. Phys. Chem.* **1964**, 68, 989.

Hildenbrand, D.L.; Kramer, W.R.; McDonald, R.A.; Stull, D.R. *J. Am. Chem. Soc.* **1958**, 80, 4129.

Hildenbrand, D.L.; Murad, E. *J. Chem. Phys.* **1966**, 44, 1524.

Hill, R.W.; Smith, P.L. *Phil. Mag.* **1953**, 44, 636.

Hill, A.E.; Willson, H.S.; Bishop, J.A. *J. Am. Chem. Soc.* **1933**, 55, 520.

Hincke, W.B. *J. Am. Chem. Soc.* **1933**, 55, 1751.

Hirschwald, W.; Stolze, F.; Stranski, J.N. *Z. Phys. Chem. (Frankfurt)* **1964**, 42, 96.

Holden, R.B.; Speiser, R.; Johnston, H.L. *J. Am. Chem. Soc.* **1948**, 70, 3897.

Holeci, I. *Chem. Průmysl* **1966**, 16, 267.

Holley, C.E., Jr.; Huber, E.J., Jr. *J. Am. Chem. Soc.* **1951**, 73, 5577.

Holmes, W.S. *Trans. Faraday Soc.* **1962**, 58, 1916.

Honig, G.; Czajkowski, M.; Stock, M.; Demtroder, W. *J. Chem. Phys.* **1979**, 71, 2138.

Hopkins, H.P., Jr.; Wulff, C.A. *J. Phys. Chem.* **1965**a, 69, 9.

Hopkins, H.P., Jr.; Wulff, C.A. *J. Phys. Chem.* **1965**b, 69, 6.

Horowitz, M.; Silvidi, A.A.; Malaker, S.F.; Daunt, J.G. *Phys. Rev.* **1952**, 88, 1182.

Horsley, J.A.; Barrow, R.F. *Trans. Faraday Soc.* **1967**, 63, 32.

Hostetler, P.B. *Am. J. Sci.* **1963**, 261, 238.

Howard, P.B.; Skinner, H.A. *J. Chem. Soc. A* **1966**, 1536.

Hu, J.-H.; White, D.; Johnston, H.L. *J. Am. Chem. Soc.* **1953**, 75, 1232.

Hubbard, S.R.; Ross, R.G. *Nature* **1982**, 295, 682.

Huber, E.J., Jr.; Holley, C.E., Jr. *J. Phys. Chem.* **1956**, 60, 498.

Huber, E.J., Jr.; Holley, C.E., Jr. *J. Chem. Thermodynamics* **1969**, 1, 267.

Huber, E.J., Jr.; Holley, C.E., Jr.; Meierkord, E.H. *J. Am Chem. Soc.* **1952**, 74, 3406.

Huber, K.P.; Herzberg, G. *Molecular Spectra and Molecular Structure. IV. Constants of Diatomic Molecules,* Van Nostrand Reinhold: New York. **1979**.

Huey, C.S.; Tartar, H.V. *J. Am. Chem. Soc.* **1940**, 62, 26.

Hull, H.; Turnbull, A.G. *Geochim. Cosmochim. Acta* **1973**, 37, 685.

Hultgren, R.; Desai, P.D.; Hawkins, D.T.; Gleiser, M.; Kelley, K.K.; Wagman, D.D. *Selected Values of the Thermodynamic Properties of the Elements.* American Society for Metals: Metals Park (Ohio). **1973**.

Humphrey, G.L. *J. Am. Chem. Soc.* **1951**, 73, 1587.

Humphrey, G.L.; O'Brien, C.J. *J. Am. Chem. Soc.* **1953**, 75, 2805.

Huntzicker, J.J.; Westrum, E.F., Jr. *J. Chem. Thermodynamics* **1971**, 3, 61.

Huston, R.; Butler, J.N. *J. Phys. Chem.* **1968**, 72, 4263.

Ioli, N.; Strumia, F.; Moretti, A. *J. Opt. Soc. Am.* **1971**, 61, 1251.

Irani, R.R.; Taulli, T.A. *J. Inorg. Nucl. Chem.* **1966**, 28, 1011.

Irving, R.J.; McKerrell, H. *Trans. Faraday Soc.* **1967**a, 63, 2582.

Irving, R.J.; McKerrell, H. *Trans. Faraday Soc.* **1967**b, 63, 2913.

Ishii, T.; Fujita, S.; *J. Chem. Eng. Data* **1978**, 23, 19.

Ishikawa F. *Sci. Repts. Tohoku Imp. Univ.* **1933**, 22, 131.

Ishikawa, F.; Kimura, G.*Osaka Lectures, Kyoto Imp. Univ.* **1927**, 252.

Ishikawa, F.; Kimura, G.; Murooka, T. *Sci. Repts. Tohoku Imp. Univ.* **1932**, 21, 455.

Ishikawa, F.; Murooka, T. *Bull. Inst. Phys. Chem. Research (Tokyo)* **1930**, 9, 781.

Ishikawa, F.; Murooka, T. *Sci. Repts. Tohoku Imp. Univ.* **1933**, 22, 138.

Ishikawa, F.; Shibata, E. *Nippon Kagaku Zasshi* **1927**, 48, 59.

Ishikawa, F.; Takai, T. *Bull. Inst. Phys. Chem. Res. (Tokyo)* **1937**, 16, 1250.

Ishikwawa, F.; Ueda, Y. *J. Chem. Soc. Japan* **1930**, 51, 634.

Ishikawa, F.; Ueda, Y. *Sci. Repts. Tohoku Imp. Univ.* **1933**, 22, 263.

Iskhakova, L.D.; Starikova, Z.A.; Morocheneta, E.P.; Trunov, V.K. *Zhur Neorg. Khim.* **1979**, 24, 1539.

Itskevich, E.S.; Strelkov, P.G. *Zhur. Fiz. Khim.* **1960**, 34, 1312.

IUPAC Commission on Atomic Weights. *Pure Appl. Chem.* **1970**, 21, 95.

Ivanova, E.F.; Aleksandrov, V.V. *Zhur. Fiz. Khim.* **1964**, 38, 878.

Ives, D.J.G.; Prasad, D. *J. Chem. Soc. B* **1970**, 1649.

Izatt, R.M.; Eatough, D.; Christensen, J.J.; Bartholomew, C.H. *J. Chem. Soc. A* **1969**a, 45.

Izatt, R.M.; Eatough, D.; Christensen, J.J.; Bartholomew, C.H. *J. Chem. Soc. A* **1969**b, 47.

Jacob, K.T.; Alcock, C.B.; Chan, J.C. *Acta Met.* **1974**, 22, 545.

Jacobson, R.L.; Langmuir, D. *Geochim. Cosmochim. Acta* **1974**, 35, 701.

Jäger, L. *Chem. Průmsyl* **1958**, 8, 136.

JANAF Thermochemical Tables, Second Edition, Stull, D.R., Prophet, H., ed. NSRD-NBS 37. Washington D.C. **1971**.

JANAF Thermochemical Tables, Second Edition, Stull, D.R., Prophet, H., ed. NSRD-NBS 37. Washington D.C. **1971**; supplements in *J. Phys. Chem. Reference Data* **1974**, 3, 311; **1975**, 4, 1; **1978**, 7, 793; **1982**, 11, 695.

Janz, G.J.; Gordon, A.R. *J. Am. Chem. Soc.* **1943**, 65, 218.

Jauch, R. *Thesis, Tech. Hochs. Stuttgart.* **1946**. See Kubaschewski, O. *Z. Metallk.* **1950**, 41, 445.

Jeapes, A.P.; Leadbetter, A.J.; Waterfield, C.G.; Wycherley, K.E. *Phil. Mag.* **1974**, 29, 803.

Jenkins, C.H.M. *Proc. Roy. Soc. (London)* **1926**, A 110, 456.

Jenkins, I.L.; Monk, C.B. *J. Am. Chem. Soc.* **1950**, 72, 2695.

Jessup, R.S. *J. Res. Nat. Bur. Stand.* **1938**, 21, 475.

Johns, J.W.C.; Barrow, R.F. *Proc. Roy. Soc. (London)* **1959**, A 251, 504.

Johnson, G.K. *J. Chem. Thermodynamics* **1977**, 9, 835.

Johnson, G.K.; Feder, H.M.; Hubbard, W.N. *J. Phys. Chem.* **1966**, 70, 1.

Johnson, G.K.; Gayer, K.H. *J. Chem. Thermodynamics* **1979**, 11, 41.

Johnson, G.K.; Hubbard, W.N. *J. Chem. Thermodynamics* **1969**, 1, 459.

Johnson, G.K.; O'Hare, P.A.G. *J. Chem. Thermodynamics* **1978**, 10, 577.

Johnson, G.K.; Smith, P.N.; Appelman, E.H.; Hubbard, W.N. *Inorg. Chem.* **1970**, 9, 119.

Johnson, G.K.; Smith, P.N.; Hubbard, W.N. *J. Chem. Thermodynamics* **1973**, 5, 793.

Johnson, G.K.; Steele, W.V. *J. Chem. Thermodynamics* **1981**, 13, 717.

Johnson, R.G.; Hudson, D.E.; Caldwell, W.C.; Spedding, F.H.; Savage, W.R. *J. Chem. Phys.* **1956**, 25, 917.

Johnson, W.H. Private communication. **1977**.

Johnson, W.H.; Ambrose, J.R. *J. Res. Nat. Bur. Stand.* **1963**, 67 A, 427.

Johnson, W.H.; Gilliland, A.A. *J. Res. Nat. Bur. Stand.* **1961**, 65 A, 63.

Johnson, W.H.; Gilliland, A.A.; Prosen, E.J. *J. Res. Nat. Bur. Stand.* **1960**, 64 A, 515.

Johnson, W.H.; Miller, R.G.; Prosen, E.J. *J. Res. Nat. Bur. Stand.* **1959**a, 62, 213.

Johnson, W.H.; Nelson, R.A.; Prosen, E.J. *J. Res. Nat. Bur. Stand.* **1959**b, 62, 49.

Johnson, W.H.; Sunner, S. *Acta Chem. Scand.* **1963**, 17, 1917.

Johnston, H.L.; Hersh, H.N.; Kerr, E.C. *J. Am. Chem. Soc.* **1951**, 73, 1112.

Johnston, H.L.; Kerr, E.C. *J. Am. Chem. Soc.* **1950**, 72, 4733.

Jolly, W.L.; Latimer, W.M. *J. Am. Chem. Soc.* **1952**, 74, 5757.

Joly, R.D.; Perachon, G. *Thermochim. Acta* **1977**, 21, 333.

Joly, R.D.; Thourey, J.; Perachon, G. *C.R. Acad. Sci. (Paris)* **1973**, 277C, 1179.

Jones, E.J. *J. Am. Chem. Soc.* **1943**, 65, 2274.

Jones, G.; Bäckstrom, S. *J. Am. Chem. Soc.* **1934**, 56, 1524.

Jones, G.E.; Gordy, W. *Phys. Rev.* **1964**, A 136, 1229.

Jones, G.H.S.; Hallett, A.C.H. *Can. J. Phys.* **1960**, 38, 696.

Jones, J.H.; Froning, H.R. *J. Am. Chem. Soc.* **1944**, 66, 1672.

Jones, W.M.; Gordon, J.; Long, E.A. *J. Chem. Phys.* **1952**, 20, 695.

Justice, B.H. *J. Chem. Eng. Data* **1969**, 14, 4.

Kaganovich, Yu. Ya.; Mishchenko, K.P. *Zhur. Obshch. Khim.* **1951**, 21, 28.

Kalishevich, G.I.; Gel'd, P.V.; Krentsis, R.P. *Zhur. Fiz. Khim.* **1965,** 39, 2999.

Kalushkina, L.A.; Shevelev, Yu. P. *Izv. Akad. Nauk SSSR, Ser. Khim.* **1969,** No.2, 11.

Kangro, W.; Groeneveld, A. *Z. Phys. Chem. (Frankfurt)* **1962,** 32, 110.

Kapustinskii, A.F.; Anvaer, B.L. *C.R. Acad. Sci. URSS* **1941,** 30, 625.

Kapustinskii, A.F.; Kan'kovskii, R.T. *Zhur. Fiz. Khim.* **1958,** 32, 2810.

Kapustinskii, A.F.; Samoilov, O.Ya. *Zhur. Fiz. Khim.* **1952,** 26, 918.

Karnaukhov, A.S. *Izvest. Sekt. Fiz. Khim. Inst. Obshch. Khim. Akud. Nauk* **1954,** 25, 334.

Katayama, I.; Shibata, J.; Kozuka, Z. *J. Jap. Inst. Metals* **1975,** 39, 990.

Katayama, I.; Shibata, J.; Kozuka, Z. *Technol. Repts., Osaka Univ.* **1979,** 29, 51.

Katzin, L.I.; Gebert, E. *J. Am. Chem. Soc.* **1955,** 77, 5814.

Kaufman, V.; Radziemskii, L.J., Jr. *J. Opt. Soc. Am.* **1969,** 59, 227.

Keenan, J.H.; Keyes, F.G.; Hill, P.G.; Moore, J.G. *Steam Tables.* John Wiley: New York. **1969**.

Keesom, P.H.; Seidel, G. *Phys. Rev.* **1959,** 113, 33.

Keesom, W.H.; van den Ende, J.N. *Proc. Koninkl. Ned. Akad. Wetenschap.* **1930,** 33, 243.

Kelley, K.K. *J. Am. Chem. Soc.* **1941,** 63, 1137.

Kelley, K.K. *U.S. Bur. Mines Rept. Inv. 3776* **1944**.

Kelley, K.K.; King, E.G. Contributions to the Data on Theoretical Metallurgy. XIV. *U.S. Bur. Mines Bull. 592* **1961**.

Kelley, K.K.; Moore, G.E. *J. Am. Chem. Soc.* **1943,** 65, 1264.

Kelley, K.K.; Southard, J.C.; Anderson, C.T. *U.S. Bur. Mines TP 625* **1941,** 73.

Kendall, J.; Andrews, J.C. *J. Am. Chem. Soc.* **1921,** 43, 1545.

Kenttämaa, J. *Suomen Kemistilehti* **1957,** 30 B, 9.

Kerr, E.C.; Hersh, H.N.; Johnston, H.L. *J. Am. Chem. Soc.* **1950,** 72, 4738.

Ketchen, E.E.; Wallace, W.E. *J. Am. Chem. Soc.* **1951,** 73, 5810.

Khamova, V.I.; Ponomareva, A.M.; Mishchenko, K.P. *Zhur. Fiz. Khim.* **1966,** 40, 1387.

Khana, V.M. *Ph.D. thesis, New York Univ.* **1963**;. *Diss. Abstr.,* **1964,** 25, 3115.

Khodakovsky, I.L.; Katorcha, L.V.; Kuyunko, N. *Geokhim.* **1980,** 11, 1606.

Khodakovsky, I.L.; Zhogina, V.V.; Ryzhenko, B.N. *Geokhim.* **1965,** 827.

Khriplovich, L.M.; Luk'yanova, I.G.; Paukov, I.E. *Zhur. Fiz. Khim.* **1975,** 49, 1337.

Khvorostin, Ya. S.; Filippov, V.K.; Reshetova, L.I. *Zhur. Fiz. Khim.* **1975,** 49, 1271.

Kilday, M.V. *J. Res. Nat. Bur. Stand.* **1980,** 85, 467.

Kilday, M.V.; Prosen, E.J. *J. Res. Nat. Bur. Stand.* **1964,** 68 A, 127.

Kilday, M.V.; Prosen, E.J. *J. Res. Nat. Bur. Stand.* **1973,** 77 A, 205.

Kilday, M.V.; Prosen, E.J.; Wagman, D.D. *J. Res. Nat. Bur. Stand.* **1973,** 77 A, 217.

Killackey, J.J. *U.S. A.E.C. Rept. BNL-756* **1962**.

Kim, J.H.; Cosgarea, A., Jr. *J. Chem. Phys.* **1966,** 44, 806.

King, E.G. *J. Am. Chem. Soc.* **1957,** 79, 2056.

King, E.G. *J. Am. Chem. Soc.* **1958,** 80, 2400.

King, R.C.; Armstrong, G.T. *J. Res. Nat. Bur. Stand.* **1968,** 72 A, 113.

King, R.C.; Armstrong, G.T. *J. Res. Nat. Bur. Stand.* **1970,** 74 A, 769.

Kirellov, P.L.; Grachev, V.S. *Inzh. Fiz. Zhur.* **1959,** 2, 53.

Kirgintsev, A.N.; Yacobi, N.Ya. *Zhur. Neorg. Khim.* **1968,** 13, 2851.

Kiriyanenko, A.A. *Issledovanie Teplofizicheskikh Svoistv.* Nauka: Novosibirsk. **1970**.

Kirkham, A.J.; Yates, B. *J. Phys.: Solid State Phys.* **1968,** C1, 1162.

Kirpichev, E.P.; Rubtsov, Yu.I.; Krivtsov, N.V.; Sorokina, T.V.; Manelis, G.B. *Zhur. Fiz. Khim.* **1975,** 49, 1975.

Kirschenbaum, A.D.; Cahill, J.A. *J. Inorg. Nucl. Chem.* **1963,** 25, 232.

Kistiakowsky, G.B. *U.S. Naval Res. Lab. Rept. NRL-2958* **1941.**

Kitchener, J.A.; Ignatowicz, S. *Trans. Faraday Soc.* **1951,** 47, 1278.

Kleboth, K. *Monatsh. Chem.* **1970,** 101, 767.

Klemenc, A.; Nagel, A. *Z. Anorg. Allg. Chem.* **1926,** 155, 257.

Knacke, O.; Prescher, K.E. *Erzbergbau u. Metallhüt.* **1964,** 28.

Kneip, G.D., Jr.; Betterton, J.O., Jr.; Scarbrough, J.O. *Phys. Rev.* **1963,** 130, 1687.

Ko, H.C.; Ahmad, N.; Chang, Y.A. *U.S. Bur. Mines Rept. Inv. 8647* **1982.**

Kobe, K.A.; Sheehy, T.M. *Ind. Eng. Chem.* **1948,** 40, 99.

Koch, R.K.; Clavert, E.D.; Thomas, C.R.; Beall, B.A. *U.S. Bur. Mines Rept. Inv. 7271* **1969.**

Kohler, K.; Zäske, P. *Z. Anorg. Allg. Chem.,* **1964,** 331, 1.

Kokovin, G.A.; Golubenko, A.M.; Chusova, J.P. *Izvest. Sib. Otdel Akad. Nauk USSR, Ser. Khim. Nauk* **1970,** 13.

Kolesov, V.P.; Popov, M.M.; Skuratov, S.M. *Zhur. Neorg. Khim.* **1959,** 4, 1233.

Kolesov, V.P.; Skuratov, S.M. *Zhur. Neorg. Khim.* **1961,** 6, 1741.

Kolthoff, I.M.; von Fischer, W. *J. Am. Chem. Soc.* **1939,** 61, 195.

Konaka, S.; Murata, Y.; Kuchitsu, K.; Morino, Y. *Bull. Chem. Soc. Japan* **1966,** 39, 1134.

Kondirev, N.V.; Berezovskii, G.V. *Zhur. Obshch. Khim.* **1935,** 5, 1246.

Kondrat'ev, N.S.; Parfent'eva, I.F. *Teplo-massoperenos v odnoi dvukhfaznykh sredakh.* Nauka: Moscow. **1971,** 19.

Königer, F.; Carter, R.O.; Müller, A. *Spectrochim. Acta* **1976** 32 A, 891.

Königer, F.; Müller, A.; Orville-Thomas, W.J. *J.Mol. Struct.* **1977,** 37, 199.

Kortüm, G.; Häussermann, W. *Ber. Bunsenges. Phys. Chem.* **1965,** 69, 594.

Kostryukov, V.N.; Morozova, G.Kh. *Zhur. Fiz. Khim.* **1960,** 34, 1833.

Kostryukov, V.N.; Samorukov, O.P.; Samorukova, N.Kh.; Nurullaev, N.G. *VINITI, Dep. No. 4458-77* M.**1977**.

Kostryukov, V.N.; Samorukov, O.P.; Samorukova, N.Kh.; Nurullaev, N.G. *Zhur. Fiz. Khim.* **1978,** 52, 1838.

Kothen, C.W. *Ph.D thesis, Ohio State Univ.* **1952.**; *Diss. Abstr.,* **1957,** 17, 2842.

Kothen, C.W.; Johnston, H.L. *J. Am. Chem. Soc.* **1953,** 75, 3101.

Kovtun, G.P.; Kruglykh, A.A.; Pavlov, V.S. *Izvest. Akad. Nauk SSSR, Met. i Gorn. Delo* **1964,** 2, 177.

Krestov, G.A.; Abrasimov, V.K. *Izvest. Vyssh. Uchebn. Zaved. Khim. i Khim. Tekhnol.* **1967,** 10, 1005.

Krestov, G.A.; Abrasimov, V.K. *Fiz. Khim. Rostvorov* **1972,** 32.

Krestov, G.A.; Egorova, I.V. *Izvest. Vyssh. Uchebn. Zaved. Khim. i Khim. Tekhnol.* **1967**a, 10, 750.

Krestov, G.A.; Egorova, I.V. *Theor. Exper. Chem.* **1967**b, 3, 71.

Krestov, G.A.; Zverev, V.A.; Krotov, V.S. *Izvest. Vyssh. Uchebn. Zaved. Khim. i Khim. Tekhnol.* **1972,** 15, 1414.

Krier, C.A.; Craig, R.S.; Wallace, W.E. *J. Phys. Chem.* **1957,** 61, 522.

Krieve, W.F.; Vango, S.P.; Mason, D.M. *J. Chem. Phys.* **1956,** 25, 519.

Krishnan, C.V.; Friedman, H.L. *J. Phys. Chem.* **1970,** 74, 2356.

Krivtsov, N.V.; Babaeva, V.P.; Rosolovskii, V.Ya. *Zhur. Neorg. Khim.* **1973**a, 18, 353.

Krivtsov, N.V.; Rosolovskii, V.Ya.; Shirokova, G.N. *Zhur. Neorg. Khim.* **1971**, 16, 2628.

Krivtsov, N.V.; Titova, K.B.; Rosolovskii, V.Ya. *Zhur. Neorg. Khim.* **1973**b, 18, 347.

Kröner, A. *Ann. Phys.* **1913**, 40, 438.

Krupenie, P.H. *J. Phys. Chem. Ref. Data* **1972**, 1, 427.

Krupkowski, A.; Golonka, J. *Bull. Acad. Polon. Sci. Cl. 3* **1964**, 12, 69.

Kryukov, P.A.; Starostina, L.I.; Tarasenko, S.Ya.; Parlyuk, L.A.; Smolyakov, B.S.; Larionov, F.G. *Mezhdunar. Geokhim. Kongr.* **1971**, 1, 186. (CA:81-69193).

Kryukov, P.A.; Starostina, L.I.; Tamsenko, S.Ya.; Primanchuk, M.P. *Geokhim.* **1974**, 1003.

Kubil, H. *Helv. Chim. Acta.* **1946**, 29, 1962.

Kuchitsu, K.; Konaka, S. *J. Chem. Phys.* **1966**, 45, 4342.

Kury, J.W.; Zielen, A.J.; Latimer, W.M. *J. Electrochem. Soc.* **1953**, 100, 468.

Kusch, P.; Hessel, M.M. *J. Mol. Spectr.* **1969**, 32, 181.

Kusch, P.; Hessel, M.M. *J. Chem. Phys.* **1978**, 68, 2591.

Kuznetsova, G.P.; Lovetskaya, G.A.; Presnyakova, V.M.; Stepin, B.D. *Zhur. Fiz. Khim.* **1974**, 48, 2141.

Kuznetsova, G.P.; Lovetskaya, G.A.; Presnyakova, V.M.; Stepin, B.D.; Smirnova, T.G. *Zhur. Neorg. Khim.* **1975**, 20, 2546.

Kybett, B.D.; Margrave, J.L. in JANAF Thermodynamic Tables, Supplement 1975. *J.Phys. Chem. Ref. Data* **1975**, 4, 1.

Lacher, J.R.; Casali, L.; Park, J.D. *J. Phys. Chem.* **1956**a, 60, 608.

Lacher, J.R.; Gottlieb, H.B.; Park, J.D. *Trans. Faraday Soc.* **1962**, 58, 2348.

Lacher, J.R.; Kianpour, A.; Oetting, F.; Park, J.D. *Trans. Faraday Soc.* **1956**b, 52, 1500.

Ladenburg, R.; Minkowski, R. *Z. Phys.* **1921**, 6, 153.

Ladenburg, R.; Thiele, E. *Z. Phys. Chem.* **1930**, B 7, 161.

Lagarde, L. *Ann. Faculté Sci. Marseille* (2), **1945**, 18, 1.

La Mer, V.K.; Parks, W.G. *J. Am. Chem. Soc.* **1931**, 53, 2040.

Lange, E.; Eichler, A. *Z. Phys. Chem.* **1927**, 129, 285.

Lange, E.; Fuoss, R.M. *Z. Phys. Chem.* **1927**, 125, 431.

Lange, E.; Martin, W. *Z. Phys. Chem.* **1937**, A 180, 233.

Lange, E.; Miederer, W. *Z. Elektrochem.* **1957**a, 61, 403.

Lange, E.; Miederer, W. *Z. Electrochem.* **1957**b, 61, 407.

Lange, E.; Monheim, J. *Z. Phys. Chem.* **1930**, A 150, 349.

Lange, E.; Monheim, J.; Robinson, A.L. *J. Am. Chem. Soc.* **1933**, 55, 4733.

Lange, E.; Shibata, Z. *Z. Phys. Chem.* **1930**, A 149, 465.

Lange, E.; Streeck, H. *Z. Phys. Chem.* **1931**, A 157, 1.

Larson, A.T. *J. Am. Chem. Soc.* **1924**, 46, 367.

Larson, A.T.; Dodge, R.L. *J. Am. Chem. Soc.* **1923**, 45, 2918.

Larson, J.W.; Cerutti, P.; Garber, H.K.; Hepler, L.G. *J. Phys. Chem.* **1968**, 72, 2902.

Larson, J.W.; Zeeb, K.G.; Hepler, L.G. *Can. J. Chem.* **1982**, 60, 2141.

Larson, W.D. *J. Am. Chem. Soc.* **1940**, 62, 764.

Latimer, W.M.; Ahlberg, J.E. *J. Am. Chem. Soc.* **1930**, 52, 549.

Latimer, W.M.; Greensfelder, B.S. *J. Am. Chem. Soc.* **1928**, 50, 2202.

Latimer, W.M.; Hicks, J.F.G., Jr.; Schutz, P.W. *J. Chem. Phys.* **1933**a, 1, 424.

Latimer, W.M.; Hicks, J.F.G., Jr.; Schutz, P.W. *J. Chem. Phys.* **1933**b, 1, 620.

Latimer, W.M.; Hoenshel, H.D. *J. Am. Chem. Soc.* **1926**, 48, 19.

Latimer, W.M.; Schutz, P.W.; Hicks, J.F.G. Jr. *J. Am. Chem. Soc.* **1934**, 56, 88.

Laurent, P.A.; Duhamel, M.J. *Bull. Soc. Chim. France* **1953,** 157.

Lavut, E.G.; Timofeyev, B.I.; Yuldasheva, V.M.; Lavut, E.A.; Gal'chenko, G.L. *J. Chem. Thermodynamics* **1981,** 13, 635.

Leadbetter, A.J. *J. Phys. C : Solid State Phys.* **1968,** 1, 1481.

Lebed', V.I.; Aleksandrov, V.V. *Zhur. Fiz. Khim.* **1964,** 38, 2608.

LeRoy, R.J. *J. Chem. Phys.* **1970**a, 52, 2678.

LeRoy, R.J. *J. Chem. Phys.* **1970**b, 52, 2683.

LeRoy, R.J.; Bernstein, R.B. *Chem. Phys. Letters* **1970**a, 5, 42.

LeRoy, R.J.; Bernstein, R.B. *J. Chem. Phys.* **1970**b, 52, 3869.

LeRoy, R.J.; Bernstein, R.B. *J. Mol. Spectr.* **1971,** 37, 109.

Leung, C.S.; Grunwald, E. *J. Phys. Chem.* **1970,** 74, 687.

Levin, I.W.; Abramowitz, S. *J. Chem. Phys.* **1965,** 43, 4213.

Levin, I.W.; Abramowitz, S. *J. Chem. Phys.* **1966,** 44, 2562.

Lewis, D.C.; Frisch, M.A.; Margrave, J.L. *Carbon* **1965,** 2, 431.

Lewis, G.N.; Argo, W.L. *J. Am. Chem. Soc.* **1915,** 37, 1983.

Lewis, G.N.; Keyes, F.G. *J. Am. Chem. Soc.* **1912,** 34, 119.

Lewis, G.N.; Randall, M. *J. Am. Chem. Soc.* **1918,** 40, 362.

Lewis, L.C. *Z. Phys.* **1931,** 69, 786.

Lietzke, M.H.; Hall, J.O. *J. Inorg. Nucl. Chem.* **1967,** 29, 1249.

Lietzke, M.H.; Stoughton, R.W.; Young, T.F. *J. Phys. Chem.* **1961,** 65, 2247.

Lietzke, M.H.; Vaughen, J.V. *J. Am. Chem. Soc.* **1955,** 77, 876.

Lilley, T.H.; Briggs, C.C. *Proc. Roy. Soc. (London)* **1976,** A 349, 355.

Lingane, J.J. *J. Am. Chem. Soc.* **1938,** 60, 724.

Linke, W.F. *Solubilities of inorganic and metal organic compounds,* Vol.I. 4th ed. Van Nostrand: Princeton. **1958.**

Linke, W.F. *Solubilities of inorganic and metal organic compounds,* Vol.II. 4th ed. Am. Chem. Soc.: Washington DC. **1965.**

Logan, J.K.; Clement, J.R.; Jeffers, H.R. *Phys. Rev.* **1957,** 105, 1435.

Loomis, F.W.; Nusbaum, R.E. *Phys. Rev.* **1932,** 39, 89.

Lovetskaya, G.A.; Kuznetsova, G.P.; Stepin, B.D. *Zhur. Fiz. Khim.* **1975,** 49, 1601.

Lovetskaya, G.A.; Kuznetsova, G.P.; Stepin, B.D.; Nikolaeva, K.I.; Starikova, Z.A. *Zhur. Neorg. Khim.* **1977,** 22, 524.

Loy, H.L.; Himmelblau, D.M. *J. Phys. Chem.* **1961,** 65, 264.

Luchinskii, G.P. *Zhur. Fiz. Khim.* **1966,** 40, 593.

Luchinskii, G.P. *Zhur. Fiz. Khim.* **1969,** 43, 1120.

Lutfullah; Dunsmore, H.S.; Paterson, R. *J. Chem. Soc. Faraday Trans. I* **1976,** 72, 495.

MacRae, D.; Van Voorhis, C.C. *J. Am. Chem. Soc.* **1921,** 43, 547.

Madgin, W.M.; Swales, D.A. *J. Chem. Soc.* **1956,** 196.

Mah, A.D. *J. Phys. Chem.* **1957,** 61, 1572.

Mah, A.D.; Adami, L.H. *U.S. Bur. Mines Rept. Inv. 6034* **1962.**

Mah, A.D.; Kelley, K.K.; Gellert, N.L.; King, E.G.; O'Brien, C.J. *U.S. Bur. Mines Rept. Inv. 5316* **1957.**

Maier, C.G.; Parks, G.S.; Anderson, C.T. *J. Am. Chem. Soc.* **1926,** 48, 2564.

Maier, C.G. *J. Am. Chem. Soc.* **1929,** 51, 194.

Maier, C.G. *J. Am. Chem. Soc.* **1930,** 52, 2159.

Makansi, M.M.; Madsen, M.; Selke, W.A.; Bonilla, C.F. *J. Phys. Chem.* **1956,** 60, 128.

Makansi, M.M.; Muendel, C.H.; Selke, W.A. *J. Phys. Chem.* **1955,** 59, 40.

Makarov, L.L.; Belousov, V.P.; Malyshev, V.N. *Zhur. Fiz. Khim.* **1967,** 41, 660.

Makarov, L.L.; Evstrop'ev, K.K.; Vlasov, Yu.G. *Zhur. Fiz. Khim.* **1958,** 32, 1618.

Makarov, L.L.; Popov, G.S. *Dokl. Akad. Nauk SSSR* **1959,** 129, 854.

Makarov, L.L.; Vlasov, Yu.G.; Azarko, V.A. *Zhur. Fiz. Khim.* **1966,** 40, 1134.

Makarov, L.L.; Vlasov, Yu.G.; Kopunets, R. *Zhur. Fiz. Khim.* **1964,** 38, 1938.

Makarov, S.Z.; Yakimov, L.S. *Zhur. Obshch. Khim.* **1933,** 3, 990.

Makolkin, I.A. *Zhur. Fiz. Khim.* **1942,** 16, 13.

Malaspina, L.; Gigli, R.; Bardi, G. *J. Chem. Thermodynamics* **1971,** 3, 827.

Männchen, W.; Bornkessel, K. *Z. Naturforsch.* **1959,** 14a, 925.

Männchen, W.; Wolf, G. *Z. Naturforsch.* **1966,** 21a, 1216.

Månsson, M.; Sunner, S. *Acta Chem. Scand.* **1963,** 17, 723.

Mar, R.W.; Bedford, R.G. *High Temp. Sci.* **1976,** 8, 365.

Mar, R.W.; Searcy, A.W. *J. Phys. Chem.* **1967,** 71, 888.

Mar, R.W.; Searcy, A.W. *J. Chem. Phys.* **1970,** 53, 3076.

Marais, E.J.; Verleger, H. *Phys. Rev.* **1950,** 80, 429.

Margrave, J.L. Private communication. **1977.**

Maronny, G. *Electrochim. Acta* **1959,** 1, 58.

Marshall, A.L.; Dornte, R.W.; Norton, F.J. *J. Am. Chem. Soc.* **1937,** 59, 1161.

Marshall, W.L.; Jones, E.V. *J. Phys. Chem.* **1966,** 70, 4028.

Marshall, W.L.; Slusher, R. *J. Phys. Chem.* **1966,** 70, 4015.

Marshall, W.L.; Slusher, R. *J. Chem. Eng. Data* **1968,** 13, 83.

Marshall, W.L.; Slusher, R.; Jones, E.V. *J. Chem. Eng. Data* **1964,** 9, 187.

Martin, D.L. *Proc. Roy. Soc. (London)* **1960a,** A 254, 433.

Martin, D.L. *Proc. Roy. Soc. (London)* **1960b,** A 254, 444.

Martin, D.L. *Proc. Phys. Soc.* **1961,** 78, 1482.

Martin, D.L. *Can. J. Phys.* **1962,** 40, 1166.

Martin, D.L. *Phys. Rev.* **1965,** 139 A, 150.

Martin, D.L. *Can. J. Phys.* **1970,** 48, 1327.

Martin, D.L. *Phys. Rev.* **1973,** 8 B, 5357.

Martynova, O.I.; Vasina, L.G.; Pozdnyakova, S.A. *Doklad. Akad. Nauk SSSR* **1971,** 201, 1022.

Mashovets, V.P.; Puchkov, L.V. *Zhur. Priklad. Khim.* **1965,** 38, 949.

Mason, C.M. *J. Am. Chem. Soc.* **1938,** 60, 1638.

Mason, C.M. *J. Am. Chem. Soc.* **1941,** 63, 220.

Mastroianni, M.; Criss, C.M. *J. Chem. Eng. Data* **1972,** 17, 222.

Matsui, M.; Oguri, S.; Kambara, S.; Kato, K. *Kogyo Kazaku Zasshi* **1929,** 32, 172 B.

Matsushita, Y.; Goto, K. *I.A.E.A. Symposium on Thermodynamics,* Vol.I, I.A.E.A: Vienna. **1965,** p. 111.

Maucherat, M. *C.R. Acad. Sci. (Paris)* **1939,** 208, 499; *J. Phys. Radium* **1939,** 10, 441.

Maxwell, L.R.; Hendricks, S.B.; Mosley, V.M. *J. Chem. Phys.* **1935,** 3, 699.

Mayer, H. *Z. Phys.* **1931,** 67, 240.

McCabe, C.L.; Birchenall, C.E. *J. Metals 3, A.I.M.E. Trans.* **1953,** 197, 707.

McCormack, J.M.; Myers, J.R.; Saxer, R.K. *J. Chem. Eng. Data* **1965,** 10, 319.

McCreary, J.R.; Thorn, R.J. *J. Chem. Phys.* **1969,** 50, 3725.

McCreary, J.R.; Thorn, R.J. *High Temp. Sci.* **1971,** 3, 300.

McCullough, J.P.; Sunner, S.; Finke, H.L.; Hubbard, W.N.; Gross, M.E.; Pennington, R.E.; Messerly, J.F.; Good, W.D.; Waddington, G. *J. Am. Chem. Soc.* **1953,** 75, 5075.

McCurdy, K.G.; Laidler, K.J. *Can. J. Chem.* **1964**, 42, 818.

Meads, P.F.; Forsythe, W.R.; Giauque, W.F. *J. Am. Chem. Soc.* **1941**, 63, 1902.

Médard, L.; Thomas, M. *Mém. Poudres* **1953**, 35, 155.

Medvedev, V.A.; Efimov, M.E.; Devina, O.A.; Klevaichuk, G.N. *Zhur. Fiz. Khim.* **1975**, 49, 1342.

Medvedev, V.A.; Efimov, M.E.; Cerutti, P.J.; McKay, R.M.; Johnson, L.H.; Hepler, L.G. *Thermochim. Acta* **1978**, 23, 87.

Medvedev, V.A. et al. *Thermal Constants of Substances,* Vol.1. Glushko, V.P.: editor. VINITI: Moscow. **1965**.

Medvedev, V.A. et al. *Thermal Constants of Substances,* Vol.2. Glushko, V.P.: editor. VINITI: Moscow. **1966**.

Medvedev, V.A. et al. *Thermal Constants of Substances,* Vol.4. Glushko, V.P.: editor. VINITI: Moscow. **1971**.

Medvedev, V.A. et al. *Thermal Constants of Substances,* Vol. 6. Glushko, V.P.: editor. VINITI: Moscow. **1972**.

Mel, H.C.; Jolly, W.L.; Latimer, W.M. *J. Am. Chem. Soc.* **1953**, 75, 3827.

Mellon, M.G.; Henderson, W.E. *J. Am. Chem. Soc.* **1920**, 42, 676.

Menzies, A.W.C. *J. Am. Chem. Soc.* **1920**, 42, 1951.

Menzies, A.W.C.; Hitchcock, C.S. *J. Phys. Chem.* **1931**, 35, 1660.

Mercer, E.E.; Farrar, D.T. *Can. J. Chem.* **1968**, 46, 2679.

Mesmer, R.E.; Baes, C.F., Jr. *J. Solution Chem.* **1974**, 3, 307.

Messer, C.E.; Fasolino, L.G.; Thalmayer, C.E. *J. Am. Chem. Soc.* **1955**, 77, 4524.

Metz, C.R.; Seifert, R.L. *J. Electrochem. Soc.* **1970**, 117, 49.

Meyer, G.; Scheffer, F.E.C. *Rec. Trav. Chim.* **1938**, 57, 604.

Miles, F.T.; Menzies, A.W.C. *J. Am. Chem. Soc.* **1937**, 59, 2392. .

Millar, R.W. *J. Am. Chem. Soc.* **1928**, 50, 2653.

Millar, R.W. *J. Am. Chem. Soc.* **1929**, 51, 207.

Mishchenko, K.P.; Pronima, M.Z. *Zhur. Obshch. Khim.* **1936**, 6, 85.

Mishchenko, K.P.; Reznikov, I.L.; Klyueva, M.L.; Sokolov, V.V.; Polyakov, Yu. A. *Zhur. Priklad. Khim.* **1965**, 38, 1939.

Mishchenko, K.P.; Shpigel, L.P. *Zhur. Obshch. Khim.* **1967**, 37, 2145.

Mishchenko, K.P.; Sukhotin, A.M. *Dokl. Akad. Nauk SSSR* **1954**, 98, 103.

Mishchenko, K.P.; Yakovlev, I.E. *Zhur. Obshch. Khim.* **1959**, 29, 1761.

Mitchell, R.E. *Ph.D Thesis, Purdue Univ.* **1964**; *Diss. Abstr.* **1965**, 26, 686.

Mitchell, R.E.; Cobble, J.W. *J. Am. Chem. Soc.* **1964**, 86, 5401.

Mitsuhashi, T.; Takahashi, Y. *J. Ceram. Soc. Jap.* **1980**, 88, 305.

Moers, K. *Z. Anorg. Allg. Chem.* **1920**, 113, 179.

Monaenkova, A.S.; Pashlova, E.B.; Vorob'ev, A.F. *Dokl. Akad. Nauk SSSR* **1971**, 199, 1332.

Monaenkova, A.S.; Vorob'ev, A.F. *Izvest. Vys. Ucheb. Zavedenii. Khim. i Khim. Tekhnol.* **1972**, 15, 2, 191.

Monk, C.B.; Amira, M.F. *J. Chem. Soc. Faraday Trans. I* **1978**, 74, 1170.

Montgomery, R.L. *U.S. Bur. Mines Rept. Inv. 6146* **1962**.

Montgomery, R.L. *Diss. Abstr.* **1975**. B36, 5063.

Montgomery, R.L.; Melaugh, R.A.; Lau, C-.C.; Meier, G.H.; Grow, R.T.; Rossini, F.D. *J. Chem. Eng. Data* **1978**, 23, 245.

Moore,C.E. *Ionization potentials and ionization limits derived from the analyses of optical spectra* NSRDS-NBS 34, **1970**.

Moore, G.E.; Kelley, K.K. *J. Am. Chem. Soc.* **1942**, 64, 2949.

Morino, Y; Uehara, H. *J. Chem. Phys.* **1966,** 45, 4543.

Morissette, B.G. *Thesis, Laval Univ.* **1952.**

Morozov, I.S.; Toptygin, D.Ya. *Zhur. Neorg. Khim.* **1956,** 1, 2601.

Morris, J.P.; Zellars, G.R. *J. Metals* **1956,** 8, 1086.

Morrison, J.A.; Patterson, D. *Trans. Faraday Soc.* **1956,** 52, 764. .

Morrison, J.A.; Patterson, D.; Dugdale, J.S. *Can. J. Chem.* **1955,** 33, 375.

Mould, H.M.; Price, W.C.; Wilkinson, G.P. *Spectrochim. Acta* **1960,** 16, 479.

Muldrow, C.N., Jr.; Hepler, L.G. *J. Phys. Chem.* **1958,** 62, 982.

Müller, F.; Reuther, H. *Z. Elektrochem.* **1941,** 47, 640.

Müller, F.; Reuther, H. *Z. Elektrochem.* **1942,** 48, 682.

Müller, W.; Meyer, W. *J. Chem. Phys.* **1984,** 80, 3311.

Munir, Z.A.; Searcy, A.W. *J. Chem. Phys.* **1965,** 42, 4223.

Muradov, V.G. *Uchen. Sapiski Ul'yanovskogo Ped. Inst.* **1964,** 18, 64.

Muradov, V.G., Geld, P.V. *Industrial Laboratory (USSR)* **1965,** 31, 1022. See Muradov, V.G. *Zhur. Fiz. Khim.* **1965,** 39, 170.

Muradova, O.N.; Fishman, I.S. *Zhur. Priklad. Spektrosk.* **1970,** 12, 971.

Muradova, O.N.; Muradov, V.G. *Zhur. Fiz. Khim.* **1972,** 46, 492.

Murata, F. *J. Chem. Soc. Japan* **1932,** 53, 574.

Murata, F. *J. Chem. Soc. Japan* **1933,** 54, 379.

Murch, L.E.; Giauque, W.F. *J. Phys. Chem.* **1962,** 66, 2052.

Murphy, G.M. *J. Chem. Phys.* **1936,** 4, 344.

Mussini, T.; Faita, G. *Ric. Sci.* **1966,** 36, 175.

Mussini, T.; Longhi, P. *Chim. Ind. (Milan)* **1977,** 59, 163.

Mussini, T.; Longhi, P.; Riva, G. *J. Chem. Thermodynamics* **1972,** 4, 591.

Mussini, T.; Maina, A.; Pagella, A. *J. Chem. Thermodynamics* **1971,** 3, 281.

Muxart, T.S. *Mezhdunarod. Geokhim. Kongr. (Dokl.)* **1971,** 4, 326.

Myl, J.; Koubova, D. *Chem. Průmysl* **1964,** 14, 38.

Myles, K.M.; Darby, J.B. *Acta Met.* **1968,** 16, 485.

Nagatani, M.; Seiyama, T.; Sakiyama, M.; Sugai, H.; Seki, S. *Bull. Chem. Soc. Japan* **1967,** 40, 1833.

Nair, V.S.K.; Nancollas, G.H. *J. Chem.Soc.* **1958,** 4144.

Nakamura, J.; Takahashi, Y.; Izumi, S.; Kanno, M. *J. Nucl. Mater.* **1980,** 88, 64.

Nakayama, F.S. *J. Inorg. Nucl. Chem.* **1971,** 33, 1287.

Nakayama, F.S.; Rasnick, B.A. *Anal. Chem.* **1967,** 39, 1022.

Natarajan, N.S. *Indian J. Pure App. Phys.* **1967,** 5, 372.

Naumov, V.N.; Nogteva, V.V.; Paukov, I.E. *Zhur. Fiz. Khim.* **1979,** 53, 497.

Nazarov, A.A.; Shul'ts, M.M.; Storonkin, A.V. *Vestnik Leningrad. Univ., Ser. Fiz. i Khim.* **1963,** (3), 94.

Neighbor, J.E.; Cochran, J.F.; Shiffman, C.A. *Phys. Rev.* **1967,** 155, 384.

Nelson, T.; Moss, C.; Hepler, L.G. *J. Phys. Chem.* **1960,** 64, 376.

Neumann, B.; Köhler, G. *Z. Elektrochem.* **1928,** 34, 218.

Neumann, B.; Kröger, C.; Kunz, H. *Z. Anorg. Allg. Chem.* **1934,** 218, 379.

Neumann, K.; Völker, E. *Z. Phys. Chem.* **1932,** A 161, 33.

Nies, N.P.; Hulbert, R.W. *J. Chem. Eng. Data* **1967,** 12, 303.

Nikolaeva, N.M.; Tolpygina, L.N. *Izvest. Sib. Otdel. Akad. Nauk SSSR, Ser. Khim.* **1969,** 3, 49.

Nims, L.F. *J. Am. Chem. Soc.* **1933,** 55, 1946.

Nosova, T.A.; Samoilov, O.Ya. *Zhur. Strukt. Khim.* **1964,** 5, 363.

Novikov, I.I.; Roshchupkin, V.V. *Tr. Vsesoyuznoi Nauchno-Tekh. Konf. po Termodinam., Sekts. Teplofiz. Svoistva Veshchestv* **1969**, 87.

Noyes, A.A.; Toabe, K. *J. Am. Chem. Soc.* **1917**, 39, 1537.

Nuñez, L.; Pilcher, G.; Skinner, H.A. *J. Chem. Thermodynamics* **1969**, 1, 31.

Nuttall, R.L.; Churney, K.; Kilday, M.V. *J. Res. Nat. Bur. Stand.* **1978**, 83, 335.

Oetting, F.L.; Rand, M.H.; Ackermann, R.J. *Chemical Thermodynamics of the Actinide Elements and Compounds. Part I.Actinide Elements*, I.A.E.A.: Vienna. **1976**.

Oguni, M.; Matsuo, T.; Suga, H.; Seki, S. *Bull. Chem. Soc. Japan* **1977**, 50, 825.

O'Hare, P.A.G.; Boerio, J. *J. Chem. Thermodynamics* **1975**, 7, 937.

O'Hare, P.A.G.; Johnson, G.K.; Appelman, E.H. *Inorg. Chem.* **1970**, 9, 332.

O'Hare, P.A.G.; Johnson, J.; Klamecki, B.; Mulvihill, M.; Hubbard, W.N. *J. Chem. Thermodynamics* **1969**, 1, 177.

Olofsson, G. *J. Chem. Thermodynamics* **1975**, 7, 507.

Olofsson, G.; Hepler, L.G. *J. Solution Chem.* **1975**, 4, 127.

Olofsson, G.; Olofsson, I. *J. Chem. Thermodynamics* **1973**, 5, 533.

Olofsson, G.; Olofsson, I. *J. Chem. Thermodynamics* **1977**, 9, 65.

Olofsson, G.; Sunner, S. *J. Chem. Thermodynamics* **1979**, 11, 605.

Onillon, M.; Olette, M. *C.R. Acad. Sci. (Paris)* **1968**, C 266, 517.

Ono, K.; Sudo, K. *Sci. Repts. Tohoku Imp. Univ.* **1955**, A7, 385.

Osborne, D.W.; Westrum, E.F., Jr. *J. Chem. Phys.* **1953**, 21, 1884.

Ots, H. *Acta Chem. Scand.* **1972**, 26, 3810.

Owen, B.B. *J. Am. Chem. Soc.* **1935**, 57, 1526.

Owen, B.B.; Brinkley, S.R., Jr. *J. Am. Chem. Soc.* **1938**, 60, 2233.

Owen, B.B.; Foering, L. *J. Am. Chem. Soc.* **1936**, 58, 1575.

Pacansky, J.; Calder, V. *J. Chem. Phys.* **1970**, 53, 4519.

Pahlman, J.E.; Smith, J.F. *Met. Trans.* **1972**, 3, 2423.

Pal'chevskii, V.V.; Phan Thi Bang *Vestnik Leningrad. Univ., Ser. Fiz. i Khim.* **1972**, (2), 144.

Pan, K.; Lin, J.L. *J. Chin. Chem. Soc. (Taiwan)* **1959**, 6, 1.

Panish, M.B. *J. Chem. Eng. Data* **1961**, 6, 592.

Paoletti, P.; Stern, J.H.; Vacca, A. *J. Phys. Chem.* **1965**, 69, 3759.

Paoletti, P.; Vacca, A. *Trans. Faraday Soc.* **1964**, 60, 50.

Papadopoulos, M.N.; Giauque, W.F. *J. Am. Chem. Soc.* **1955**, 77, 2740.

Papadopoulos, M.N.; Giauque, W.F. *J. Phys. Chem.* **1962**, 66, 2049.

Parker, V.B. *Thermal properties of aqueous uni-univalent electrolytes* NSRDS-NBS 2, **1965**.

Parker, V.B. *J. Res. Nat. Bur. Stand.* **1973**, 77 A, 227.

Parks, G.A. *Am.Mineralogist* **1972**, 57, 1163.

Partington, J.R.; Soper, W.E. *Phil. Mag.* **1929**, 7, 209.

Pattoret, A.; Drowart, J.; Smoes, S. *Trans. Faraday Soc.* **1969**, 65, 98.

Paukov, I.E. *Zhur. Fiz. Khim.* **1969**, 43, 2021.

Paukov, I.E.; Anisimov, M.P.; Luk'yanova, I.G. *Zhur. Fiz. Khim.* **1974**, 48, 1602.

Paukov, I.E.; Khriplovich, L.M. *Zhur. Fiz. Khim.* **1969**, 43, 2678.

Paukov, I.E.; Khriplovich, L.M.; Korotikikh, A.M. *Zhur. Fiz. Khim.* **1968**, 42, 1297.

Paukov, I.E.; Lavrent'eva, M.N. *Zhur. Fiz. Khim.* **1968**, 42, 1842.

Paukov, I.E., Lavrent'eva, M.N. *Zhur. Fiz. Khim.* **1969a**, 43, 1390.

Paukov, I.E., Lavrent'eva, M.N.; Anisimov, M.P. *Zhur. Fiz. Khim.* **1969b**, 43, 2120.

Paule, R.C.; Mandel, J. *Nat. Bur. Stand. Special Publication 260-19* **1970**.

Paule, R.C.; Mandel, J. *Nat. Bur. Stand. Special Publication 260-21* **1971**.

Paule, R.C.; Margrave, J.L. *J. Phys. Chem.* **1963**, 67, 1368.

Pavlyuk, L.A.; Kryukov, P.A. *Izv. Sib. Otdel. Akad. Nauk SSSR, Ser. Khim. Nauk* **1976**, 14, 25.

Payne, J.H. *J. Am. Chem. Soc.* **1937**, 59, 947.

Pearce, J.N. *J. Am. Chem. Soc.* **1936**, 58, 376.

Pearce, J.N.; Eckstrom, H.C. *J. Am. Chem. Soc.* **1937**, 59, 2689.

Pearce, J.N.; Hopson, H. *J. Phys. Chem.* **1937**, 41, 535.

Pearce, J.N.; Nelson, A.F. *J. Am. Chem. Soc.* **1932**, 54, 3544.

Pearce, J.N.; Pumplin, G.G. *J. Am. Chem. Soc.* **1937**, 59, 1219.

Pearce, J.N.; Taylor, M.D.; Bartlett, R.M. *J. Am. Chem. Soc.* **1928**, 50, 2951.

Peiper, J.C.; Pitzer, K.S. *J. Chem. Thermodynamics* **1982**, 14, 613.

Penciner, J.; Marcus, Y. *J. Chem. Eng. Data* **1965**, 10, 105.

Peppler, R.B.; Newman, E.S. *J. Res. Nat. Bur. Stand.* **1951**, 46, 121.

Perelygin, B.G.; Byval'tsev, Y.A.; Vorob'ev, A.F. *Zhur. Fiz. Khim.* **1977**, 51, 2398.

Perelygin, B.G.; Byval'tsev, Y.A.; Vorob'ev, A.F. *Zhur. Fiz. Khim.* **1978**a, 52, 484.

Perelygin, B.G.; Byval'tsev, Y.A.; Vorob'ev, A.F. *Zhur. Fiz. Khim.* **1978**b, 52, 1558.

Perelygin, B.G.; Byval'tsev, Y.A.; Vorob'ev, A.F. *Zhur. Fiz. Khim.* **1978**c, 52, 1836.

Perman, E.P.; Urry, W.D. *Trans. Faraday Soc.* **1928**, 28, 337.

Perret, R. *Bull. Soc. Chim. France* **1966**, 755.

Peters, H.; Möbius, H.H. *Z. Phys. Chem. (Leipzig)* **1958**, 209, 298.

Phillips, R.C.; George, P.; Rutman, R.J. *Biochemistry* **1963**, 2, 501.

Piacente, V.; De Maria, G. *Ric. Sci.* **1969**, 39, 549.

Piesbergen, U. *Z. Naturforsch.* **1963**, 18a, 141.

Pike, F.P.; Foster, C.T., Jr. *J. Chem. Eng. Data* **1959**, 4, 305.

Pitzer, K.S. *J. Am. Chem. Soc.* **1937**, 59, 2365.

Pitzer, K.S. *J. Am. Chem. Soc.* **1938**, 60, 1828.

Pitzer, K.S. *J. Am. Chem. Soc.* **1941**, 63, 516.

Pitzer, K.S.; Coulter, L.V. *J. Am. Chem. Soc.* **1938**, 60, 1310.

Pitzer, K.S.; Mayorga, G. *J. Solution Chem.* **1974**, 3, 539.

Pitzer, K.S.; Roy, R.N.; Silvester, L.F. *J. Am. Chem. Soc.* **1977**, 99, 4930.

Pitzer, K.S.; Smith, W.V. *J. Am. Chem. Soc.* **1937**, 59, 2633.

Pitzer, K.S.; Smith, W.V.; Latimer, W.M. *J. Am. Chem. Soc.* **1938**, 60, 1826.

Platford, R.F. *Can. J. Chem.* **1969**, 47, 2271.

Plummer, N.L.; Busenberg, E. *Geochim. Cosmochim. Acta* **1982**, 46, 1011.

Plyler, E.K. *J. Res. Nat. Bur. Stand.* **1960**, 64 A, 377.

Pollitzer, F. *Z. Anorg. Chem.* **1909**, 64, 121.

Pollitzer, F. *Z. Elektrochem.* **1911**, 17, 5.

Pollitzer, F. *Z. Elektrochem.* **1913**, 19, 513.

Ponslet, A.; Bariaux, D. *Bull. Cl. Sci. Acad. Roy. Belg.* **1966**, 52, 248.

Popov, M.M.; Gagarinskii, Yu. V.; Stepanenko, N.N. *Zhur. Neorg. Khim.* **1957**, 2, 1457.

Popov, M.M.; Ivanov, M.I. *Atomnaya Energ.* **1957**, 2, 360.

Porter, R.F.; Schissel, P.; Inghram, M.G. *J. Chem. Phys.* **1955**, 23, 339.

Potier, A. *Mém. Services Chim. État (Paris)* **1953**, 38, 391.

Potter, N.D.; Boyer, M.H.; Ju, F.; Hildenbrand, D.L.; Murad, E.; Hall, W.F. *Philco Corp. Tech. Report AD-715567* **1970**.

Potter, R.W.; Clynne, M.A. *J. Res. U.S. Geological Survey* **1978**, 6, 701.

Power, W.H.; Fabuss, B.M.; Satterfield, C.N. *J. Chem. Eng. Data* **1964**, 9, 437.

Prasad, R.; Venugopal, V.; Iyer, P.N.; Sood, D.D. *J. Chem. Thermodynamics* **1978,** 10, 135.

Preuner, G. *Z. Anorg. Chem.* **1909,** 55, 279.

Preuner, G.; Schupp, W. *Z. Phys. Chem.* **1909,** 68, 157.

Price, G.H.; Stuart, W.I. *J. Chem. Soc. Faraday Trans. I* **1973,** 69, 1498.

Priepke, R.J.; Vosburgh, W.C. *J. Am. Chem. Soc.* **1930,** 52, 4831.

Priselkov, Yu. A. *Dissertation,* **1954**. Moscow State University; cited by A.N. Nesmeyanov, in *Vapor Pressures of the Chemical Elements.* Moscow. **1961.**

Priselkov, Yu.A.; Nesmeyanov, A.N. *Dokl. Akad. Nauk SSSR* **1954,** 95, 1207.

Prosen, E.J.; Johnson, W.H.; Pergiel, F.Y. *J. Res. Nat. Bur. Stand.* **1958,** 61, 247.

Prosen, E.J.; Johnson, W.H.; Pergiel, F.Y. *J. Res. Nat. Bur. Stand.* **1959,** 62, 43.

Prosen, E.J.; Rossini, F.D. *J. Res. Nat. Bur. Stand.* **1944,** 33, 439.

Provost, R.H.; Wulff, C.A. *J. Chem. Thermodynamics* **1970,** 2, 655.

Prue, J.E.; Read, A.J.; Romeo, G. *Trans. Faraday Soc.* **1971,** 67, 420.

Prutton, C.F.; Brosheer, J.C.; Maron, S.H. *J. Am. Chem. Soc.* **1935,** 57, 1656.

Puschin, G.I.; Subbotin, V.A.; Kharaturov, E.Kh. *Teplofiz. Vys. Temp.* **1975,** 13, 747.

Quintin, M. *J. Chim. Phys.* **1936,** 33, 111.

Randall, M.; Bichowsky, F.R. *J. Am. Chem. Soc.* **1918,** 40, 368.

Randall, M.; Bisson, C.S. *J. Am. Chem. Soc.* **1920,** 42, 347.

Randall, M.; Spencer, H.M. *J. Am. Chem. Soc,* **1928,** 50, 1572.

Rank, D.H.; Baldwin, W.M. *J. Chem. Phys.* **1951,** 19, 1210.

Rank, D.H.; Fink, U.; Wiggins, T.A. *J. Mol. Spectr.* **1965**b, 18, 170.

Rank, D.H.; Rao, B.S.; Wiggins, T.A. *J. Mol. Spectr.* **1965**a, 17, 122. .

Rao, D.B.; Motzfeldt, K. *Acta Chem. Scand.* **1970,** 24, 2796.

Rao, S.R.; Hepler, L.G. *Hydrometallurgy,* **1977,** 2, 293.

Rao, Y.V.; Venkateswarlu, P. *J. Mol. Spectr.* **1964,** 13, 288.

Raridon, R.J.; Baldwin, W.H.; Kraus, K.A. *J. Phys. Chem.* **1968,** 72, 925.

Rashkovskaya, E.A.; Chernen'kaya, E.I. *Zhur. Priklad. Khim.* **1967,** 40, 301.

Ratner, A.P.; Makarov, L.L. *Zhur. Fiz. Khim.* **1958,** 32, 1809.

Rau, H. *J. Chem. Thermodynamics* **1975,** 7, 903.

Rayne, J.A. *J. Phys. Chem. Solids* **1958,** 7, 268.

Read, A.J. *J. Solution Chem.* **1975,** 4, 53.

Readnour, J.M.; Cobble, J.W. *Inorg. Chem.* **1969,** 8, 2174.

Rengade, E. *C.R. Acad. Sci. (Paris)* **1908,** 129, 131.

Reshetnikov, N.A. *Zhur. Neorg. Khim.* **1961,** 6, 682.

Rezukhina, T.N.; Sisoeva, T.F.; Kholokhonova, L.T. *Sixth All-Union Conference on Calorimetry.* Tbilisi, U.S.S.R. **1973,** p. 324.

Ricci, J.E.; Fischer, J. *J. Am. Chem. Soc.* **1952,** 74, 1607.

Ricci, J.E.; Linke, W.F. *J. Am. Chem. Soc.* **1951,** 73, 3601.

Ricci, J.E.; Selikson, B. *J. Am. Chem. Soc.* **1952,** 74, 1956.

Ricci, J.E.; Skarulis, J.A. *J. Am. Chem. Soc.* **1951,** 73, 3618.

Richards, G.W.; Woolf, A.A. *J. Chem. Soc. A,* **1968,** 470.

Richards, T.W.; Yngve, V. *J. Am. Chem. Soc.* **1918,** 40, 164.

Ricks, J.M.; Barrow, R.F. *Can. J. Phys.* **1969,** 47, 2423.

Rigney, D.V.; Kapelner, S.M.; Cleary, R.E. *U.S. A.E.C. Rept. TIM 810* **1965**.

Rittenberg, D.; Urey, H.C. *J. Chem. Phys.* **1934**a, 2, 106.

Rittenberg, D.; Urey, H.C. *J. Am. Chem. Soc.* **1934**b, 56, 1885.

Roberts, L.M. *Proc. Phys. Soc.* **1957,** 70 B, 738.

Robie, R.A.; Hemingway, B.S.; Wilson, W.H. *J. Research U.S. Geol. Survey* **1976,** 4, 631.

Robinson, R.A. *Trans. Faraday Soc.* **1940,** 36, 1135.

Robinson, R.A.; Jones, R.S. *J. Am. Chem. Soc.* **1936,** 58, 959.

Robinson, R.A.; Macaskill, J.B. *J. Solution Chem.* **1979,** 8, 35.

Robinson, R.A.; Stokes, R.H. *Trans. Faraday Soc.* **1940,** 36, 740.

Robinson, R.A.; Stokes, R.H. *Electrolyte Solutions,* 2nd ed. Butterworths: London. **1959**.

Robinson, R.A.; Tait, D.A. *Trans. Faraday Soc.* **1941,** 37, 569.

Robson, H.E.; Gilles, P.W. *J. Phys. Chem.* **1964,** 68, 983.

Rodebush, W.H.; DeVries, T. *J. Am. Chem. Soc.* **1925,** 47, 2488.

Rodebush, W.H.; Dixon, A.L. *J. Am. Chem. Soc.* **1925**a, 47, 1036.

Rodebush, W.H.; Dixon, A.L. *Phys. Rev.* **1925**b, 26, 851.

Rodebush, W.H.; Henry, W.F. *J. Am. Chem. Soc.* **1930,** 52, 3159.

Rodebush, W.H.; Walters, E.G. *J. Am. Chem. Soc.* **1930,** 52, 2654.

Roeder, A.; Morawietz, W. *Z. Elektrochem.* **1956,** 60, 431.

Romanova, I.I.; Samoilov, O.Ya. *Zhur. Neorg. Khim.* **1969,** 14, 252.

Rosen, B. *Spectroscopic Constants for Diatomic Molecules.* Pergamon: New York. **1970**.

Rosenberg, T. *Acta Chem. Scand.* **1949,** 3, 50.

Rossini, F.D. *J. Res. Nat. Bur. Stand.* **1931**a, 6, 1.

Rossini, F.D. *J. Res. Nat. Bur. Stand.* **1931**b, 6, 37.

Rossini, F.D. *J. Res. Nat. Bur. Stand.* **1932,** 9, 679.

Rossini, F.D. *J. Res. Nat. Bur. Stand.* **1939,** 22, 407.

Rossini, F.D.; Wagman, D.D.; Evans, W.H.; Levine, S.; Jaffe, I. *Nat. Bur. Stand. Circ. 500* Washington, D.C. **1952**.

Roth, W.A.; Becker, G. *Z. Phys. Chem.* **1932,** A 159, 1.

Roth, W.A.; Berendt, H.; Wirths, G. *Z. Electrochem.* **1941,** 47, 185.

Roth, W.A.; Bertram, A. *Z. Elektrochem.* **1937,** 43, 376.

Roth, W.A.; Grau, R.; Meichsner, A. *Z. Anorg. Allg. Chem.* **1930,** 193, 161.

Roth, W.A.; Richter, H. *Z. Phys. Chem.* **1934,** A 170, 123.

Roth, W.A.; Wolf, U.; Fritz, O. *Z. Elektrochem.* **1940,** 46, 42.

Roth, W.A.; Zeumer, H. *Z. Angew. Chem.* **1931,** 44, 559.

Rubin, T.R.; Giauque, W.F. *J. Am. Chem. Soc.* **1952,** 74, 800.

Rudzitis, E.; Feder, H.M.; Hubbard, W.N. *J. Phys. Chem.* **1964,** 68, 2978.

Rudzitis, E.; Feder, H.M.; Hubbard, W.N. *Inorg. Chem.* **1967,** 6, 1716.

Ruff, O.; Johannsen, O. *Ber. Dtsch. Chem. Ges.* **1905,** 38, 3601.

Rupert, J.P.; Hopkins, H.P., Jr.; Wulff, C.A. *J. Phys. Chem.* **1965,** 69, 3059.

Rush, R.M.; Johnson, J.S. *J. Phys. Chem.* **1968,** 72, 767.

Ryabchikov, J.V.; Mikulinskii, A.S. *Tsvet. Met.* **1963,** 1, 95.

Ryss, I.G.; Slutskaya, M.M.; Vitukhnovskaya, B.S. *Zhur. Priklad. Khim.* **1952,** 25, 148.

Saegusa, F. *J. Chem. Soc. Japan* **1948,** 69, 24.

Saegusa, F. *Sci. Repts. Tohoku Imp. Univ., Ser.I* **1950,** 34, 147.

Saheki, Y.; Funaki, K. *Nippon Kagaku Zasshi* **1957,** 78, 754.

Samoilov, O.Ya. *Izvest. Akad. Nauk SSSR* **1956,** 1415.

Samoilov, O.Ya.; Tsvetkov, V.G. *Zhur. Strukt. Khim.* **1968,** 9, 193.

Samuseva, R.G.; Egorova, R.S.; Plyushchev, V.E. *Zhur. Neorg. Khim.* **1962**b, 7, 1666.

Samuseva, R.G.; Plyushchev, V.E.; Egorova, R.S. *Zhur. Neorg. Khim.* **1962**a, 7, 1415.

Sandin, T.R.; Keesom, P.H. *Phys. Rev.* **1969,** 177, 1370.

Sano, K. *J. Chem. Soc. Japan* **1936,** 57, 1025.

Scatchard, G.; Breckenridge, R.C. *J. Phys. Chem.* **1954,** 58, 596.

Scatchard, G.; Tefft, R.F. *J. Am. Chem. Soc.* **1930,** 52, 2272.

Schäfer, H.; Breil, G.; Pfeffer, G. *Z. Anorg. Allg. Chem.* **1954,** 276, 325.

Schäfer, H.; Zeppernick, F. *Z. Anorg. Allg. Chem.* **1953,** 272, 274.

Schindler, P.; Althaus, H.; Feitknecht, W. *Helv. Chim. Acta* **1964,** 47, 982.

Schins, H.E.J.; van Wijk, R.W.M.; Dorpema, B. *Z. Metallkunde* **1971,** 62, 330.

Schissel, P.O.; Trulson, O.C. *J. Phys. Chem.* **1962,** 66, 1492.

Schmahl, N.G.; Sieben, P. in *The Physical Chemistry of Metallic Solutions and Intermetallic Compounds,* Volume I. H.M.S.O.: London. **1960.**

Schmidt, E.A.; Zalukaev, V.L.; Sharpataya, G.A.; Golushina, L.N. *Thesis, Ivanova, USSR* **1979.**

Schmidt, H.G.; Wolf, G. *Solid State Commun.* **1975,** 16, 1085.

Schneider, A.; Gattow, G. *Z. Anorg. Allg. Chem.* **1954,** 277, 41.

Scholle, S.; Brunclikova, Z. *Chem. Prümysl.* **1968,** 18, 533.

Schuffenecker, L.; Balesdent, D.; Houriez, J. *Bull. Soc. Chim. France* **1970,** 1239.

Schulz, D.A.; Searcy, A.W. *Thesis, Univ. California* **1961** - see Hultgren et al. **1973.**

Schulz, G.; Schaefer, H. *Ber. Bunsenges. Phys. Chem.* **1966,** 70, 21.

Schumm, R.H.; Prosen, E.J.; Wagman, D.D. *J. Res. Nat. Bur. Stand.* **1974,** 78 A, 375.

Schwiete, H.E.; Pranschke, A. *Zement* **1935,** 24, 593.

Scott, A.F.; Durham, E.J. *J. Phys. Chem.* **1930,** 34, 531.

Scott, D.H. *Phil. Mag.* **1924,** 47, 32.

Searcy, A.W. *J. Am. Chem. Soc.* **1952,** 74, 4789.

Searcy, A.W.; Freeman, R.D. *J. Am. Chem. Soc.* **1954,** 76, 5229.

Searcy, A.W.; Freeman, R.D. *J. Chem. Phys.* **1955,** 23, 88.

Seryakov, G.V.; Vaks, S.A.; Sidorina, S.L. *Zhur. Obshch. Khim.* **1960,** 30, 2130.

Sharma, L.; Prasad, B. *J. Indian Chem. Soc.* **1969,** 46, 241.

Sharma, L.; Prasad, B. *J. Indian Chem. Soc.* **1970**a, 47, 193.

Sharma, L.; Prasad, B. *J. Indian Chem. Soc.* **1970**b, 47, 379.

Sharma, L.; Sahu, G.; Prasad, B. *J. Indian Chem. Soc.* **1968,** 45, 580.

Shchukarev, S.A.; Lilich, L.S.; Latysheva, V.A. *Dokl. Akad. Nauk SSSR* **1953,** 91, 273.

Shchukarev, S.A.; Vasilkova, I.V.; Sharunun, B.H. *Vestnik Leningrad. Univ., Ser. Fiz. i Khim.* **1960,** 2, 112.

Shearman, R.W.; Menzies, A.W.C. *J. Am. Chem. Soc.* **1937,** 59, 185.

Shibata, F.L.E. *J. Sci. Hiroshima Univ.* **1931,** A1, 215.

Shibata, F.L.E.; Kobayashi, Y.; Furukawa, S. *J. Chem. Soc. Japan* **1931**a, 52, 404.

Shibata, F.L.E.; Murata, F. *J. Chem. Soc. Japan* **1931**a, 52, 393.

Shibata, F.L.E.; Murata, F. *J. Chem. Soc. Japan* **1931**b, 52, 645.

Shibata, F.L.E.; Murata, F. *J. Sci. Hiroshima Univ.* **1937,** A7, 335.

Shibata, F.L.E.; Murata, F.; Toyada, Y. *J. Chem. Soc. Japan* **1931**b, 52, 639.

Shibata, F.L.E.; Oda, S.; Furukawa, S. *J. Chem. Soc. Japan* **1930,** 51, 71.

Shibata, F.L.E.; Taketa, T. *J. Sci. Hiroshima Univ.* **1932,** A2, 243.

Shidlovskii, A.A.; Voskresenskii, A.A. *Zhur. Fiz. Khim.* **1963,** 37, 2062.

Shin, C.; Criss, C.M. *J. Chem. Thermodynamics* **1979,** 11, 663.

Shirley, D.A. *J. Am. Chem. Soc.* **1960,** 82, 3841.

Shirley, D.A.; Giauque, W.F. *J. Am. Chem. Soc.* **1959,** 81, 4778.

Shiu, D.H.; Munir, Z.A. *Metallurg. Trans.* **1971**a, 2, 2953.

Shiu, D.H.; Munir, Z.A. *High Temp. Sci.* **1971**b, 3, 381.

Shklovskaya, R.M.; Arkhipov, S.M.; Kuzina, V.A.; Tsibulevskaya, K.A. *Zhur. Neorg. Khim.* **1976,** 21, 2868.

Shomate, C.H. *J. Am. Chem. Soc.* **1945,** 67, 1096.

Shomate, C.H. *J. Am. Chem. Soc.* **1947,** 69, 218.

Shomate, C.H.; Huffman, E.H. *J. Am. Chem. Soc.* **1943,** 65, 1625.

Shomate, C.H.; Kelley, K.K. *J. Am. Chem. Soc.* **1944,** 66, 1490.

Shpil'rain, E.E.; Belova, A.M. *Nauch. Issled. Inst. Vys. Temp., Report No.264* **1962.**

Shpil'rain, E.E.; Belova, A.M. *Teplofiz. Vys. Temp.* **1967,** 5, 477.

Shpil'rain, E.E.; Belova, A.M. *Teplofiz. Vys. Temp.* **1968,** 6, 342.

Shpil'rain, E.E.; Nikanorov, E.V. *Teplofizicheskie Svoistva Gasov.* Nauka: Moscow. **1970,** p. 141.

Shpil'rain, E.E.; Nikanorov, E.V. *Teplofiz. Vys. Temp.* **1971,** 9, 434.

Shpil'rain, E.E.; Nikanorov, E.V. *Teplofiz. Vys. Temp.* **1972,** 10, 297.

Shpil'rain, E.E.; Totskii, E.E.; Karmyshin, Yu.V. *Trudy Mosk. Energ. Inst.* **1970**a, 75, 62.

Shpil'rain, E.E.; Totskii, E.E.; Shereshevskii, V.A. *Teplofiz. Vys. Temp.* **1968,** 6, 924.

Shpil'rain, E.E; Yakomovich, K.A.; Totskii, E.E.; Timrot, D.L.; Fomin, V.A. *Thermophysical Properties of the Alkali Metals.* Standards: Moscow **1970**b.

Shpil'rain, E.E.; Zvereva, E.M. *Inzh.-Fiz. Zhur.* **1963,** 6, 74.

Shrawder, J., Jr.; Cowperthwaite, I.A. *J. Am. Chem. Soc.* **1934,** 56, 2340.

Shuiman, G. *Thesis, Columbia Univ. School of Engineering and Appl. Sci., N.Y.* **1963.**

Shul'ts, M.M.; Simanova, S.A. *Zhur. Fiz. Khim.* **1966,** 40, 462.

Shul'ts, M.M.; Makarov, L.L.; Su Yu-jeng *Zhur. Fiz. Khim.* **1962,** 36, 2194.

Siemens, P.R.; Giauque, W.F. *J. Phys. Chem.* **1969,** 73, 149.

Simanova, S.A.; Shul'ts, M.M. *Zhur. Neorg. Khim.* **1967,** 12, 223.

Singh, D. *J. Sci. Res. Bandaras Hindu Univ.* **1955,** 6, 131.

Skarulis, J.A.; Ricci, J.E. *J. Am. Chem. Soc.* **1941,** 63, 3429.

Skinner, H.A; Bennett, J.E.; Pedley, J.B. *Pure Appl. Chem.* **1961,** 2, 17.

Skinner, H.A.; Smith, N.B. *Trans. Faraday Soc.* **1953,** 49, 60.

Skinner, G.B.; Ruehrwein, R.A. *J. Phys. Chem.* **1955,** 59, 113.

Skinner, G.B.; Searcy, A.W. *J. Phys. Chem.* **1968,** 72, 3375.

Slansky, C.M. *J. Am. Chem. Soc.* **1940,** 62, 2430.

Slonim, C.; Hüttig, G.F. *Z. Anorg. Allg. Chem.* **1929,** 181, 55.

Smets, J.; Coppens, P.; Drowart, J. *Chem. Phys.* **1977,** 20, 243.

Smisko, J.; Mason, L.S. *J. Am. Chem. Soc.* **1950,** 72, 3679.

Smith, A.; Eastlack, H.E. *J. Am. Chem. Soc.* **1916,** 38, 1261.

Smith, D.F.; Woods, H.K. *J. Am. Chem. Soc.* **1923,** 45, 2632.

Smith, E.R.; Taylor, J.K. *J. Res. Nat. Bur. Stand.* **1940,** 25, 731.

Smith, J.F. *Symposium on Thermodynamics of Nuclear Materials.* I.A.E.A.: Vienna. **1962.**

Smith, M.B.; Bass, G.E., Jr. *J. Chem. Eng. Data* **1963,** 8, 342.

Smith, P.L. *Phil. Mag.* **1955,** 46, 744.

Smith, P.L.; Wolcott, N.M. *Bull. Internat. Inst. Refrig.* **1955,** Suppl.3, 283.

Smith, P.L.; Wolcott, N.M. *Phil. Mag.* **1956,** 1, 854.

Smith, R.P. *J. Am. Chem. Soc.* **1946,** 68, 1163.

Smith, W.V.; Brown, O.L.I.; Pitzer, K.S. *J. Am. Chem. Soc.* **1937,** 59, 1213.

Snyder, L.E.; Edwards, T.H. *J. Mol. Spectr.* **1969,** 31, 347.

Snyder, P.E.; Seltz, H. *J. Am. Chem. Soc.* **1945,** 67, 683.

Sorai, M.; Suga, H.; Seki, S. *Bull. Chem. Soc. Japan* **1965**, 38, 1125.

Sorai, M.; Suga, H.; Seki, S. *Bull. Chem. Soc. Japan* **1968**, 41, 312.

Soulier, J.P.; Gautier, J. *C. R. Acad. Sci. (Paris)* **1966**, 263C, 1485.

Southard, J.C. *Ind. Eng. Chem.* **1940**, 32, 442.

Southard, J.C. *J. Am. Chem. Soc.* **1941**, 63, 3147.

Southard, J.C.; Nelson, R.A. *J. Am. Chem. Soc.* **1933**, 55, 4865.

Sowa, E.S. *Nucleonics* **1963**, 21, 76.

Spencer, H.M. *J. Chem. Phys.* **1946**, 14, 729.

Spencer, H.M.; Mote, J.H. *J. Am. Chem. Soc.* **1932**, 54, 4618.

Speros, D.M.; Woodhouse, R.L. *J. Phys. Chem.* **1968**, 72, 2846.

Sproesser, W.C.; Taylor, G.B. *J. Am. Chem. Soc.* **1921**, 43, 1782.

Sreedharan, O.M.; Athiappan, E.; Pankajavalli, R.; Gnanamoorthy, J.B. *J. Less-Common Met.* **1979**, 68, 143.

Sretenskaya, N.G. *Geokhim.* **1974**, 996.

Stalinski, B.; Bieganski, Z. *Rocz. Chem.* **1961**, 35, 273.

Staples, B.S. *J. Phys. Chem. Ref. Data* **1981**, 10, 779.

Starck, S.; Bodenstein, M. *Z. Elektrochem.* **1910**, 16, 961.

Starck, S.; Bodenstein, M. *Z. Elektrochem.* **1916**, 22, 327.

Staveley, L.A.K.; Linford, R.G. *J. Chem. Thermodynamics* **1969**, 1, 1.

Ste. Marie, J.; Torma, A.E.; Gübeli, A.O. *Can. J. Chem.*, **1966**, 42, 662.

Stepin, B.D.; Serebrennikova, G.M.; Chicherina, G.P.; Trunov, V.K.; Oboznenko, Y.V. *Zhur. Neorg. Khim.* **1976**, 21, 3148.

Stephens, H.P.; Cobble, J.W. *Inorg. Chem.* **1971**, 10, 619.

Stephenson, C.C. Private communication. 1975.

Stephenson, C.C.; Abajian, P.G.; Provost, R.; Wulff, C.A. *J. Chem. Eng. Data* **1968**, 13, 191.

Stephenson, C.C.; Bentz, D.R.; Stevenson, D.A. *J. Am. Chem. Soc.* **1955**, 77, 2161.

Stephenson, C.C.; Hooley, J.G. *J. Am. Chem. Soc.* **1944**, 66, 1397.

Stephenson, C.C.; Hopkins, H.P.; Wulff, C.A. *J. Phys. Chem.* **1964**, 68, 1427.

Stephenson, C.C.; McMahon, H.O. *J. Am. Chem. Soc.* **1939**, 61, 437.

Stephenson, C.C.; Potter, R.L.; Maple, T.G.; Morrow, J.C. *J. Chem. Thermodynamics* **1969**, 1, 59.

Stephenson, C.C.; Zettlemoyer, A.C. *J. Am. Chem. Soc.* **1944**, 66, 1405.

Stern, J.H.; Lowe, E.; O'Connor, M.E. *J. Solution Chem.* **1974**, 3, 823.

Stern, J.H.; Passchier, A.A. *J. Chem. Eng. Data* **1962**, 7, 73.

Stern, T.H. *Ph.D. thesis, Univ. of Washington* **1958**.

Stimson, H.F. *J. Res. Nat. Bur. Stand.* **1969**, 73 A, 493.

Stock, A.; Gibson, G.E.; Stamm, E. *Ber. Dtsch. Chem. Ges.* **1912**, 45, 3527.

Stokes, R.H. *J. Am. Chem. Soc.* **1945**a, 67, 1689.

Stokes, R.H. *Trans. Faraday Soc.* **1945**b, 41, 642.

Stokes, R.H. *Trans. Faraday Soc.* **1948**, 44, 295.

Stokes, R.H.; Robinson, R.A. *J. Am. Chem. Soc.* **1948**, 70, 1870.

Stokes, R.H.; Stokes, J.M. *Trans. Faraday Soc.* **1945**, 41, 688.

Stone, J.P.; Ewing, C.T.; Spann, J.R.; Steinkuller, E.W.; Williams, D.D.; Miller, R.R. *J. Chem. Eng. Data* **1966**, 11, 315.

Storms, E.K. *Refractory Materials,* Academic Press: New York. **1967**.

Storms, E.K. *Thermodynamics,* Vol I. I.A.E.A.: Vienna. **1966**, 309.

Storms, E.K.; Mueller, B. *J. Phys. Chem.* **1977**, 81, 318.

Stoughton, R.W.; Lietzke, M.H. *J. Phys. Chem.* **1960**, 64, 133.

Stout, J.W. *J. Chem. Phys.* **1941**, 9, 285.

Stout, J.W.; Chisholm, R.C. *J. Chem. Phys.* **1962**, 36, 979.

Strassmair, H.; Stark, D. *Z. Angew. Phys.* **1967**, 23, 40.

Stull, D.R.; Hildenbrand, D.L.; Oetting, F.L.; Sinke, G.C. *J. Chem. Eng. Data* **1970**, 15, 52.

Stwalley, W.C. *Chem. Phys. Letters* **1970**, 7, 600.

Sugawara, S.; Sato, T.; Minamiyama, T. *Bull. Japan Soc. Mech. Eng.* **1962**, 5, 711.

Sunner, S.; Thorén, S. *Acta Chem. Scand.* **1964**, 18, 1528.

Sunner, S.; Wadsö, I. *Trans. Faraday Soc.* **1957**, 53, 455.

Sweeton, F.H.; Mesner, R.E; Baes, C.E.,Jr. *J. Solution Chem.* **1974**, 3, 191.

Taft, R.; Krishnan, V.R. *Trans. Kansas Acad. Sci.* **1954**, 57, 101.

Takahashi, S. Private communication to M.W. Chase. **1981**.

Tamas, J.; Kosza, G. *Magyar Kem. Folyoirat* **1964**, 70, 148.

Tanaka, Y.; Ogawa, M.; Jursa, A.S. *J. Chem. Phys.* **1964**, 40, 3690.

Taniewska-Osinska, S.; Piestrzynska, B.; Logwinienko, R. *Can. J. Chem.* **1980**, 58, 1584.

Tanji, K.K. *Environmental Sci. Technol.* **1969**, 3, 656.

Taylor, A.H., Jr.; Crist, R.H. *J. Am. Chem. Soc.* **1941**, 63, 1377.

Taylor, A.R.; Gardner, T.E.; Smith, D.R. *U.S. Bur. Mines Rept. Inv. 6157* **1963**.

Taylor, G.B. *Ind. Eng. Chem.* **1925**, 17, 633.

Taylor, H.S.; Perrott, G.S. *J. Am. Chem. Soc.* **1921**, 43, 484.

Taylor, J.B.; Langmuir, I. *Phys. Rev.* **1937**, 51, 753.

Taylor, K.; Wells, L.S. *J. Res. Nat. Bur. Stand.* **1938**, 21, 133.

Tepper, F.; Murchison, A.; Zelenak, J.; Roelich, F. *U.S. Air Force Materials Lab. Rept. RTD-TDR-63-4018.* Pt.1. Wright-Patterson AFB: Ohio. **1963**.

Thakker, M.T.; Chi, C.W.; Peck, R.E.; Wasan, D.T. *J. Chem. Eng. Data* **1968**, 13, 553.

Thiele, E. *Ann. Phys.* **1932**, 14, 937.

Thompson, C.J.; Sinke, G.C.; Stull, D.R. *J. Chem. Eng. Data* **1962**, 7, 380.

Thomsen, J. *Thermochemische Untersuchungen,* J. Barth: Leipzig. **1883-1886**.

Thorvaldson, T.; Brown, W.G.; Peaker, C.R. *J. Am. Chem. Soc.* **1930**, 52, 910.

Thourey, J.; Perachon, G. *Thermochim. Acta* **1980**, 39, 243.

Tilford, S.G.; Vanderslice, J.T.; Wilkinson, P.G. *Astrophys. J.* **1965**, 142, 1203.

Todd, S.S. *J. Am. Chem. Soc.* **1949**, 71, 4115.

Todd, T.R.; Clayton, C.M.; Telfair, W.B.; McCubbin, T.K.; Pliva, J. *J. Mol. Spectr.* **1976**, 62, 201.

Tomassi, W.; Wroblowa, H. *Roczn. Chem.* **1956**, 30, 873.

Topol, L.E. *Inorg. Chem.* **1968**, 7, 451.

Torgeson, D.R.; Sahama, T.G. *J. Am. Chem. Soc.* **1948**, 70, 2156.

Torgeson, D.R.; Shomate, C.H. *J. Am. Chem. Soc.* **1947**, 69, 2103.

Toriumi, T.; Hara, R. *J. Chem. Soc. Japan* **1934**, 55, 1051.

Tret'yakov, G.D.; Geiderikh, V.A. *Zhur. Fiz. Khim.* **1968**, 42, 1786.

Treumann, W.B.; Ferris, L.M. *J. Am. Chem. Soc.* **1958**, 80, 5048.

Treverton, J.A.; Margrave, J.L. *J. Chem. Thermodynamics* **1971**, 3, 473.

Trumpler, G.; Schuler, D.; Ibl, N. *Helv. Chim. Acta* **1950**, 33, 790.

Tseplyaeva, A.V.; Severin, V.I.; Levchuk, G.K.; Prisekov, Yu.A.; Ryabtseva, L.P. *High Temp. (USSR)* **1982**, 20, 395.

Tsvetkov, V.G.; Rabinovich, I.B. *Zhur. Fiz. Khim.* **1969**, 43, 1213.

Tsvetkov, Yu.V.; Edel'stein, V.M. *Trudy Inst. Metallurg. Akad. Nauk SSSR* **1963**, 95.

Turmanova, A.A.; Mishchenko, K.P.; Flis, I.E. *Zhur. Neorg. Khim.* **1957**, 2, 1990.

Ueda, Y. *J. Chem. Soc. Japan* **1931**, 52, 740.

Ueda, Y. *J. Chem. Soc. Japan* **1932**, 53, 559.

Ueda, Y. *Sci. Repts. Tohoku Imp. Univ.* **1933**a, 22, 879.

Ueda, Y. *Sci. Repts. Tohoku Imp. Univ.* **1933**b, 22, 448.

Ueno, K. *J. Chem. Soc. Japan* **1941**, 62, 990.

Van Artsdalen, E.R.; Anderson, K.P. *J. Am. Chem. Soc.* **1951**, 73, 579.

Vanderryn, J. *J. Chem. Phys.* **1959**, 30, 331.

Vanderzee, C.E. *J. Chem. Thermodynamics* **1982**, 14, 219.

Vanderzee, C.E.; Gier, L.J. *J. Chem. Thermodynamics* **1974**, 6, 441.

Vanderzee, C.E.; King, D.L. *J. Chem. Thermodynamics* **1972**, 4, 675.

Vanderzee, C.E.; King, D.L.; Wadsö, I. *J. Chem. Thermodynamics* **1972**a, 4, 685.

Vanderzee, C.E.; Månsson, M.; Sunner, S. *J. Chem. Thermodynamics* **1972**b, 4, 533.

Vanderzee, C.E.; Månsson, M.; Wadsö, I.; Sunner, S. *J. Chem. Thermodynamics* **1972**c, 4, 541.

Vanderzee, C.E.; Nutter, J.D. *J. Phys. Chem.* **1963**, 67, 2521.

Vanderzee, C.E.; Quist, A.S. *J. Phys. Chem.* **1961**, 65, 118.

Vanderzee, C.E.; Rhodes, D.E. *J. Am. Chem. Soc.* **1952**, 74, 3552.

Vanderzee, C.E.; Rodenburg, M.L.N.; Berg, R.L. *J. Chem. Thermodynamics* **1974**, 6, 17.

Vanderzee, C.E.; Rodenburg, W.W. *J. Chem. Thermodynamics* **1971**, 3, 267.

Vanderzee, C.E.; Swanson, J.A. *J. Phys. Chem.* **1963**a, 67, 285.

Vanderzee, C.E.; Swanson, J.A. *J. Phys. Chem.* **1963**b, 67, 2608.

Vanderzee, C.E.; Swanson, J.A. *J. Chem. Thermodynamics* **1974**, 6, 827.

Vanderzee, C.E.; Waugh, D.H.; Haas, N.C. *J. Chem. Thermodynamics* **1980**a, 12, 21.

Vanderzee, C.E.; Waugh, D.H.; Haas, N.C.; Wigg, D.A. *J. Chem. Thermodynamics* **1980**b, 12, 27.

Vandoni, R.; Laudy, M. *J. Chim. Phys. Phys.-Chim. Biol.* **1952**, 49, 99.

Vasil'ev, V.P.; Glavina, S.R. *Elektrokhim.* **1971**, 7, 1395.

Vasil'ev, V.P.; Glavina, S.R. *Izvest. Vys. Ucheb. Zaved., Khim. i Khim. Tekhnol.* **1973**, 16, 39.

Vasil'ev, V.P.; Kochergina, L.A. *Zhur. Fiz. Khim.* **1967**, 41, 2133.

Vasil'ev, V.P.; Kochergina, L.A. *Zhur. Fiz. Khim.* **1968**, 42, 373.

Vasil'ev, V.P.; Kokurin, N.I.; Vasil'eva, V.N. *Zhur. Neorg. Khim.* **1973**a, 18, 300.

Vasil'ev, V.P.; Kokurin, N.I.; Vasil'eva, V.N. *Zhur. Neorg. Khim.* **1976**, 21, 407.

Vasil'ev, V.P.; Kozlovskii, E.V. *Zhur. Neorg. Khim.* **1974**, 19, 267.

Vasil'ev, V.P.; Kozlovskii, E.V.; Kunin, B.T. *Izvest. Vyssh. Uchebn. Zaved. Khim. i Khim. Tekhnol.* **1973**c, 16, 365.

Vasil'ev, V.P.; Kozlovskii, E.V.; Shitova, V.V. *Zhur. Fiz. Khim.* **1971**, 45, 191.

Vasil'ev, V.P.; Kunin, B.T. *Zhur. Fiz. Khim.* **1972**a, 46, 2957.

Vasil'ev, V.P.; Kunin, B.T. *Zhur. Neorg. Khim.* **1972**b, 17, 2169.

Vasil'ev, V.P.; Vasil'eva, V.N.; Dmitrieva, N.G.; Kokurin, N.I. *Zhur. Neorg. Khim.* **1981**, 26, 30.

Vasil'ev, V.P.; Vasil'eva, V.N.; Kokurin, N.I. *Zhur. Neorg. Khim.* **1973**b, 18, 1465.

Vasil'kova, I.V.; Barvinok, G.M. *Issled. Obl. Khim. i Tekhnol. Mineralnikh Solei i Okislov,* Nauka: Moscow. **1965**.

Vdovenko, V.M.; Lazarev, L.M.; Khvorostin, S. *Zhur. Neorg. Khim.* **1967**, 12, 1152.

Venkataswaran, C.S. *Proc. Indian Acad. Sci.* **1935**, 2A, 260.

Venkataswaran, C.S. *Proc. Indian Acad. Sci.* **1936**, 4A, 345.

Verdonk, A.H.; Nedermeijer, J.; Laverman, J.W. *J. Chem. Thermodynamics* **1975**, 7, 1047.

Verma, R.D. *J. Chem. Phys.* **1960**, 32, 738.

Vichutinski, A.A; Golikov, A.G.; *Dokl. Akad. Nauk SSSR* **1978**, 238, 127.

Vidale, G.L. *Gen. Electric Space Sci. Lab. Rept. T60SD330* **1960**, 43.

Vinal, G.W.; Brickwedde, L.H. *J.Res. Nat. Bur. Stand.* **1941**, 26, 455.

Vinogradov, Yu.K.; Volyak, L.D. *Teplofiz. Vys. Temp.* **1966**, 4, 50.

Vlasov, Y.S.; Seleznev, B.L. *Zhur. Fiz. Khim.* **1972**, 46, 2387.

Vlasova, I.V.; Stepina, S.B.; Stancheva, L.I.; Plyushchev, V.E. *Zhur. Neorg. Khim.* **1966**, 11, 1424.

Volokhov, Yu.A.; Pavlov, L.N.; Eremin, N.I.; Mironov, V.E. *Zhur. Priklad. Khim.* **1971**, 44, 246.

Volyak, L.D.; Vinogradov, Yu.K.; Anisimov, V.M. *Teplofiz. Vys. Temp.* **1968**a, 6, 754.

Volyak, L.D.; Vinogradov, Yu.K.; Anisimov, V.M. *Teplofiz. Vys. Temp.* **1968**b, 6, 521.

Volyak, L.D.; Vinogradov, Yu.K.; Anisimov, V.M.; Tarlakov, Y.V. *Mosk. Aviats. In-Ta* **1975**, 91.

Vorob'ev, A.F.; Broier, A.F. *Zhur. Fiz. Khim.* **1971**a, 45, 1307.

Vorob'ev, A.F.; Broier, A.F. *Zhur. Fiz. Khim.* **1971**b, 45, 2390.

Vorob'ev, A.F.; Broier, A.F.; Skuratov, S.M. *Dokl. Akad. Nauk SSSR* **1967**a, 173, 385.

Vorob'ev, A.F.; Ibrahim, N.A.; Skuratov, S.M. *Vestnik Mosk. Univ., Ser. Khim.* **1965**, 5, 3.

Vorob'ev, A.F.; Ibrahim, N.A.; Skuratov, S.M. *Zhur. Neorg. Khim.* **1966**, 11, 25.

Vorob'ev, A.F.; Kolesov, V.P.; Skuratov, S.M. *Zhur. Neorg. Khim.* **1960**a, 5, 1402.

Vorob'ev, A.F.; Monaenkova, A.S. *Vestnik Mosk. Univ., Ser. Khim.* **1972**, 13, 182.

Vorob'ev, A.F.; Monaenkova, A.S.; Skuratov, S.M. *Vestnik. Mosk. Univ., Ser. Khim.* **1967**b, 22(6), 3.

Vorob'ev, A.F.; Perelygin, B.G.; Byval'tsev, Y.A. *Thermodynamic Properties of Solutions.* Moscow. **1980**, p. 3.

Vorob'ev, A.F.; Privalova, N.M. *Vestnik Mosk. Univ., Ser.II, Khim.* **1963**, 18(6), 22.

Vorob'ev, A.F.; Privalova, N.M.; Monaenkova, A.S.; Skuratov, S.M. *Dokl. Akad. Nauk SSSR* **1960**b, 135, 1388.

Vorob'ev, A.F.; Skuratov, S.M. *Zhur. Fiz. Khim.* **1958**, 32, 2580.

Vorob'ev, A.F.; Umyarova, R.S.; Urusov, V.S. *Zhur. Obshch. Khim.* **1974**, 44, 979.

Voronin, G.F.; Shiu, N.T.; Gerasimov, Ya.I. *Zhur. Fiz. Khim.* **1967**, 41, 1468.

Vosburgh, W.C.; Craig, D.N. *J. Am. Chem. Soc.* **1929**, 51, 2009.

Vosburgh, W.C.; McClure, R.S. *J. Am. Chem. Soc.* **1943**, 65, 1060.

Voskresenskaya, N.K.; Ponomareva, K.S. *Dokl. Akad. Nauk SSSR* **1944**, 45, 188.

Voskresenskaya, N.K.; Ponomareva, K.S. *Zhur. Fiz. Khim.* **1946**, 20, 433.

Wagman, D.D.; Evans, W.H.; Parker, V.B.; Halow, I.; Bailey, S.M.; Schumm, R.H. *Selected Values of Chemical Thermodynamic Properties.* NBS TN 270-3, Washington DC, **1968**.

Wagman, D.D.; Kilday, M.V. *J. Res. Nat. Bur. Stand.* **1973**, 77 A, 569.

Wallace, W.E. *J. Am. Chem. Soc.* **1949**, 71, 2485.

Wallace, W.E. *J. Phys. Chem.* **1946,** 50, 152.

Wallace, W.E.; Robinson, A.L. *J.Am. Chem. Soc.* **1941,** 63, 958.

Walling, J.F. *J. Phys. Chem.* **1963,** 67, 1380.

Ward, G.K.; Millero, F.J. *J. Chem. Thermodynamics* **1973,** 5, 591.

Warren, T.E. *J. Am. Chem. Soc.* **1927,** 49, 1904.

von Wartenberg, H. *Z. Elektrochem.* **1909,** 15, 866.

von Wartenberg, H. *Z. Elektrochem.* **1914,** 20, 443.

von Wartenberg, H.; Hanisch, K. *Z. Phys. Chem.* **1932,** A 161, 463.

Waterfield, C.G.; Linford, R.G.; Goalby, B.B.; Bates, T.R.; Elyard, C.A.; Staveley,
    L.A.K. *Trans. Faraday Soc.* **1968,** 64, 868.

Waterfield, C.G.; Staveley, L.A.K. *Trans. Faraday Soc.* **1967,** 63, 2349.

Webb, D.U.; Rao, K.N. *J. Mol. Spectr.* **1968,** 28, 121.

Webb, T.J. *J. Phys. Chem* **1925,** 29, 816.

Weed, H.C., Jr. *Ph.D. thesis, Ohio State Univ.* **1957**; *Diss. Absts.* **1958,** 18, 843.

Wegscheider, R.; Mehl, J. *Monatsh. Chem.* **1928,** 49, 283.

Weiler, J. *Ann. Phys.* **1929,** 1, 361.

Weintraub, R.; Apelblat, A.; Tamir, A. *J. Chem. Thermodynamics* **1982,** 14, 887.

Weiwad, F.; Kehlen, H.; Kuschei, F.; Sackmann, H. *Z. Phys. Chem. (Leipzig)* **1973,**
    253, 114.

Weller, W.W. *U.S. Bur. Mines Rept. Inv. 6669* **1965**.

Wells, L.S.; Taylor, K. *J. Res. Nat. Bur Stand.* **1937,** 19, 215.

Wendrow, B.; Kobe, K.A. *Ind. Eng. Chem.* **1952,** 44, 1439.

West, W.A.; Menzies, A.W.C. *J. Phys. Chem.* **1929,** 33, 1880.

Westrum, E.F., Jr. *Proc. IV Internat. Cong. Verre.* **1956,** 396.

Westrum, E.F., Jr. Private communication. **1960**.

Westrum, E.F., Jr. *Thermodynamics.* Vol. I. I.A.E.A.: Vienna. **1966,** 497.

Westrum, E.F., Jr.; Grønvold, F. *J. Am. Chem. Soc.* **1959,** 81, 1777.

Westrum, E.F., Jr.; Justice, B.H. *J. Chem. Phys.* **1969,** 50, 5083.

Westrum, E.F., Jr.; Pitzer, K.S. *J. Am. Chem. Soc.* **1949,** 71, 1940.

Wetmore, F.E.W.; Gordon, A.R. *J. Chem. Phys.* **1937,** 5, 60.

Wilson, G.L.; Miles, F.D. *Trans. Faraday Soc.* **1940,** 36, 356.

Wilson, R.E. *J. Am. Chem. Soc.* **1921,** 43, 704.

Winkler, L.W. *Z. Physik. Chem.* **1906a,** 55, 30.

Winkler, L.W. *Z. Physik. Chem.* **1906b,** 55, 350.

Wise, S.S.; Margrave, J.L.; Feder, H.M.; Hubbard, W.N. *J. Phys. Chem.* **1963,** 67,
    815.

Wise, S.S.; Margrave, J.L.; Feder, H.M.; Hubbard, W.N. *J. Phys. Chem.* **1966,** 70,
    7.

Wishaw, B.F.; Stokes, R.H. *Trans. Faraday Soc.* **1953,** 49, 27.

Wishaw, B.F.; Stokes, R.H. *Trans. Faraday Soc.* **1954,** 50, 952.

Wittig, F.E.; Schmatz, W. *Z. Elektrochem.* **1959,** 63, 470.

Wolcott, N.M. *Phil. Mag.* **1957,** 2, 1246.

Woolf, P.L.; Zellars, G.R.; Foerster, E.; Morris, J.P. *U.S. Bur. Mines Rept. Inv. 5634*
    1960.

Worswick, R.D.; Dunn, A.G.; Staveley, L.A.K. *J. Chem. Thermodynamics* **1974,** 6,
    565.

Wright, R.H.; Maass, O. *Can. J. Res.* **1932,** 6, 94.

Wu, C.H. *J. Chem. Phys.* **1976,** 65, 3181.

Wu, C.H.; Birky, M.M.; Hepler, L.G. *J. Phys. Chem.* **1963,** 67, 1202.

Wu, C.H.; Witonsky, R.J.; George, P.; Rutman, R.J. *J. Am. Chem. Soc.* 89, **1967**, 1987.

Wu, H.Y.; Wahlbeck, P.G. *High Temp. Sci.* **1971**, 3, 469.

Wu, Y.C.; Friedman, H.L. *J. Phys. Chem.* **1966**, 70, 501.

Wu, Y.C.; Hamer, W.J. *Nat. Bur. Stand. Rept. 10002, Parts XI, XIII, XIV.* **1969**.

Wu, Y.C.; Hamer, W.J. *J. Phys. Chem. Ref. Data* **1980**, 9, 513.

Wu, Y.C.; Rush, R.M.; Scatchard, G. *J. Phys. Chem.* **1969**, 73, 2047.

Wu, Y.C.; Young, T.F. *J. Res. Nat. Bur. Stand.* **1980**, 85, 11.

Wüst, J.; Lange, E. *Z. Phys. Chem.* **1925**, 116, 161.

Yakimov, M.A.; Guzhavina, E.I.; Lazeeva, M.S. *Zhur. Neorg. Khim.* **1969**, 14, 1927.

Yatlov, V.S.; Polyakova, E.M. *Zhur. Obshch. Khim.* **1938**, 8, 774.

Yokoyama, H.; Yamatera, H. *Bull. Chem. Soc. Japan* **1975**, 48, 2708.

Yost, D.M.; DeVault, D.; Anderson, T.E.; Lassettre, E.N. *J Chem. Phys.* **1938**, 6, 424.

Young, F.E. *J. Am. Chem. Soc.* **1944**, 66, 773.

Young, T.F.; Irish, D.E. *Ann. Rev. Phys. Chem.* **1962**, 13, 435.

Yui, N. *Sci. Repts. Tohoku Imp. Univ.* **1951**, 1, 35, 53.

Zalubas, R. *J. Opt. Soc. Am.* **1968**, 58, 1195.

Zalubas, R. *J. Res. Nat. Bur. Stand.* **1976**, 80 A, 223.

Zavitsanos, P.D. *Rev. Sci. Instr.* **1964**, 35, 1061.

Zavodnov, S.S.; Kryukov, P.A. *Izv. Akad. Nauk SSSR, Otdel. Khim. Nauk* **1960**, 1704.

Zdanovskii, A.B.; Spiridonov, F.P. *Zhur. Priklad. Khim.* **1967**, 40, 1152.

Zeumer, H.; Roth, W.A. *Z. Elektrochem.* **1934**, 40, 777.

Zhogin, D.Yu.; Kosarukina, E.A.; Kolesov, V.P. *Zhur. Fiz. Khim.* **1980**, 54, 916.

Ziegler, W.T.; Messer, C.E., *J. Am. Chem. Soc.* **1941**, 63, 2694.

Zielen, A.J. *J. Am. Chem. Soc.* **1959**, 81, 5022.

Zmbov, K.F.; Ames, L.L.; Margrave, J.L. *High Temp. Sci.* **1973**, 5, 235.

Zukowsky, G.J. *Z. Anorg. Chem.* **1911**, 71, 403.

HISTORY OF THE CODATA TASK GROUP ON KEY VALUES FOR THERMODYNAMICS

This is a brief account of the more significant events in the life of the Task Group.

April 1968

Professor F.D. Rossini, President of CODATA, invited individuals with experience in thermodynamic data assessment to join a Task Group on Key Values for Thermodynamics (TGKVT).

June 30 – July 2, 1968

TGKVT held its first meeting in Arnoldshain, Federal Republic of Germany, in connection with the First International CODATA Conference. The initial membership was established, as follows:

Prof S. Sunner (Sweden)	- Chairman
Dr J.D. Cox (UK)	- Member
Prof L.V. Gurvich (USSR)	- Member
Prof D.M. Newitt (UK)	- Member
Mr D.D. Wagman (USA)	- Member
Prof L. Waldman (FRG)	- Member
Prof F.D. Rossini (USA)	- Member, Ex Officio
Dr C. Schäfer (FRG)	- Member, Ex Officio

The following account of the Task Group's first meeting appeared in CODATA Newsletter, No. 1, October 1968: "The Task Group met during the First International CODATA Conference at Arnoldshain, (Federal Republic of Germany), and approved the following report:

1. Certain compounds and certain properties shall be designated as key compounds and key properties with respect to thermodynamic tabulations.
2. The properties at this particular time so designated are $\Delta_f H°(298.15$ K), $S°(298.15$ K), and $H°(298.15$ K) - $H°(0)$.
3. The key substances shall include all the elements in their standard reference states and standard monatomic gaseous states; and such additional compounds as the Task Group may determine.
4. The Task Group proposes to recommend a list of key property values for the key substances for consideration by the scientific community over a one year period. This may be accomplished by publication in the Bulletin of Thermodynamics and Thermochemistry and in other appropriate journals. Submission of this list to the public provides an opportunity for comments from the scientific community to the Task Group.
5. After the lapse of the prescribed period, the Task Group will proceed to recommend final values for the properties involved. The final approval rests on

CODATA. The recommended values will be published according to the procedures laid down by CODATA.

6. After the final publication of the first list of approved key values, the Task Group will from time to time make additions to the list and such revisions as are deemed necessary."

September 2 – 4, 1970
The second formal meeting of TGKVT was held in Teddington, UK. Tentative values for key thermodynamic properties of 32 chemical species were agreed upon; a report on the first two years' work was drafted. Membership for the period 1970-1972 was established as: Chairman, S. Sunner; Members: J.D. Cox, L.V. Gurvich, D.M. Newitt, D.D. Wagman; Consultants: C.B. Alcock (Canada), W.H. Evans (USA), V.A. Medvedev (USSR), I. Wadsö (Sweden).

September 7 – 10, 1970
The Task Group met informally during the Second International CODATA Conference, held at St. Andrews, UK. The draft report prepared during the previous week was finalized for publication, and a work programme for the ensuing five years was outlined.

November 1970
TGKVT's first technical report was published as CODATA Bulletin, No. 2, November 1970, with the title "Tentative Set of Key Values for Thermodynamics - Part I".

July 19, 1971
The third formal meeting of TGKVT was held in Washington DC, USA. Some values in Tentative Set I were revised, values for Tentative Set II were agreed upon, and the data-assessment program for Tentative Set III was outlined.

December 1971
TGKVT's second technical report was published as CODATA Bulletin, No. 5, December 1971, with the title "Final Set of Key Values for Thermodynamics - Part I". The Task Group's third technical report was published as CODATA Bulletin, No 6, December 1971, with the title "Tentative Set of Key Values for Thermodynamics - Part II".

June 27 - 28, 1972
The fourth formal meeting of TGKVT was held in Le Creusot, France, during the Third International CODATA Conference. The text for a publication on the next set of tentative key values was prepared. Membership for the period 1972-1974 was established as: Chairman, S. Sunner; Members: J.D. Cox, L.V. Gurvich, D.D. Wagman; Consultants: W.H. Evans, V.A. Medvedev, J.B. Pedley (UK).

August 1972
The Task Group's fourth technical report was published as CODATA Bulletin, No. 7, August 1972, with the title "Tentative Set of Key Values for Thermodynamics - Part III".

August 26 – 29, 1973
TGKVT's fifth formal meeting was held in Munich, Federal Republic of Germany. Comments received from the scientific public on Tentative Sets II and III were reviewed, and some revisions made. It was decided to conjoin and publish the data of Tentative Sets II and III, whether revised or unchanged, in a document that used the word "Recommended" in its title, rather than the word "Final" (cf. December 1971). Data assessments for Tentative Set IV were discussed, and plans laid for Tentative Set V.

December 1973
The Task Group's fifth technical report was published as CODATA Bulletin, No 10, December 1973, with the title "CODATA Recommended Key Values for Thermodynamics, 1973".

January 1974
TGKVT's sixth technical report was published under the title "Tentative Set of Key Values for Thermodynamics - Part IV". The publication was not part of any regular CODATA series, but was nevertheless considered to be an official CODATA publication.

June 19 – 26, 1974
TGKVT met informally in Moscow, USSR, and in Tsakhcadzor, USSR, during the Fourth International CODATA Conference. The species to be included in Tentative Set V were agreed upon. It was decided to terminate the data-assessment programme by about 1978. Membership for the period 1974-1976 was established as: Chairman, J.D. Cox; Members: J. Drowart (Belgium), L.V. Gurvich, L.G. Hepler (Canada), J.B. Pedley, D.D. Wagman; Consultants: W.H. Evans, V.A. Medvedev, S. Sunner.

April 21 – 25, 1975
The sixth formal meeting of TGKVT was held in Moscow, USSR. Comments received on Tentative Set IV were reviewed, values for Tentative Set V were agreed upon, and the programme for Tentative Set VI was outlined.

September 1975
The Task Group's seventh technical report was published as CODATA Special Report, No. 3, September 1975, with the title "Tentative Set of Key Values for Thermodynamics - Part V".

January 1976
The Task Group's eighth technical report was published as CODATA Bulletin, No. 17, January 1976, with the title "CODATA Recommended Key Values for Thermodynamics, 1975".

June 24 – 26, 1976
The seventh formal meeting of TGKVT was held in Boulder, Colorado, USA. Comments received on Tentative Set V were reviewed, values for Tentative Set VI were agreed upon, and the program for Tentative Set VII was outlined. Membership for the period 1976-1978 was kept unchanged from that for 1974-1976.

March 1977
The Task Group's ninth technical report was published as CODATA Special Report, No. 4, March 1977, with the title "Tentative Set of Key Values for Thermodynamics - Part VI".

May 1977
The Task Group's tenth technical report was published as CODATA Bulletin, No. 22, May 1977, with the title "CODATA Recommended Key Values for Thermodynamics, 1976".

August 29 – 30, 1977
TGKVT held its eighth formal meeting in Lund, Sweden. Comments received on Tentative Set VI were discussed, and values for Tentative Set VII were agreed. It was decided to prepare only one more Tentative Set, and to include auxiliary tabulations of heat capacities and enthalpies of dilution in the publication programme.

April 1978
TGKVT's eleventh technical report was published as CODATA Special Report, No 7, April 1978, with the title "Tentative Set of Key Values for Thermodynamics - Part VII". The Task Group's twelfth technical report was published as CODATA Bulletin, No 28, April 1978, with the title "CODATA Recommended Key Values for Thermodynamics, 1977".

September 4 – 5, 1978
The ninth formal meeting of TGKVT was held in Harwell, UK. Plans were laid for writing a final report on the Task Group's activities, covering key values for about 150 chemical species. Contact was made with CODATA's newly created Task Group on Internationalization and Systematization of Thermodynamic Tables, which also met at Harwell, concerning matters of mutual interest. Membership for the period 1978 - 1980 was established as: Chairman: J.D. Cox; Members: J. Drowart, L.G. Hepler, V.A. Medvedev, D.D. Wagman; Consultants: W.H.Evans, L.V. Gurvich, J.B. Pedley, S. Sunner.

September 14 – 15, 1979
The tenth formal meeting of TGKVT was held in Grenoble, France. Values for Tentative Set VIII were agreed upon.

March 14, 1980
The death occurred of Prof D.M. Newitt (Member 1968-1972).

April 1980
TGKVT's thirteenth technical report was published as CODATA Special Report, No. 8, April 1980, with the title "Tentative Set of Key Values for Thermodynamics - Part VIII".

June 3, 1980
The death occurred of Prof S. Sunner (Chairman, 1968-1974, Consultant, 1974-1980).

August 26, 1980
The work of the Task Group was explained in a Main Lecture by J.D. Cox at the Sixth International Conference on Thermodynamics, held in Merseburg, German Democratic Republic. A silent tribute was paid at the Conference to the memory of the late Stig Sunner.

October 13 – 14, 1980
Renewal of the Task Group was not sought at the Twelfth CODATA General Assembly, held in Kyoto, Japan. Hence, TGKVT formally came to an end. However, permission was granted for a "shadow" Task Group, with the same membership as for 1978-1980, to continue in being until a final report on the Task Group's work had been written.

June 1981
Revised values for the fundamental constants utilized by TGKVT were received from the Chairman of CODATA's Task Group on Fundamental Constants. This information was crucial for the drafting of TGKVT's final report.

August 22 – 28, 1981
An informal meeting of TGKVT was held in Louvain/Leuven, Belgium, in association with CODATA's Task Group on Internationalization and Systematization of Thermodynamic Tables. The work of revising key values published during the previous eleven years was launched, and plans were laid for writing a final report under the editorship of J.D. Cox and D.D. Wagman.

December 1984
Revision of key values and drafting of a final report were completed.

June 24, 1985
The death occurred of Dr W.H. Evans (Consultant 1970-1980).

February 24, 1988
The death occurred of Prof. V.A. Medvedev (Consultant and Member of the Task Group since its inception; editor of this Final Report).

KEY VALUES FOR AQUEOUS IONS:
A LISTING OF THE REACTIONS CONSIDERED

This Annex contains a listing, Table II-1, of all of the thermodynamic data used to evaluate the enthalpies of formation and entropies of the aqueous ions, as indicated in Chapter 2. It is followed by an index, Table II-2, to the listing.

A simultaneous least squares solution was made for the properties of the following ions:

$$OH^-, \ F^-, \ Cl^-, \ Br^-, \ I^-, \ ClO_4^-, \ SO_4^{2-}, \ NO_3^-,$$
$$NH_4^+, \ Hg_2^{2+}, \ Ag^+, \ Li^+, \ Na^+, \ K^+, \ Rb^+, \ and \ Cs^+$$

all at the hypothetical one molal standard state at 298.15 K and 0.101325 MPa. For F^-, Cl^-, SO_4^{2-}, NO_3^-, and Ag^+, the data are primarily those used to derive the entropies. Their enthalpies of formation are covered in the Notes in Chapter 2. The ion of mercury, listed above, is also discussed in the Notes.

Data are not given here for other ions discussed in Chapter 2. These ions are:

$$HSO_4^-, \ HS^-, \ HPO_4^{2-}, \ H_2PO_4^-, \ CO_3^{2-}, \ HCO_3^-,$$
$$Sn^{2+}, \ Pb^{2+}, \ Al^{3+}, \ Zn^{2+}, \ Cd^{2+}, \ Cu^{2+}, \ Hg^{2+},$$
$$UO_2^{2+}, \ Mg^{2+}, \ and \ Ca^{2+}.$$

Table II-1 contains the following information: The chemical reaction or substance measured, the property and value obtained (with the estimated overall uncertainty of the value), and a brief alphanumeric bibliographic reference code, which is described below. Comments on a reaction, such as ranges of concentration measured, temperatures, etc., are given immediately underneath the reaction.

In the column headed REFERENCE the two initial digits are the last two digits of the year of publication of the article (numbers greater than 86 refer to nineteenth century publications; a few references to 82THO are references to the nineteenth century measurements by Julius Thomsen published in his Collected Works (1882-1886)). Following the two digits are the first three letters of the last names of the first and second authors, in upper case letters and separated by a slash. A lower-case letter at the end of the code indicates that there is more than one reference having the same first two author codes and year of publication. These reference codes are keyed to the BIBLIOGRAPHY in Chapter 3, in which the full reference, including all authors, journal, volume, pages, and year of publication, is given in an alphabetical arrangement.

Table II-2 contains the following information: The formulas of the substances in the reactions in Table II-1 for which thermodynamic property values were determined in the analysis, and the numbers of the reactions in which these substances occur. (The values of the properties of other substances in the reactions were held fixed as indicated in Table 5 or in the notes to it in Chapter 2.)

Values for thermochemical properties in Table II-1 are all molar values. For the chemical reactions and physical processes ΔH, ΔG, EMF's and pressures all apply to the reactions and processes as written, with each substance being taken in the requisite (molar) amounts. Symbols are defined in paragraph 1.3.1, and follow the IUPAC conventions. These are extended by "ai" meaning "aqueous, ionized", and "ao" meaning "aqueous, not ionized further", both at the standard state. The subscript "m", meaning molar, is not appended to thermodynamic symbols.

Table II-1. Measurements on Aqueous Thermochemistry at 298.15 K and 0.101325 MPa.

NO.	REACTION OR SUBSTANCE	PROP. TYPE	OBSVD. VALUE kJ/mol or J/(mol·K)	± UNC.	REFERENCE
1	$HCl(ai) + NaOH(ai) = NaCl(ai) + H_2O(l)$ Used HCl(0.01-0.04 mol/kg) and NaOH(0.01-0.04 mol/kg)	$\Delta H=$	-55.789	0.080	70LEU/GRU
2	$HCl(ai) + KOH(ai) = KCl(ai) + H_2O(l)$ 0.098 in KOH and 0.1 mol/kg HCl. Apparently used enthalpies of diln. from Rossini et al. (1952)	$\Delta H=$	-55.86	0.03	71AHR/KULa
3	$HCl(ai) + NaOH(ai) = NaCl(ai) + H_2O(l)$ 0.10 mol/kg HCl + 0.0099 mol/kg NaOH. Diln. corr. = -0.594 kJ/mol, calc from Rossini et al. (1952)	$\Delta H=$	-55.835	0.020	72OTS
4	$HCl(ai) + NaOH(ai) = NaCl(ai) + H_2O(l)$ Used HCl(0.504 mol/kg) and NaOH(0.10 mol/kg). Diln. corr. = -1.287 kJ/mol calc. from Parker (1965)	$\Delta H=$	-55.840	0.140	77OLO/OLO
5	$HClO_4(ai) + NaOH(ai) = NaClO_4(ai) + H_2O(l)$ Used HClO$_4$(0.01 mol/kg). + NaOH(0.013-0.311 mol/kg). Diln. corr. from +0.25 to 0.51 kJ/mol	$\Delta H=$	-55.798	0.070	63VAN/SWAb

Table II-1. Measurements on Aqueous Thermochemistry at 298.15 K and 0.101325 MPa (cont'd).

NO.	REACTION OR SUBSTANCE	PROP. TYPE	OBSVD. VALUE kJ/mol or	± UNC. J/(mol·K)	REFERENCE
6	H^+(ao) + OH^-(ao) = H_2O(l) From soln. of I_2O_5(cr) in H_2O(l) and alkali	$\Delta H=$	-55.81	0.20	78CER/KO
7	HCl(ai) + KOH(ai) = H_2O(l) + KCl(ai) HCl(0.111 mol/kg) and KOH(0.00989 mol/kg). Diln. corr. calc. from Rossini et al.(1952) = 0.55 to 0.70 kJ/mol	$\Delta H=$	-55.840	0.050	70GRE/OTS
8	H^+(ao) + OH^-(ao) = H_2O(l)	$\Delta H=$	-55.92	0.11	73OLO/OLO
9	HCl(ai) + NaOH(ai) = NaCl(ai) + H_2O(l) HCl(0.507 mol/kg), NaOH(0.020 mol/kg). Total diln. corr. = +1.20 kJ/mol calc. from Parker (1965)	$\Delta H=$	-55.72	0.30	79OLO/SUN
10	H^+(ao) + OH^-(ao) = H_2O(l) Review of literature	$\Delta H=$	-55.815	0.050	75OLO/HEP
11	H^+(ao) + OH^-(ao) = H_2O(l) Review of literature	$\Delta G=$	-79.906	0.040	75OLO/HEP
12	H^+(ao) + OH^-(ao) = H_2O(l) Review of calorimetric meas.	$\Delta H=$	-55.835	0.100	65PAR
13	NaOH(ai) + HCl(ai) = NaCl(ai) + H_2O(l) Used NaOH(2.0 mol/kg)+HCl(0.098 mol/kg). Diln. corr.from 0 to -0.070 kJ/mol calc. from Parker (1965)	$\Delta H=$	-55.856	0.100	68GOL/HEP
14	H^+(ao) + OH^-(ao) = H_2O(l) Used HCl(0.1 mol/kg) with NaOH(0.03 mol/kg) and KOH(0.1 mol/kg)	$\Delta H=$	-55.77	0.15	72CIA/PAL
15	KOH(ai) + HCl(ai) = KCl(ai) + H_2O(l) Used KOH(0.10 mol/kg), HCl(0.50 mol/kg). Diln. corr. = +1.55 kJ/mol, calc. from Rossini et (1952)	$\Delta H=$	-55.780	0.045	63GER/LED
16	H^+(ao) + OH^-(ao) = H_2O(l) Recalc. of data given for 278-323 K in Harned and Owen (1958)	$\Delta G=$	-79.877	0.010	84COD
17	H^+(ao) + OH^-(ao) = H_2O(l) From conductance meas. on H_2O(l), using ionic equiv. conductances	$\Delta G=$	-79.907	0.100	62DUE/HAL

Table II-1. Measurements on Aqueous Thermochemistry at 298.15 K and 0.101325 MPa (cont'd).

NO.	REACTION OR SUBSTANCE	PROP. TYPE	OBSVD. VALUE kJ/mol or J/(mol·K)	± UNC.	REFERENCE
18	$H^+(ao) + OH^-(ao) = H_2O(l)$ Our recalc. of H_2|OH-, NaCl|AgCl|Ag cells from 278-358 K. Authors take ΔG, -79.895 kJ/mol as "best"	$\Delta G=$	-79.986	0.050	73BEZ/CAM
19	$H^+(ao) + OH^-(ao) = H_2O(l)$ Meas. on tetramethylammonium salts as function of molality	$\Delta G=$	-79.923	0.040	71PRU/REA
20	$H^+(ao) + OH^-(ao) = H_2O(l)$ From cell meas. as func. of T and conc.	$\Delta G=$	-79.872	0.050	74SWE/MES
21	$H^+(ao) + OH^-(ao) = H_2O(l)$ Conductance meas. from 370 K plus data of Harned and Owen (1958) at 298, 323 K	$\Delta G=$	-79.895	0.040	72FIS/BAR
22	$H_2O_2(l) = H_2O(l) + 0.5 O_2(g)$	$\Delta H=$	-98.07	0.08	55GIG/MOR
23	$H_2O_2(l) = H_2O_2(20H_2O)$	$\Delta H=$	-3.347	0.040	52MOR
24	$H_2O_2(l) = H_2O(l) + 0.5 O_2(g)$	$\Delta H=$	-98.15	0.10	30ROT/GRA
25	$0.5 H_2(g) + 0.5 Cl_2(g) = HCl(ai)$	$\Delta H=$	-167.24	0.30	70KIN/ARM
26	$0.5 H_2(g) + 0.5 Cl_2(g) = HCl(ai)$ EMF meas. from 298 - 353 K	$\Delta G=$	-131.080	0.080	68CER/LON
27	$0.5 H_2(g) + 0.5 Cl_2(g) = HCl(ai)$ ΔS from temp. depend. of EMF	$\Delta S=$	-120.03	0.50	68CER/LON
28	$HCl(g) = HCl(ai)$ Diln. corr. from $HCl(2400H_2O) = -0.272$ kJ/mol; corr. for non-ideal $HCl(g) = -0.047$ kJ/mol	$\Delta H=$	-74.790	0.050	63VAN/NUT
29	$HCl(g) = HCl(ai)$ See corr. in Gunn (1964)	$\Delta H=$	-74.827	0.040	63GUN/GRE
30	$HCl(g) = HCl(ai)$	$\Delta H=$	-74.52	0.30	37ROT/BER
31	$HCl(g) = HCl(ai)$ Partial pressure of $HCl(g)$ over $HCl(aq)$ as function of conc.	$\Delta G=$	-35.960	0.080	63HAA/NAA
32	$HCl(g) = HCl(ai)$ Combination of EMF meas.	$\Delta G=$	-36.009	0.050	55AST/GIT

Table II-1. Measurements on Aqueous Thermochemistry at 298.15 K and 0.101325 MPa (cont'd).

NO.	REACTION OR SUBSTANCE	PROP. TYPE	OBSVD. VALUE kJ/mol or J/(mol·K)	± UNC.	REFERENCE
33	HCl(g) = HCl(ai) Partial pressure of HCl(g) over HCl(aq) as function of conc.	$\Delta G=$	-36.015	0.050	19BAT/KIR
34	NH_4ClO_4(cr) Heat cap. meas. from 5-350 K. S(5 K) = 0.091 J/(mol·K)	$S=$	184.16	0.15	69WES/JUS
35	$NaClO_4$(cr) Details of meas. not available	$S=$	143.97	0.40	79SCH/ZAL
36	$KClO_4$(cr) Heat cap. meas. from 12-298 K. S(12 K) = 1.21 J/(mol·K)	$S=$	151.25	0.60	30LAT/AHL
37	$CsClO_4$(cr) Heat cap. meas. from 15-293 K. S(15 K) = 3.43 J/(mol·K)	$S=$	175.05	0.50	38PIT/SMI
38	NH_4ClO_4(cr) = NH_4ClO_4(ai) Soly. = 2.12 mol/kg from Ayerst and Phillips (1966) and Simanova and Shul'ts (1967); act. coef. = 0.393 from Esval and Tyree (1962)	$\Delta G=$	0.905	0.050	84COD
39	NH_4ClO_4(cr) = NH_4ClO_4(ai) Meas at 0.001-0.009 mol/kg; aver. corr. for diln. = -0.080 kJ/mol (from $NaClO_4$ data)	$\Delta H=$	33.51	0.20	60BIR/HEP
40	NH_4ClO_4(cr) = NH_4ClO_4(ai) Meas. at 0.14 mol/kg, corr. for diln. = 0 (from $NaClO_4$)	$\Delta H=$	33.41	0.20	63VOR/PRI
41	$AgClO_4$(cr) = AgCl(cr) + 2 O_2(g) Thermal decomp. in a bomb with benzoic acid	$\Delta H=$	-97.45	0.55	66GIL/WAG
42	$AgClO_4$(cr) = $AgClO_4$(ai) Diln. corr. from $AgClO_4$(6500H_2O) = -0.080 kJ/mol (estimated)	$\Delta H=$	7.100	0.050	66GIL/WAG
43	$AgClO_4$(cr) = $AgClO_4$(ai) Corr. from $AgClO_4$(5000H_2O) = 0.08 kJ/mol	$\Delta H=$	8.60	0.25	64DRA/CHA
44	$LiClO_4$(cr) = $LiClO_4$(ai)	$\Delta H=$	-26.55	0.20	60BIR/HEP
45	$LiClO_4$(cr) = $LiClO_4$(ai)	$\Delta H=$	-26.80	0.20	63VOR/PRI

Table II-1. Measurements on Aqueous Thermochemistry at 298.15 K and 0.101325 MPa (cont'd).

NO.	REACTION OR SUBSTANCE	PROP. TYPE	OBSVD. VALUE kJ/mol or	± UNC. J/(mol·K)	REFERENCE
46	$NaClO_4(cr) = NaCl(cr) + 2\ O_2(g)$ Thermal decomposition in bomb using benzoic acid	$\Delta H=$	-29.41	0.30	66GIL/WAG
47	$NaClO_4(cr) = NaCl(cr) + 2\ O_2(g)$ Thermal decomposition in bomb by electric heating	$\Delta H=$	-27.95	0.40	75KIR/RUB
48	$NaClO_4(cr) = NaClO_4(ai)$	$\Delta H=$	13.870	0.030	63VAN/SWAa
49	$NaClO_4(cr) = NaClO_4(ai)$	$\Delta H=$	13.82	0.05	72MAS/CRI
50	$NaClO_4(cr) = NaClO_4(ai)$	$\Delta H=$	13.81	0.05	63VOR/PRI
51	$AgClO_4(cr) + NaCl(cr) = NaClO_4(cr) + AgCl(cr)$ Corr. for enthalpies of soln. to 0.017 mol/kg for $NaClO_4$(14.045 kJ/mol) and $AgClO_4$(7.185 kJ/mol)	$\Delta H=$	-68.41	0.50	66GIL/WAG
52	$KClO_4(cr) = KClO_4(ai)$ Soly. = 0.149 mol/kg from Guenther (1969) Biktimirov and Kuzovkina (1970) and from Kirgintsev and Yakobi (1968). Act. coef. = 0.73, estim. by comparison of data for Na and K salts. Latimer and Ahlberg (1930) estimated activity coef. to be 0.70, which gives ΔG = 11.21 kJ/mol	$\Delta G=$	11.00	0.40	84COD
53	$KClO_4(cr) = KClO_4(ai)$	$\Delta H=$	51.48	0.25	60BIR/HEP
54	$KClO_4(cr) = KClO_4(ai)$	$\Delta H=$	50.92	0.20	69TSV/RAB
55	$KClO_4(cr) = KClO_4(ai)$ From soly. as func. of ionic strength obtains $\log K_{sp}$ = -1.9444	$\Delta G=$	11.10	0.05	69GUE
56	$KClO_4(cr) = KClO_4(ai)$	$\Delta H=$	51.450	0.050	75EFI/MED
57	$KClO_4(cr) = KClO_4(ai)$	$\Delta H=$	50.85	0.30	66GIL/WAG
58	$KClO_4(cr) = KClO_4(ai)$	$\Delta H=$	50.74	0.25	30LAT/AHL
59	$KClO_4(cr) = KClO_4(ai)$	$\Delta H=$	50.75	0.20	63VOR/PRI
60	$KClO_4(cr) = KCl(cr) + 2\ O_2(g)$ Thermal decomp. in a bomb with benzoic acid	$\Delta H=$	-4.02	0.40	61JOH/GIL

Table II-1. Measurements on Aqueous Thermochemistry at 298.15 K and 0.101325 MPa (cont'd).

NO.	REACTION OR SUBSTANCE	PROP. TYPE	OBSVD. VALUE kJ/mol or J/(mol·K)	± UNC.	REFERENCE
61	$KClO_4(cr) = KCl(cr) + 2 O_2(g)$ Thermal decomposition in bomb by electric heating	$\Delta H=$	-4.14	0.40	75KIR/RUB
62	$AgClO_4(cr) + KCl(cr) = KClO_4(cr) + AgCl(cr)$ Corr. for enthalpies of soln. to 0.017 mol/kg for $KClO_4$(50.85 kJ/mol) and $AgClO_4$(7.185 kJ/mol). If Efimov and Medvedev (1975) value for $KClO_4$ soln. is used, $\Delta H = -92.46$ kJ/mol	$\Delta H=$	-91.84	0.50	66GIL/WAG
63	$KClO_4(cr) + 1.3333\ KBr(cr) = 1.3333$ $\qquad KBrO_3(cr) + KCl(cr)$ From combination of react. of $KClO_4(cr)$ and $KBrO_3(cr)$ with $TiCl_3$ soln.	$\Delta H=$	40.12	0.70	75EFI/MED
64	$LiClO_4(cr) + KCl(cr) = LiCl(cr) + KClO_4(cr)$ From combination of enthalpies of soln. in satd. $KClO_4$ solns.	$\Delta H=$	-24.50	0.26	61GIL/JOH
65	$NaClO_4(cr) + KCl(cr) = NaCl(cr) + KClO_4(cr)$ From combination of enthalpies of soln. in satd. $KClO_4$ solns.	$\Delta H=$	-24.24	0.26	61GIL/JOH
66	$NH_4ClO_4(cr) + KCl(cr) = NH_4Cl(cr) + KClO_4(cr)$ From combination of enthalpies of soln. in satd. $KClO_4$ solns.	$\Delta H=$	-15.40	0.20	61GIL/JOH
67	$CsClO_4(cr) = CsClO_4(ai)$ Meas. from 0.002-0.009 mol/kg. Corr. for diln. = -0.060 kJ/mol	$\Delta H=$	55.44	0.40	38PIT
68	$CsClO_4(cr) = CsClO_4(ai)$ Meas. in NaOH solutions $\quad$ LISTED FOR INFORMATION ONLY.	$\Delta H=$	57.03	0.80	73KRI/BAB
69	$CsClO_4(cr) = CsClO_4(ai)$ Meas. of soly. as func. of conc. extrap. to $\mu = 0$, $\log K_{sp} = -2.380$. Soly. = 0.0850 mol/kg and estim. act. coef. (0.78) yields $\Delta G = 13.45 \pm 0.30$ kJ/mol	$\Delta G=$	13.59	0.20	69GUE
70	$HBr(g)$ Result is weighted average for $H^{79}Br$ and $H^{81}Br$. Molecular consts. from Rank et al. (1965), Plyler (1960), Mould et al. (1960), Barrow and Stamper (1961), Ginter and Tilford (1971), and Jones and Gordy (1964)	$S=$	198.591	0.004	84COD

Table II-1. Measurements on Aqueous Thermochemistry at 298.15 K and 0.101325 MPa (cont'd).

NO.	REACTION OR SUBSTANCE	PROP. TYPE	OBSVD. VALUE kJ/mol or J/(mol·K)	± UNC.	REFERENCE
71	$0.5\ H_2(g) + 0.5\ Br_2(g) = HBr(g)$ Corr. from 376 K = +0.27 kJ/mol	$\Delta H=$	-51.92	0.40	56LAC/CAS
72	$Cl_2(g) + 2\ Br^-(ao) = Br_2(l) + 2\ Cl^-(ao)$ Combination of reactions of $Cl_2(g)$ and $Br_2(l)$ with SO_2 ($2500H_2O$). Corr. to std. state = -0.067 kJ/mol	$\Delta H=$	-91.29	0.40	63JOH/SUN
73	$Cl_2(g) + 2\ Br^-(ao) = Br_2(l) + 2\ Cl^-(ao)$ Combination of reactions with As_2O_3($12000H_2O$). Corr. to std. state negligible	$\Delta H=$	-91.29	0.80	64SUN/THO
74	$Cl_2(g) + 2\ Br^-(ao) = Br_2(l) + 2\ Cl^-(ao)$ Corr. from 292 K = +0.12 kJ/mol; corr. for $Br_2(ao)$ = 4.52 kJ/mol from Thomsen (1882)	$\Delta H=$	-91.55	2.00	82THO
75	$HBr(g) = HBr(ai)$ Corr. to std. state from HBr($2640H_2O$) = -0.268 kJ/mol; corr. for non-ideality of HBr(g) = -0.065 kJ/mol	$\Delta H=$	-85.144	0.060	63VAN/NUT
76	$HBr(g) = HBr(ai)$	$\Delta H=$	-85.06	0.60	82THO
77	$HBr(g) = HBr(ai)$	$\Delta H=$	-85.23	0.40	37ROT/BER
78	$0.5\ H_2(g) + 0.5\ Br_2(l) = HBr(ai)$ Combination of cell meas. against $Ag|AgBr|Br^-(ao)$ electrode	$\Delta G=$	-102.81	0.80	34JON/BAE
79	$Br_2(l) = Br_2(ao)$	$\Delta H=$	-0.87	0.15	63WU/BIR
80	$Br_2(l) = Br_2(ao)$ Calorimetric meas. in dil. $HClO_4$	$\Delta H=$	-2.17	0.20	73VAS/KOZ
81	$HBr(g) = HBr(ai)$	$\Delta G=$	-51.38	1.00	63HAA/NAA
82	$Br_2(ao) + H_2(g) = 2\ HBr(ai)$ From meas. on $Ag(cr)|AgBr(cr)|Br^-(aq)|Br_2(ao)$ cell from 273 to 323 K, combined with data from Hetzer et al.(1962)	$\Delta H=$	-240.94	0.15	66MUS/FAI
83	$Br_2(ao) + H_2(g) = 2\ HBr(ai)$ From meas. on $Ag(cr)|AgBr(cr)|Br^-(aq)|Br_2(ao)$ cell from 273 to 323 K, combined with data from Hetzer et al.(1962)	$\Delta G=$	-209.83	0.04	66MUS/FAI

Table II-1. Measurements on Aqueous Thermochemistry at 298.15 K and 0.101325 MPa (cont'd).

NO.	REACTION OR SUBSTANCE	PROP. TYPE	OBSVD. VALUE kJ/mol or J/(mol·K)	± UNC.	REFERENCE
84	Br_2(ao) + Br^-(ao) = Br_3^-(ao) From meas. on Ag(cr)|AgBr(cr)|Br^-(aq)|Br_2(l) cell from 273-323 K	$\Delta G=$	-7.150	0.025	66MUS/FAI
85	Br_2(ao) + Br^-(ao) = Br_3^-(ao) From calorimetric meas. in dilute $HClO_4$	$\Delta H=$	-7.53	0.30	73VAS/KOZ
86	HI(g) Molec. consts. from Boyd and Thompson (1952), Haeusler et al. (1963), and Cowan and Gordy (1956)	$S=$	206.481	0.004	84COD
87	H_2O_2(l) + 2 HI(ai) = I_2(cr) + 2 H_2O(l) Corr. from I_2(ao) = -1.23 ± 0.20 kJ/mol; corr. from H_2O_2($20H_2O$) = -3.45 kJ/mol; corr. for diln. from HI($250H_2O$) = +0.586 kJ/mol	$\Delta H=$	-271.60	1.00	67VOR/BRO
88	Br_2(l) + 2 I^-(ao) = I_2(cr) + 2 Br^-(ao) Corr. to I_2(cr) = -5.61 kJ/mol from authors' meas.	$\Delta H=$	-128.62	1.00	63WU/BIR
89	HI(g) = HI(ai)	$\Delta H=$	-83.283	0.085	74VAN/GIE
90	HI(g) = HI(ai) Pressure of HI(g) over HI(aq) from 5.9-9.2 mol/kg	$\Delta G=$	-53.93	0.50	19BAT/KIR
91	I_2(cr) + 0.5 N_2H_4($300H_2O$) = 0.5 N_2(g) + 2 HI($50H_2O$)	$\Delta H=$	-130.21	0.25	66HOW/SKI
92	N_2(g) + 2 H_2(g) = N_2H_4($300H_2O$) Adjusted to concentration	$\Delta H=$	34.52	0.20	68WAG/EVA
93	HI($50H_2O$) = HI(ai)	$\Delta H=$	-1.021	0.020	74VAN/GIE
94	0.5 H_2(g) + 0.5 I_2(g) = HI(g) Equil. meas. between 695-780 K	$\Delta H=$	-4.64	0.15	47BRI/HAG
95	0.5 H_2(g) + 0.5 I_2(g) = HI(g) Equil. meas. between 667-765 K	$\Delta H=$	-4.970	0.080	41TAY/CRI
96	0.5 H_2(g) + 0.5 I_2(g) = HI(g) Equil. meas. between 671-741 K	$\Delta H=$	-4.510	0.060	34RIT/UREb
97	H_2SO_4(l) Meas. from 16-305 K. S(15 K) = 1.13 J/(mol·K). See Rubin and Giauque (1952)	$S=$	156.90	0.20	60GIA/HOR

Table II-1. Measurements on Aqueous Thermochemistry at 298.15 K and 0.101325 MPa (cont'd).

NO.	REACTION OR SUBSTANCE	PROP. TYPE	OBSVD. VALUE kJ/mol or J/(mol·K)	± UNC.	REFERENCE
98	$H_2SO_4(115H_2O) = H_2SO_4(ai)$	$\Delta H=$	-21.455	0.020	80WU/YOU
99	$H_2SO_4(l) = H_2SO_4(ai)$ Diln. corr. from Wu and Young (1980)	$\Delta H=$	-95.265	0.080	60GIA/HOR
100	$H_2SO_4(l) = H_2SO_4(ai)$ Corr. from 1 mol/kg using act. coef. = 0.1215 from Pitzer et al. (1977)	$\Delta G=$	-53.846	0.060	60GIA/HOR
101	$H_2SO_4(l) = H_2SO_4(ai)$ Used act. coef. = 0.1247 at 1 mol/kg from Staples (1981); value at 1 mol/kg from Giauque et al. (1960)	$\Delta G=$	-54.103	0.060	84COD
102	$NH_4Cl(cr)$	$S=$	95.02	0.20	75STE
103	$NH_4Cl(cr)$ Meas. from 8-300 K. ΔS(trs, 242.5 K) = 4.816 J/(mol·K)	$S=$	95.27	0.20	72CHI/NAK
104	$NH_4Cl(cr)$ Meas. from 18-320 K. No points given	$S=$	94.81	0.20	74AMI/KOV
105	$NH_4Br(cr)$ Meas. from 15-305 K. S(15 K) = 0.950 J/(mol·K)	$S=$	112.84	0.40	65SOR/SUG
106	$NH_4Br(cr)$ Calculated by authors from combination of own and other published data	$S=$	111.21	0.40	80BAR/CAL
107	$NH_4Br(cr)$ Value taken from unpublished thesis	$S=$	111.38	0.40	68STE/ABA
108	$(NH_4)_2SO_4(cr)$ S(52.8 K) = 21.00 J/(mol·K) by Debye-Einstein extrap.	$S=$	220.50	2.00	45SHO
109	$NH_4NO_3(cr)$ Meas. from 15-314 K; S(15 K) = 0.615 J/(mol·K).	$S=$	151.07	0.20	55STE/BEN
110	$0.5\ N_2(g) + 1.5\ H_2(g) + H^+(ao) = NH_4^+(ao)$ From enthalpy of comb. of succinic acid and mono-ammonium succinate, enthalpy of soln. of acid and salt	$\Delta H=$	-133.82	0.85	72VAN/MANc

Table II-1. Measurements on Aqueous Thermochemistry at 298.15 K and 0.101325 MPa (cont'd).

NO.	REACTION OR SUBSTANCE	PROP. TYPE	OBSVD. VALUE kJ/mol or J/(mol·K)	± UNC.	REFERENCE
111	$0.5\ N_2(g) + 1.5\ H_2(g) + H^+(ao) = NH_4^+(ao)$ From enthalpy of comb. of succinic acid and di-ammonium succinate, enthalpy of soln. of acid and salt	$\Delta H=$	-133.15	0.85	72VAN/MANc
112	$0.5\ N_2(g) + 1.5\ H_2(g) + H^+(ao) = NH_4^+(ao)$ From enthalpy of comb. and soln. of oxalic acid and ammonium oxalate	$\Delta H=$	-133.30	1.00	34BEC/ROT
113	$NH_3(g) = NH_3(ao)$	$\Delta H=$	-35.346	0.060	72VAN/KIN
114	$NH_3(g) = NH_3(ao)$ Priv. commun. from L. A. K. Staveley	$\Delta H=$	-35.380	0.120	72VAN/KIN
115	$NH_4^+(ao) = H^+(ao) + NH_3(ao)$ See Olofsson (1975) for corr.	$\Delta G=$	52.732	0.050	54EVE/LAN
116	$NH_3(ao) + H_2O(l) = NH_4^+(ao) + OH^-(ao)$ From enthalpies of soln. of NH_4Cl in dil. $H^+(aq)$ and NaOH(aq)	$\Delta H=$	3.849	0.025	72VAN/KINa
117	$NH_3(ao) + H_2O(l) = NH_4^+(ao) + OH^-(ao)$ Meas. of pK from 278-323 K	$\Delta G=$	27.119	0.040	50BAT/PIN
118	$NH_3(ao) + H_2O(l) = NH_4^+(ao) + OH^-(ao)$ Meas. of pK from 278-323 K	$\Delta H=$	4.42	0.60	50BAT/PIN
119	$N_2(g) + 0.5\ O_2(g) + 2\ H_2O(l) = NH_3(g) + HNO_3(ai)$ Combin. of enthalpy of decomp. and soln. of $NH_4NO_3(cr)$ and enthalpy of neut. of $HNO_3(ai)$, all by authors	$\Delta H=$	319.30	0.52	34BEC/ROT
120	$NH_4NO_3(cr) = N_2(g) + 0.5\ O_2(g) + 2\ H_2O(l)$ Thermal decomp. in bomb with paraffin oil	$\Delta H=$	-206.67	0.60	34BEC/ROT
121	$NH_4NO_3(cr) = N_2(g) + 0.5\ O_2(g) + 2\ H_2O(l)$ Thermal decomp. in bomb with nitrobenzene	$\Delta H=$	-206.08	1.00	53MED/THO
122	$NH_4NO_3(cr) = N_2(g) + 0.5\ O_2(g) + 2\ H_2O(l)$ Thermal decomp. with paraffin oil. Used dry sample of pure crystal form (IV)	$\Delta H=$	-206.05	0.30	79COX/HAR
123	$NH_4NO_3(cr) = NH_3(g) + HNO_3(g)$ Meas. on $NH_4NO_3(l)$ betw. 463-543 K by boiling point method	$\Delta H=$	185.31	0.15	54FEI

Table II-1. Measurements on Aqueous Thermochemistry at 298.15 K and 0.101325 MPa (cont'd).

NO.	REACTION OR SUBSTANCE	PROP. TYPE	OBSVD. VALUE kJ/mol or J/(mol·K)	± UNC.	REFERENCE
124	$NH_4NO_3(cr) = NH_3(g) + HNO_3(g)$ Third law average of data on solid (349-438 K) and liq. (443-513 K)	$\Delta H=$	185.20	0.60	62BRA/JUN
125	$HNO_3(l) = HNO_3(ai)$ Corr. from 0.13 mol/kg is -0.400 kJ/mol from Parker (1965)	$\Delta H=$	-33.480	0.080	42FOR/GIA
126	$HNO_3(l) = HNO_3(g)$ Calorim.; corr. from 293 K = -0.28 kJ/mol	$\Delta H=$	39.16	0.13	40WIL/MIL
127	$NH_3(g) + HNO_3(ai) = NH_4NO_3(ai)$ Recalc. and corr. by Vanderzee and King (1972a)	$\Delta H=$	-87.23	0.25	34BEC/ROT
128	$NH_4NO_3(cr) = NH_4NO_3(ai)$	$\Delta H=$	25.544	0.030	80VAN/WAU
129	$NH_4NO_3(cr) = NH_4NO_3(ai)$	$\Delta H=$	25.418	0.030	78MED/EFI
130	$NH_4NO_3(cr) = NH_4NO_3(ai)$ Cryst. form not specified, but probably (IV)	$\Delta H=$	25.86	0.40	72KRE/ABR
131	$NH_4NO_3(cr) = NH_4NO_3(ai)$ Corr. from $NH_4NO_3(1000H_2O) = -0.22$ kJ/mol	$\Delta H=$	25.49	0.20	34BEC/ROT
132	$NH_4NO_3(cr) = NH_4NO_3(ai)$ Soly. = 25.96 mol/kg from Wishaw and Stokes (1953) (25.95 mol/kg), Prutton et al. (1935) (26.22 mol/kg), and Flatt and Fritz (1950) (25.96 mol/kg). Act. coef. = 0.131 from Hamer and Wu (1972)	$\Delta G=$	-6.068	0.080	84COD
133	$NH_4Cl(cr) = NH_4Cl(ai)$	$\Delta H=$	14.820	0.020	72VAN/KINa
134	$NH_4Cl(cr) = NH_4Cl(ai)$ Corr. from $NH_4Cl(37H_2O) = -0.556$ kJ/mol from Parker (1965)	$\Delta H=$	14.765	0.080	66KHA/PON
135	$NH_4Cl(cr) = NH_4Cl(ai)$ Corr. from $NH_4Cl(500H_2O) = -0.385$ kJ/mol from Parker (1965)	$\Delta H=$	14.979	0.080	69TSV/RAB
136	$NH_4Cl(cr) = NH_4Cl(ai)$ Meas. from 0.92 to 0.292 mol/kg. Corr. to std. state from Parker (1965)	$\Delta H=$	14.98	0.20	67MAK/BEL

Table II-1. Measurements on Aqueous Thermochemistry at 298.15 K and 0.101325 MPa (cont'd).

NO.	REACTION OR SUBSTANCE	PROP. TYPE	OBSVD. VALUE kJ/mol or J/(mol·K)	± UNC.	REFERENCE
137	$NH_4Cl(cr) = NH_4Cl(ai)$ Review of literature	$\Delta H=$	14.78	0.06	65PAR
138	$NH_4Cl(cr) = NH_4Cl(ai)$ Soly. = 7.400 mol/kg from Wishaw and Stokes (1953), Ricci and Skarulis (1951), Pearce & Pumplin (1937), Prutton et al. (1935), Shul'ts et al. (1962). Act. coef. = 0.565 from Hamer and Wu (1972)	$\Delta G=$	-7.092	0.020	84COD
139	$NH_3(ao) + HCl(ai) = NH_4Cl(ai)$ Diln. corr. = -0.02 kJ/mol. Meas. from 278-418 K	$\Delta H=$	-51.92	0.11	75OLO
140	$NH_4Br(cr) = NH_4Br(ai)$ Meas. as function of concentration; $\Delta H = 16.765 + 1.648\ m^{0.5}$ kJ/mol	$\Delta H=$	16.765	0.030	68STE/ABA
141	$NH_4Br(cr) = NH_4Br(ai)$ Review of literature	$\Delta H=$	16.78	0.40	65PAR
142	$NH_4Br(cr) = NH_4Br(ai)$ Soly. = 8.05 mol/kg from Smith and Eastlock (1916), Flatt and Burkhardt (1944), and Scott and Durham (1930). Act. coef. = 0.605 from Shul'ts and Simonova (1966)	$\Delta G=$	-7.849	0.040	84COD
143	$(NH_4)_2SO_4(cr) = (NH_4)_2SO_4(ai)$ Diln. corr. from $(NH_4)_2SO_4(500H_2O)= -0.928$ kJ/mol from data for $Na_2SO_4(ai)$	$\Delta H=$	6.603	0.200	69TSV/RAB
144	$(NH_4)_2SO_4(cr) = (NH_4)_2SO_4(ai)$ Corr. from 293 K = -1.36 kJ/mol; corr. from $(NH_4)_2SO_4\ (400H_2O)$ = -0.83 kJ/mol	$\Delta H=$	6.59	0.30	31ROT/ZEU
145	$(NH_4)_2SO_4(cr) = (NH_4)_2SO_4(ai)$ Soly. = 5.83 mol/kg from Wishaw and Stokes (1954) (5.843 mol/kg) and Ricci and Selikson (1952) (5.810 mol/kg). Act. coef. (0.105) recalculated from data of Wishaw and Stokes (1954)	$\Delta G=$	0.123	0.080	84COD
146	$Hg_2Cl_2(cr)$ S calc. from cell data on $Hg\|Hg_2Cl_2\|Cl^-(ao)$ electrode by Ives and Prasad (1970), Gupta et al. (1963), Covington et al. (1967), and Gerke (1922)	$S=$	191.6	1.5	84COD

Table II-1. Measurements on Aqueous Thermochemistry at 298.15 K and 0.101325 MPa (cont'd).

NO.	REACTION OR SUBSTANCE	PROP. TYPE	OBSVD. VALUE kJ/mol or J/(mol·K)	± UNC.	REFERENCE
147	$Hg_2SO_4(cr)$ Meas. from 16-298 K. See also Brackett et al. (1960) (5-21 K). $S(5\ K) = 0.114\ J/(mol·K)$.	$S=$	200.66	0.10	62PAP/GIA
148	$H_2(g) + Hg_2^{+2}(ao) = 2\ H^+(ao) + 2\ Hg(l)$ Review of literature	$\Delta G=$	-153.60	0.02	74VAN/SWA
149	$Hg(l) + Hg^{+2}(ao) = Hg_2^{+2}(ao)$	$\Delta G=$	-11.10	0.10	56HIE/SIL
150	$HgO(cr) + 2\ H^+(ao) = Hg^{+2}(ao) + H_2O(l)$ Red. From enthalpy of soln. in $HClO_4$(1 mol/kg). Diln. corr. = -2.18 kJ/mol	$\Delta H=$	-24.83	0.16	74VAN/ROD
151	$H_2(g) + Hg_2Cl_2(cr) = 2\ HCl(ai) + 2\ Hg(l)$	$\Delta G=$	-51.747	0.020	70IVE/PRA
152	$H_2(g) + Hg_2Cl_2(cr) = 2\ HCl(ai) + 2\ Hg(l)$	$\Delta S=$	-57.41	0.80	70IVE/PRA
153	$H_2(g) + Hg_2Cl_2(cr) = 2\ Hg(l) + 2\ HCl(ai)$ Meas. from 278-318 K. Our eqn.: $E/V =$ $0.26816-2.9660\ \text{x}\ 10^{-4}\ (T/K\ -298.15) - 3.003\ \text{x}$ $10^{-6}\ (T/K\ -298.15)^2$	$\Delta G=$	-51.747	0.020	62DAS/IVE
154	$H_2(g) + Hg_2Cl_2(cr) = 2\ Hg(l) + 2\ HCl(ai)$ Meas. from 278-318 K. Our eqn.: $E/V =$ $0.26816-2.9660\ \text{x}\ 10^{-4}\ (T/K\ -298.15) - 3.003\ \text{x}$ $10^{-6}\ (T/K\ -298.15)^2$	$\Delta S=$	-57.24	0.80	62DAS/IVE
155	$H_2(g) + Hg_2Cl_2(cr) = 2\ HCl(ai) + 2\ Hg(l)$ Meas. against quinhydrone electrode (278-308 K). Values for QHE taken from Bates (1964)	$\Delta G=$	-51.768	0.040	68SHA/SAH
156	$H_2(g) + Hg_2Cl_2(cr) = 2\ HCl(ai) + 2\ Hg(l)$ Meas. against quinhydrone electrode (278-308 K). Values for QHE taken from Bates (1964)	$\Delta S=$	-57.86	0.40	68SHA/SAH
157	$H_2(g) + Hg_2Cl_2(cr) = 2\ HCl(ai) + 2\ Hg(l)$	$\Delta G=$	-51.741	0.040	67COV/DOBa
158	$H_2(g) + Hg_2Cl_2(cr) = 2\ HCl(ai) + 2\ Hg(l)$	$\Delta S=$	-56.93	0.80	67COV/DOBb
159	$H_2(g) + Hg_2Cl_2(cr) = 2\ HCl(ai) + 2\ Hg(l)$ Refit of his meas. from 283-323 K	$\Delta G=$	-51.739	0.040	58GRZa
160	$H_2(g) + Hg_2Cl_2(cr) = 2\ HCl(ai) + 2\ Hg(l)$ Refit of his meas. from 283-323 K	$\Delta S=$	-60.67	0.80	58GRZa

Table II-1. Measurements on Aqueous Thermochemistry at 298.15 K and 0.101325 MPa (cont'd).

NO.	REACTION OR SUBSTANCE	PROP. TYPE	OBSVD. VALUE kJ/mol or J/(mol·K)	± UNC.	REFERENCE
161	$H_2(g) + Hg_2Cl_2(cr) = 2 HCl(ai) + 2 Hg(l)$	$\Delta G=$	-51.747	0.020	63GUP/HILa
162	$H_2(g) + Hg_2Cl_2(cr) = 2 HCl(ai) + 2 Hg(l)$	$\Delta S=$	-57.33	0.40	63GUP/HILa
163	$2 Hg(l) + Cl_2(g) = Hg_2Cl_2(cr)$	$\Delta G=$	-210.452	0.050	22GER
164	$2 Hg(l) + Cl_2(g) = Hg_2Cl_2(cr)$	$\Delta S=$	-182.4	1.0	22GER
165	$Hg_2Cl_2(cr) = Hg_2^{+2}(ao) + 2 Cl^-(ao)$ From soly. meas.	$\Delta G=$	102.04	0.10	55DRY/GLE
166	$Hg_2Cl_2(cr) = Hg_2^{+2}(ao) + 2 Cl^-(ao)$ From EMF on $Hg\|HgNO_3(aq)\|\|KCl(aq)\|Hg_2Cl2(cr)\|Hg$ LISTED FOR INFORMATION ONLY.	$\Delta G=$	102.40	0.50	29BRO
167	$Hg_2Cl_2(cr) = Hg_2^{+2}(ao) + 2 Cl^-(ao)$	$\Delta H=$	98.08	0.20	74VAN/SWA
168	$Hg_2Cl_2(cr) = Hg_2^{+2}(ao) + 2 Cl^-(ao)$	$\Delta G=$	101.86	0.20	74VAN/SWA
169	$H_2(g) + Hg_2Br_2(cr) = 2 HBr(ai) + 2 Hg(l)$ Calc. from HBr(0.10015 mol/kg) with act. coef. = 0.806	$\Delta G=$	-26.878	0.040	27GER/GED
170	$H_2(g) + Hg_2Br_2(cr) = 2 HBr(ai) + 2 Hg(l)$ Meas. cells from 278-318 K	$\Delta G=$	-26.867	0.020	63GUP/HILb
171	$H_2(g) + Hg_2Br_2(cr) = 2 HBr(ai) + 2 Hg(l)$ Meas. cells from 278-318 K	$\Delta S=$	-30.13	0.40	63GUP/HILb
172	$Hg_2Br_2(cr) = Hg_2^{+2}(ao) + 2 Br^-(ao)$ From EMF on $Hg\|HgNO_3(aq)\|\|KBr(aq)\|Hg_2Br_2(cr)\|Hg$	$\Delta G=$	127.06	0.60	29BRO
173	$Hg_2Cl_2(cr) + 2 Br^-(ao) = Hg_2Br_2(cr) + 2Cl^-(ao)$ Meas. from 283-313 K. Assume effect of ion conc. cancels	$\Delta G=$	-24.534	0.050	26BRO
174	$Hg_2Cl_2(cr) + 2 Br^-(ao) = Hg_2Br_2(cr) + 2Cl^-(ao)$ Meas. from 283-313 K. Assume effect of ion conc. cancels LISTED FOR INFORMATION ONLY.	$\Delta S=$	-36.28	0.80	26BRO
175	$2 Hg(l) + H_2SO_4(ai) = Hg_2SO_4(cr) + H2(g)$ Reextrap. by Pitzer et al. (1977). Meas. from 278-328 K, and from 0.1-8.0 mol/kg	$\Delta G=$	118.180	0.050	60BEC/DOB

Table II-1. Measurements on Aqueous Thermochemistry at 298.15 K and 0.101325 MPa (cont'd).

NO.	REACTION OR SUBSTANCE	PROP. TYPE	OBSVD. VALUE kJ/mol or J/(mol·K)	± UNC.	REFERENCE
176	$2 \ Hg(l) + H_2SO_4(ai) = Hg_2SO_4(cr) + H2(g)$ Reextrap. by Pitzer et al.(1977). Meas. from 278-328 K, and from 0.1-8.0 mol/kg	$\Delta S=$	161.75	0.40	60BEC/DOB
177	$H_2(g) + Hg_2SO_4(cr) = H_2SO_4(ai) + 2 \ Hg(l)$ Meas. from 278-308 K against quinhydrone electrode (0.6998 V at 298 K)	$\Delta G=$	-118.348	0.050	70SHA/PRAb
178	$H_2(g) + Hg_2SO_4(cr) = H_2SO_4(ai) + 2 \ Hg(l)$ Meas. from 278-308 K against quinhydrone electrode (0.6998 V at 298 K)	$\Delta S=$	-163.0	1.0	70SHA/PRAb
179	$H_2(g) + Hg_2SO_4(cr) = H_2SO_4(ai) + 2 \ Hg(l)$	$\Delta G=$	-118.745	0.100	72HAM
180	$H_2(g) + Hg_2SO_4(cr) = H_2SO_4(ai) + 2 \ Hg(l)$	$\Delta G=$	-118.19	0.08	65COV/DOBa
181	$Hg_2SO_4(cr) = Hg_2^{+2}(ao) + SO_4^{+2}(ao)$ Sol. product meas. analytically between 288 - 308 K	$\Delta G=$	34.78	0.40	70SHA/PRAa
182	$AgCl(cr)$ Calc. from meas. by Berg (1976). (2-20 K) and Eastman and Milner (1933) (20-310 K)	$S=$	96.36	0.20	84COD
183	$AgBr(cr)$ Meas. from 24-290 K; $S(11 \ K) = 1.13 \ J/(mol·K)$. See also Eucken et al. (1931) (11-270 K)	$S=$	107.24	0.40	33EAS/MIL
184	$AgI(cr)$ Meas. from 15-302 K. $S(15 \ K) = 6.3 \ J/(mol·K)$	$S=$	115.06	0.80	41PIT
185	$Ag_2SO_4(cr)$ Meas. from 14-297 K. $S(15 \ K) = 4.06 \ J/(mol·K)$	$S=$	200.96	0.06	33LAT/HICa
186	$AgNO_3(cr)$ Meas. from 13-297 K. $S(13 \ K) = 0.80 \ J/(mol·K)$	$S=$	141.04	0.40	37SMI/BRO
187	$Ag(cr) + H^+(ao) = Ag^+(ao) + 0.5 \ H_2(g)$ Meas. from 278-318 K. $E/V = -0.79910 + 9.930 \times 10^{-4} \ (T/K -298.15) + 4.057 \times 10^{-7} \ (T/K -298.15)^2$	$\Delta G=$	77.102	0.020	38OWE/BRI
188	$Ag(cr) + H^+(ao) = Ag^+(ao) + 0.5 \ H_2(g)$ Meas. from 278-318 K. $E/V = -0.79910 + 9.930 \times 10^{-4} \ (T/K -298.15) + 4.057 \times 10^{-7} \ (T/K -298.15)^2$	$\Delta S=$	95.81	0.20	38OWE/BRI

Table II-1. Measurements on Aqueous Thermochemistry at 298.15 K and 0.101325 MPa (cont'd).

NO.	REACTION OR SUBSTANCE	PROP. TYPE	OBSVD. VALUE kJ/mol or J/(mol·K)	± UNC.	REFERENCE
189	$Ag_2O(cr) + 2 H^+(ao) = 2 Ag^+(ao) + H_2O(l)$ From enthalpies of soln. in dilute HNO_3	$\Delta H=$	-43.64	0.25	37PIT/SMI
190	$Ag_2O(cr) + H_2O(l) + 2 Cl^-(ao) = 2$ $AgCl(cr) + 2 OH^-(ao)$	$\Delta G=$	-23.415	0.100	37PIT/SMI
191	$0.5 H_2(g) + AgCl(cr) = Ag(cr) + HCl(ai)$ Meas. from 273-368 K	$\Delta G=$	-21.459	0.020	54BAT/BOW
192	$0.5 H_2(g) + AgCl(cr) = Ag(cr) + HCl(ai)$ Meas. from 273-368 K	$\Delta S=$	-62.30	0.40	54BAT/BOW
193	$Ag(cr) + 0.5 Cl_2(g) = AgCl(cr)$	$\Delta G=$	-109.603	0.040	67FAI/LON
194	$Ag(cr) + 0.5 Cl_2(g) = AgCl(cr)$	$\Delta S=$	-57.70	0.20	67FAI/LON
195	$AgNO_3(ai) + KCl(ai) = AgCl(cr) + KNO_3(ai)$ Corr. from 296 K = 0.40 kJ/mol; diln. corr. from -0.37 to +0.22 kJ/mol	$\Delta H=$	-65.690	0.080	27LAN/FUO
196	$AgNO_3(ai) + KCl(ai) = AgCl(cr) + KNO_3(ai)$ Assume authors included diln. corr.	$\Delta H=$	-65.56	0.75	68BRO/DHAb
197	$Ag^+(ao) + Cl^-(ao) = AgCl(cr)$ Calorim. enthalpy of pptn.; ΔG from review of literature	$\Delta H=$	-65.722	0.080	73WAG/KIL
198	$Ag^+(ao) + Cl^-(ao) = AgCl(cr)$ Calorim. enthalpy of pptn.; ΔG from review of literature	$\Delta G=$	-55.651	0.020	73WAG/KIL
199	$Ag^+(ao) + Cl^-(ao) = AgCl(cr)$ From soly. by conduct. meas. from 278-328 K; $\log K_{sp} = 1.9470 - 3487.9(K/T)$	$\Delta G=$	-55.662	0.050	54GLE/MAL
200	$Ag^+(ao) + Cl^-(ao) = AgCl(cr)$ Soly. meas. at 298.25 and 318.15 K using radioactive Ag	$\Delta G=$	-55.68	0.10	68BRO/DHAa
200a	$Ag^+(ao) + Cl^-(ao) = AgCl(cr)$ Radiochemical method, correcting for AgCl(ao)	$\Delta G=$	-55.75	0.10	77BRO/HAY
201	$Ag^+(ao) + Cl^-(ao) = AgCl(cr)$ From study of complexing constants for AgCl	$\Delta G=$	-55.65	0.10	71AND/BUT

Table II-1. Measurements on Aqueous Thermochemistry at 298.15 K and 0.101325 MPa (cont'd).

NO.	REACTION OR SUBSTANCE	PROP. TYPE	OBSVD. VALUE kJ/mol or	± UNC. J/(mol·K)	REFERENCE
202	Ag(cr) + 0.5 Cl$_2$(g) = AgCl(cr) High temp. cells (649-973 K). Third law calc.	$\Delta H=$	-126.82	0.40	63NAZ/SHU
203	Ag(cr) + 0.5 Cl$_2$(g) = AgCl(cr) Ag|AgCl|Cl$_2$(g) cell from 288-308 K	$\Delta G=$	-109.66	0.08	22GER
204	Ag(cr) + 0.5 Cl$_2$(g) = AgCl(cr) Ag|AgCl|Cl$_2$(g) cell from 288-308 K	$\Delta S=$	-57.4	0.5	22GER
205	0.5 H$_2$(g) + AgCl(cr) = Ag(cr) + HCl(ai) From cell meas. (288-328 K), HCl conc. from 0.005-0.1 mol/kg	$\Delta G=$	-21.455	0.010	81ELS/FOU
206	0.5 H$_2$(g) + AgCl(cr) = Ag(cr) + HCl(ai) From cell meas. (288-328 K), HCl conc. from 0.005-0.1 mol/kg	$\Delta S=$	-62.2	0.2	81ELS/FOU
207	Ag(cr) + 0.5 Br$_2$(l) = AgBr(cr) High temp. (700-1000 K) cell meas. Third Law calc.	$\Delta H=$	-99.75	0.40	70MET/SEI
208	Ag(cr) + 0.5 Br$_2$(l) = AgBr(cr) From cell meas. at 298 K	$\Delta G=$	-95.935	0.020	34JON/BAE
209	AgBr(cr) + Cl$^-$(ao) = AgCl(cr) + Br$^-$(ao) Cell meas., Ag|AgCl vs Ag|AgBr	$\Delta G=$	14.584	0.010	34JON/BAE
210	Ag(cr) + HBr(ai) = AgBr(cr) + 0.5 H$_2$(g) Cell meas. from 273 - 333 K	$\Delta G=$	6.862	0.040	36HAR/KES
211	Ag(cr) + HBr(ai) = AgBr(cr) + 0.5 H$_2$(g) Cell meas. from 273 - 333 K	$\Delta S=$	47.61	0.80	36HAR/KES
212	Ag(cr) + HBr(ai) = AgBr(cr) + 0.5 H$_2$(g)	$\Delta G=$	6.883	0.040	36OWE/FOE
213	Ag(cr) + HBr(ai) = AgBr(cr) + 0.5 H$_2$(g)	$\Delta S=$	48.66	0.80	36OWE/FOE
214	Ag(cr) + HBr(ai) = AgBr(cr) + 0.5 H$_2$(g) From cell meas. (288-328 K) as func. of HBr concentration	$\Delta G=$	6.855	0.010	82ELS/FOU
215	Ag(cr) + HBr(ai) = AgBr(cr) + 0.5 H$_2$(g) From cell meas. (288-328 K) as func. of HBr concentration	$\Delta S=$	47.3	0.2	82ELS/FOU

Table II-1. Measurements on Aqueous Thermochemistry at 298.15 K and 0.101325 MPa (cont'd).

NO.	REACTION OR SUBSTANCE	PROP. TYPE	OBSVD. VALUE kJ/mol or J/(mol·K)	± UNC.	REFERENCE
216	$Ag(cr) + HBr(ai) = AgBr(cr) + 0.5 H_2(g)$ Meas. fitted from 273-373 K; $E/V = -0.07109 + (4.867 \times 10^{-4})(T/K)$	$\Delta G=$	6.870	0.020	62HET/ROB
217	$Ag(cr) + HBr(ai) = AgBr(cr) + 0.5 H_2(g)$ Meas. fitted from 273-323 K; $E/V = -0.07109 + (4.867 \times 10^{-4})(T/K)$	$\Delta S=$	47.47	0.40	62HET/ROB
218	$Ag^+(ao) + Br^-(ao) = AgBr(cr)$ Soly. meas. by conductance at 298 K. Temp. dependence appears to be in error	$\Delta G=$	-70.161	0.050	54GLE/MAL
219	$Ag^+(ao) + Br^-(ao) = AgBr(cr)$ Review of literature	$\Delta G=$	-70.206	0.080	73WAG/KIL
220	$Ag^+(ao) + Br^-(ao) = AgBr(cr)$ Corr. for std. state and demixing = 0.12-0.44 kJ/mol	$\Delta H=$	-84.828	0.080	73WAG/KIL
221	$AgCl(cr) + HBr(ai) = AgBr(cr) + Br^-(ao)$	$\Delta G=$	-14.62	0.08	32SHI/TAK
222	$AgNO_3(ai) + HBr(ai) = AgBr(cr) + HNO_3(ai)$ Calorim. diln. corr. = +3.93 kJ/mol	$\Delta H=$	-84.77	0.40	49GEL/SKI
223	$Ag(cr) + 0.5 I_2(cr) = AgI(cr)$ Meas. $E/V = 0.6629 \pm 0.0010$ at 298 K	$\Delta G=$	-63.960	0.080	66SOU/GAU
224	$Ag(cr) + HI(ai) = AgI(cr) + 0.5 H_2(g)$ Calc. from data at 273-323 K; $E/V = +0.15237 + 3.1898 \times 10^{-4}(T/K -298.15) + 3.024 \times 10^{-6} (T/K -298.15)^2$	$\Delta G=$	-14.702	0.030	64HET/ROB
225	$Ag(cr) + HI(ai) = AgI(cr) + 0.5 H_2(g)$ Calc. from data at 273-323 K; $E/V = +0.15237 + 3.1898 \times 10^{-4}(T/K -298.15) + 3.024 \times 10^{-6} (T/K -298.15)^2$	$\Delta S=$	30.78	0.80	64HET/ROB
226	$Ag(cr) + HI(ai) = AgI(cr) + 0.5 H_2(g)$	$\Delta G=$	-14.660	0.050	40GOU/VOS
227	$Ag(cr) + HI(ai) = AgI(cr) + 0.5 H_2(g)$	$\Delta S=$	31.63	0.40	40GOU/VOS
228	$Ag(cr) + HI(ai) = AgI(cr) + 0.5 H_2(g)$ Cell meas. from 278-313 K; obtained equation (corrrected to absolute V): $E/V = 0.15244 + 3.28 \times 10^{-4} (T/K -298.15) + 3.6 \times 10^{-6} (T/K -298.15)^2$	$\Delta G=$	-14.695	0.020	35OWE

Table II-1. Measurements on Aqueous Thermochemistry at 298.15 K and 0.101325 MPa (cont'd).

NO.	REACTION OR SUBSTANCE	PROP. TYPE	OBSVD. VALUE kJ/mol or J/(mol·K)	± UNC.	REFERENCE
229	$Ag(cr) + HI(ai) = AgI(cr) + 0.5\ H_2(g)$ Cell meas. from 278-313 K; obtained equation (corrrected to absolute V): $E/V = 0.15244 + 3.28 \times 10^{-4}\ (T/K -298.15) + 3.6 \times 10^{-6} (T/K -298.15)^2$	$\Delta S=$	31.6	1.0	35OWE
230	$AgNO_3(ai) + KI(ai) = AgI(cr) + KNO_3(ai)$ Corr. to std. state = -0.12 kJ/mol	$\Delta H=$	-111.21	0.20	30LAN/SHI
231	$Ag(cr) + HI(ai) = AgI(cr) + 0.5\ H_2(g)$	$\Delta G=$	-14.670	0.040	65KOR/HAU
232	$Ag(cr) + HI(ai) = AgI(cr) + 0.5\ H_2(g)$	$\Delta S=$	30.50	0.80	65KOR/HAU
233	$Ag(cr) + HI(ai) = AgI(cr) + 0.5\ H_2(g)$	$\Delta G=$	-14.686	0.030	63FEA/WAT
234	$Ag(cr) + 0.5\ I_2(cr) = AgI(cr)$ Cell data; $E/V = 0.6830 + 1.50 \times 10^{-4}$ $(T/K -273.15)$ from 293-373 K	$\Delta G=$	-66.26	0.08	68TOP
235	$Ag(cr) + 0.5\ I_2(cr) = AgI(cr)$ Cell data; $E/V = 0.6830 + 1.50 \times 10^{-4}$ $(T/K -273.15)$ from 293-373 K	$\Delta S=$	14.48	0.80	68TOP
236	$Ag(cr) + 0.5\ I_2(cr) = AgI(cr)$ Reaction meas. in concentrated KI solutions	$\Delta H=$	-62.47	0.20	25WEB
237	$Ag^+(ao) + I^-(ao) = AgI(cr)$ Corr. to std. states = 0.38 kJ/mol	$\Delta H=$	-111.13	0.10	73WAG/KIL
238	$Ag^+(ao) + I^-(ao) = AgI(cr)$ Review of literature	$\Delta G=$	-91.827	0.080	73WAG/KIL
239	$AgNO_3(ai) + KI(ai) = AgI(cr) + KNO_3(ai)$ Corr. to std. states = -0.12 kJ/mol	$\Delta H=$	-111.21	0.20	30LAN/SHI
240	$AgNO_3(ai) + HI(ai) = AgI(cr) + HNO_3(ai)$ Corr. to std. states = +5.2 kJ/mol	$\Delta H=$	-109.91	0.80	49GEL/SKI
241	$Ag_2SO_4(cr) = 2\ Ag^+(ao) + SO_4^{-2}(ao)$ Meas. from 0.005 - 0.014 mol/kg. Corr. for formation of $(AgSO_4)^-(ao)$	$\Delta H=$	17.32	0.20	65HOP/WULa
242	$Ag_2SO_4(cr) = 2\ Ag^+(ao) + SO_4^{-2}(ao)$ Calculated from temperature dependence of EMF meas. on $Ag|Ag_2SO_4(cr)|SO_4^{-2}(ao)$ electrode	$\Delta H=$	17.63	0.40	59PAN/LIN

Table II-1. Measurements on Aqueous Thermochemistry at 298.15 K and 0.101325 MPa (cont'd).

NO.	REACTION OR SUBSTANCE	PROP. TYPE	OBSVD. VALUE kJ/mol or J/(mol·K)	± UNC.	REFERENCE
243	$Ag_2SO_4(cr) = 2\ Ag^+(ao) + SO_4^{-2}(ao)$ Calculated from meas. by Pan and Lin (1959) (28.06 kJ/mol) and Hopkins and Wulff (1965) (28.20 kJ/mol) including corr. for $AgSO_4^-(aq)$). See also Lietzke and Hall (1967), Stoughton and Lietzke (1960), and Kenttamaa (1957).	$\Delta G=$	28.12	0.20	84COD
244	$AgNO_3(cr) = AgNO_3(ai)$ Includes data from Graham and Hepler (1958)	$\Delta H=$	22.732	0.040	73WAG/KIL
245	$AgNO_3(cr) = AgNO_3(ai)$ Enthalpy of soln. meas. at 0.125 mol/kg, diln. corr. = 0.17 kJ/mol	$\Delta H=$	22.38	0.40	63SHI/VOS
246	$AgNO_3(cr) = AgNO_3(ai)$ Enthalpy of soln. meas. at 0.00854 mol/kg; diln. corr. = -0.105 kJ/mol	$\Delta H=$	22.43	0.20	37SMI/BRO
247	$AgNO_3(cr) = AgNO_3(ai)$ Enthalpy of soln. meas. at 0.116 mol/kg; diln. corr. = 0.15 kJ/mol	$\Delta H=$	22.91	0.20	37LAN/MAR
248	$AgNO_3(cr) = AgNO_3(ai)$ Soly. = 15.01 mol/kg from Skarulis and Ricci (1941) (15.010 mol/kg), Ricci and Linke (1951) (15.003 mol/kg). Act. coef. (0.0822) from Kangro and Groeneveld (1962)	$\Delta G=$	-1.042	0.040	73WAG/KIL
249	$Ag(cr) + 0.5\ Hg_2Cl_2(cr) = Hg(l) + AgCl(cr)$ Cell as func. of HCl conc. and T (298-525 K)	$\Delta G=$	-4.44	0.16	55LIE/VAU
250	$Ag(cr) + 0.5\ Hg_2Cl_2(cr) = Hg(l) + AgCl(cr)$ Cell as func. of HCl conc. and T (298-525 K)	$\Delta S=$	33.8	0.4	55LIE/VAU
251	$Ag(cr) + 0.5\ Hg_2Cl_2(cr) = Hg(l) + AgCl(cr)$	$\Delta G=$	-4.398	0.040	22GER
252	$Ag(cr) + 0.5\ Hg_2Cl_2(cr) = Hg(l) + AgCl(cr)$	$\Delta S=$	32.56	0.40	22GER
253	$2\ Ag(cr) + Hg_2Br_2(cr) = 2\ AgBr(cr) + 2\ Hg(l)$ Meas. between 293-303 K	$\Delta G=$	-13.200	0.050	40LAR
254	$2\ Ag(cr) + Hg_2Br_2(cr) = 2\ AgBr(cr) + 2\ Hg(l)$ Meas. between 288-308 K	$\Delta G=$	-13.130	0.050	40DAK/EWI

Table II-1. Measurements on Aqueous Thermochemistry at 298.15 K and 0.101325 MPa (cont'd).

NO.	REACTION OR SUBSTANCE	PROP. TYPE	OBSVD. VALUE kJ/mol or J/(mol·K)	± UNC.	REFERENCE
255	2 Ag(cr) + Hg$_2$Br$_2$(cr) = 2 AgBr(cr) + 2 Hg(l) Meas. between 288-308 K LISTED FOR INFORMATION ONLY.	$\Delta S=$	60.2	1.0	40DAK/EWI
256	Ag(cr) + Br$^-$(ao) + 0.5 Hg$_2$Cl$_2$(cr) = Hg(l) + Cl$^-$(ao) + AgBr(cr) Meas. of Ag|AgBr vs. Hg|Hg$_2$Cl$_2$	$\Delta G=$	-18.975	0.020	34JON/BAE
257	LiOH(cr) Meas. from 16-302 K. S(16 K) = 0.12 J/(mol·K)	$S=$	42.76	0.20	50BAU/JOH
258	LiOH:H$_2$O(cr) Meas. from 15-302 K. S(15 K) = 0.12 J/(mol·K)	$S=$	71.42	0.20	50BAU/JOH
259	LiF(cr) Meas. from 18-272 K. See also Clusius and Eichenauer (1956) (10-111 K)	$S=$	35.66	0.12	49CLU/GOL
260	LiCl(cr) Meas. from 14-320 K. S(14 K) = 0.06 J/(mol·K). See also Stull et al. (1970) (15-298 K)	$S=$	59.27	0.10	60SHI
261	LiBr(cr) Meas. from 12-318 K. S(12 K) = 0.15 J/(mol·K)	$S=$	74.01	0.20	74PAU/ANI
262	LiI(cr) Meas. from 5-315 K. See also Berg (1976) (2-20 K). S(5 K) = 0.022 J/(mol·K)	$S=$	86.77	0.20	75KHR/LUK
263	Li$_2$SO$_4$(cr) Heat capacity meas. from 12 - 311 K. S(12 K) = 0.17 J/(mol·K)	$S=$	113.97	0.20	69PAU
264	Li(cr) + H$_2$O(l) = LiOH(ai) + 0.5 H$_2$(g) Diln. corr. from 0.055 mol/kg is -0.427 kJ/mol	$\Delta H=$	-222.63	0.16	67GUN
265	Li(cr) + H$_2$O(l) = LiOH(ai) + 0.5 H$_2$(g) Corr. from 0.617 mol/kg is -1.293 kJ/mol	$\Delta H=$	-221.90	0.80	55MES/FAS
266	Li(cr) + H$_2$O(l) = LiOH(ai) + 0.5 H$_2$(g) Corr. from 0.0278 mol/ kg and 291 K is -1.51 kJ/mol	$\Delta H=$	-222.63	0.40	20MOE
267	Li(cr) + H$^+$(ao) = Li$^+$(ao) + 0.5 H$_2$(g) Corr. for Li amalgam electrode = 0.8438 V. Corr. act. coef. of LiOH from Hamer and Wu (1972)	$\Delta G=$	-292.83	0.40	68HUS/BUT

Table II-1. Measurements on Aqueous Thermochemistry at 298.15 K and 0.101325 MPa (cont'd).

NO.	REACTION OR SUBSTANCE	PROP. TYPE	OBSVD. VALUE kJ/mol or J/(mol·K)	± UNC.	REFERENCE
268	Li(cr) + AgCl(cr) = LiCl(ai) + Ag(cr) Corr. for amalgam electrode (=1.017 V) calc. from Cogley and Butler (1968). Extrap. to zero ionic strength LiCl(ai)	$\Delta G=$	-312.32	0.50	64LEB/ALE
269	LiOH(cr) = LiOH(ai) Corr. from 0.139 mol/kg is -0.640 kJ/mol from Parker (1965)	$\Delta H=$	-21.09	0.20	33UED2
270	LiOH(cr) = LiOH(ai) Corr. from 0.139 mol/kg is -0.640 kJ/mol from Parker (1965)	$\Delta H=$	-23.58	0.20	61RES
271	LiOH:H$_2$O(cr) = LiOH(cr) + H$_2$O(l) From difference in enthalpies of solution	$\Delta H=$	16.820	0.080	33UEDb
272	LiOH:H$_2$O(cr) = LiOH(ai) + H$_2$O(l) Soly. = 5.14 mol/kg, act. coef. = 0.495, activity of H$_2$O = 0.841. See also Ueda (1933 b)	$\Delta G=$	-4.201	0.080	62KAN/GRO
273	LiOH:H$_2$O(cr) = LiOH(cr) + H$_2$O(l) Vap. pres. of H$_2$O(g) = 0.00515 atm at 298 K; log P/atm = 8.172-3118.66(K/T)	$\Delta G=$	4.470	0.050	33UEDb
274	LiOH:H$_2$O(cr) = LiOH(cr) + H$_2$O(l) Vap. pres. of H$_2$O(g) = 0.00467 atm at 298 K; log P/atm = 8.7681 - 3309.91(K/T)	$\Delta G=$	4.712	0.050	64BAC/BOA
275	LiOH:H$_2$O(cr) = LiOH(cr) + H$_2$O(l) Meas. from 303-323 K with H$_2$O manometer. Calc. P(298 K) = 0.00489 atm	$\Delta G=$	4.601	0.100	68THA/CHI
276	LiF(cr) = LiF(ai) Debye-Hueckel extrap. from 0.00925 mol/kg	$\Delta H=$	4.418	0.100	65COX/HAR
277	LiF(cr) = LiF(ai) Meas. to 0.02 mol/kg. No individual data	$\Delta H=$	4.48	0.20	64STE/HOP
278	LiF(cr) = LiF(ai) Corr. from 0.005 mol/kg from Parker (1965)	$\Delta H=$	4.30	0.20	79GER/PER
279	LiF(cr) = LiF(ai) Corr. from 0.015 mol/kg at 294.6 K = -0.053 kJ/mol	$\Delta H=$	4.70	0.20	61KOL/SKU

Table II-1. Measurements on Aqueous Thermochemistry at 298.15 K and 0.101325 MPa (cont'd).

NO.	REACTION OR SUBSTANCE	PROP. TYPE	OBSVD. VALUE kJ/mol or J/(mol·K)	± UNC.	REFERENCE
280	LiF(cr) = LiF(ai) Soly. = 0.0513 mol/kg from Payne (1937). Act. coef. = 0.82, estim. from NaF, KF, LiCl, NaCl	ΔG=	15.71	0.20	84COD
281	LiCl(cr) = LiCl(ai) Extrap. from 0.053 mol/kg from Parker (1965)	ΔH=	-35.955	0.080	77JOL/PER
282	LiCl(cr) = LiCl(ai) Meas. to 0.079 mol/kg; extrap. from Parker (1965)	ΔH=	-36.99	0.10	64DRA/CHA
283	LiCl(cr) = LiCl(ai) Meas. to 0.111 mol/kg; extrap. from Parker (1965)	ΔH=	-37.58	0.30	69TSV/RAB
284	LiCl:H$_2$O(cr) = LiCl(ai) + H$_2$O(l) Soly. = 19.88 mol/kg, act. coef. = 61.7 from Clynne and Potter (1979); Hildago and Orr (1968). Water activity = 0.110 from Hamer and Wu (1972). See also Gibbard and Scatchard (1973)	ΔG=	-29.740	0.100	84COD
285	LiCl:H$_2$O(cr) = LiCl(cr) + H$_2$O(l) From difference in enthalpies of solution at 292 K. Corr. to 298 K = 0.226 kJ/mol	ΔH=	16.980	0.100	29SLO/HUT
286	LiCl:H$_2$O(cr) = LiCl(cr) + H$_2$O(l) log P/atm = 20.4256 - (4100.7)(K/T) - 4.026 log T/K	ΔG=	10.188	0.200	29SLO/HUT
287	LiCl:H$_2$O(cr) = LiCl(cr) + H$_2$O(l) Meas. 298-308 K; log P/atm = 7.4024 - (3148.46)(K/T)	ΔG=	9.432	0.200	31UED
288	LiBr(cr) = LiBr(ai) Review of literature	ΔH=	-48.83	0.20	65PAR
289	LiBr(cr) = LiBr(ai)	ΔH=	-48.37	0.20	77JOL/PER
290	LiI(cr) = LiI(ai) Review of literature	ΔH=	-63.30	0.20	65PAR
291	LiI(cr) = LiI(ai) Extrap. from 0.017 mol/kg from Parker (1965)	ΔH=	-61.95	0.15	77JOL/PER
292	Li$_2$SO$_4$(cr) = Li$_2$SO$_4$(ai) Extrap. from 0.055 mol/kg is -1.76 kJ/mol	ΔH=	-29.71	0.20	52KAP/SAM

Table II-1. Measurements on Aqueous Thermochemistry at 298.15 K and 0.101325 MPa (cont'd).

NO.	REACTION OR SUBSTANCE	PROP. TYPE	OBSVD. VALUE kJ/mol or J/(mol·K)	± UNC.	REFERENCE
293	$Li_2SO_4(cr) = Li_2SO_4(ai)$ Extrap. from 0.278 mol/kg is -2.63 kJ/mol	$\Delta H=$	-30.00	0.40	64MCC/LAI
294	$Li_2SO_4(cr) = Li_2SO_4(ai)$ Extrap. from 0.111 mol/kg is -2.13 kJ/mol	$\Delta H=$	-30.00	0.30	69TSV/RAB
295	$Li_2SO_4(cr) = Li_2SO_4(ai)$ Corr. from Li_2SO_4(0.018 mol/kg) = -1.25 kJ/mol	$\Delta H=$	-30.20	0.50	69ROM/SAM
296	$Li_2SO_4:H_2O(cr) = Li_2SO_4(ai) + H_2O(l)$ Soly. = 3.14 mol/kg, act. coef. = 0.2993, act. of H_2O = 0.8435. See also Pearce and Eckstrom (1937), Kangro and Groeneveld (1962), and Ueda (1933a)	$\Delta G=$	-2.553	0.040	81GOL
297	$Li_2SO_4:H_2O(cr) = Li_2SO_4(cr) + H_2O(l)$ Average eqn. from 2 sets of meas: log P/atm = 7.466 - (2992)(K/T)	$\Delta G=$	6.073	0.050	73PRI/STU
298	$Li_2SO_4:H_2O(cr) = Li_2SO_4(cr) + H_2O(l)$ Average eqn. from 2 sets of meas: log P/atm = 7.466 - (2992)(K/T)	$\Delta H=$	13.28	1.00	73PRI/STU
299	$Li_2SO_4:H_2O(cr) = Li_2SO_4(cr) + H_2O(l)$ From difference in enthalpies of solution (corr. to infinite diln)	$\Delta H=$	12.55	0.50	69ROM/SAM
300	$Li_2SO_4:H_2O(cr) = Li_2SO_4(cr) + H_2O(l)$ From difference in enthalpies of solution	$\Delta H=$	12.20	0.20	33UEDa
301	$Li_2SO_4:H_2O(cr) = Li_2SO_4(cr) + H_2O(l)$ Meas. vap. pres. from 294-363 K; log P/atm = 7.8779 - (3149.9)(K/T)	$\Delta G=$	6.745	0.200	64BAC/BOA
302	$Li_2SO_4:H_2O(cr) = Li_2SO_4(cr) + H_2O(l)$ Meas. vap. pres. from 294-363 K; log P/atm = 7.8779 - (3149.9)(K/T)	$\Delta H=$	16.3	4.0	64BAC/BOA
303	$Li_2SO_4:H_2O(cr) = Li_2SO_4(cr) + H_2O(l)$ Meas. vap. pres. from 293-318 K; log P/atm = 7.347 - (2967)(K/T)	$\Delta G=$	6.27	0.20	57POP/GAG
304	$Li_2SO_4:H_2O(cr) = Li_2SO_4(cr) + H_2O(l)$ Meas. vap. pres. from 293-318 K; log P/atm = 7.347 - (2967)(K/T)	$\Delta H=$	12.80	2.00	57POP/GAG

Table II-1. Measurements on Aqueous Thermochemistry at 298.15 K and 0.101325 MPa (cont'd).

NO.	REACTION OR SUBSTANCE	PROP. TYPE	OBSVD. VALUE kJ/mol or J/(mol·K)	± UNC.	REFERENCE
305	Li_2SO_4:H_2O(cr) = Li_2SO_4(cr) + H_2O(l) Meas. vap. pres. at 298.15 K; enthalpies of soln. of hydrate and anhydrous salt in 400 H_2O	$\Delta G=$	5.84	0.20	33UEDa
306	Li_2SO_4:H_2O(cr) = Li_2SO_4(cr) + H_2O(l) Meas. vap. pres. at 298.15 K; enthalpies of soln. of hydrate and anhydrous salt in 400 H_2O	$\Delta H=$	12.63	0.80	33UEDa
307	NaOH(cr) S(15 K) = 0.083 J/(mol·K). Corrected from Murch and Giauque (1962)	$S=$	64.454	0.080	69SIE/GIA
308	NaOH:H_2O(cr) Corrected from meas. of Murch and Giauque (1962). S(15 K) = 0.234 J/(mol·K)	$S=$	99.52	0.20	69SIE/GIA
309	NaF(cr) From the meas. by Kirkham and Yates (1968) (25-300 K), Harrison et al. (1968) (0.05-15 K), Birch et al. (1979) (2.3-20 K), and King (1957) (54-296 K)	$S=$	51.13	0.10	84COD
310	NaCl(cr) Meas. from 3-267 K. See also Morrison and Patterson (1956) and Morrison et al. (1955)	$S=$	72.13	0.20	64BAR/LEA
311	NaBr(cr) Meas. from 7-302 K. See also Birch et al. (1979) (3-19 K)	$S=$	86.93	0.02	64GAR/TAY
312	NaI(cr) From the meas. by Berg and Morrison (1957) (3-270 K) and Gardner and Taylor (1964) (56-301 K)	$S=$	98.53	0.02	84COD
313	Na_2SO_4(cr) S(15 K) = 0.32 J/(mol·K). See also Pitzer and Coulter (1938) (14-300 K)	$S=$	149.58	0.20	72BRO/GIA
314	Na_2SO_4:10H_2O(cr) Includes corr. of 5.73 J/(mol·K) for zero point entropy	$S=$	591.28	0.40	58BRO/GIA
315	$NaNO_3$(cr) Meas. from 16-287 K. S(16 K) = 0.60 J/(mol·K)	$S=$	116.27	0.20	33SOU/NEL

Table II-1. Measurements on Aqueous Thermochemistry at 298.15 K and 0.101325 MPa (cont'd).

NO.	REACTION OR SUBSTANCE	PROP. TYPE	OBSVD. VALUE kJ/mol or J/(mol·K)	± UNC.	REFERENCE
316	$Na(cr) + H_2O(l) = NaOH(ai) + 0.5 H_2(g)$ Extrap. from 0.0555 mol/kg is -0.350 kJ/mol	$\Delta H=$	-184.615	0.030	67GUN
317	$Na(cr) + H_2O(l) = NaOH(ai) + 0.5 H_2(g)$ Extrap. from 0.370 mol/kg is -0.490 kJ/mol	$\Delta H=$	-184.548	0.080	55MES/FAS
318	$Na(cr) + HCl(ai) = NaCl(ai) + 0.5 H_2(g)$ Corr. for sat. calomel electrode = +23.287 kJ/mol	$\Delta G=$	-261.98	0.20	50TRU/SCH
319	$Na(cr) + H^+(ao) = Na^+(ao) + 0.5 H_2(g)$ From recalc. of meas. of Mussini et al. (1971)	$\Delta G=$	-262.179	0.050	77MUS/LON
320	$Na(cr) + H_2O(l) = NaOH(ai) + 0.5 H_2(g)$ Extrap. to infinite diln. range from -0.42 to -0.50 kJ/mol	$\Delta H=$	-184.32	0.40	51KET/WAL
321	$NaOH(cr) = NaOH(ai)$ Corr. for mol. wt. and for enthalpy of diln. Extrap. to anhyd. NaOH(cr). See Siemens and Giauque (1969)	$\Delta H=$	-44.493	0.050	62MUR/GIA
322	$NaOH(cr) = NaOH(ai)$ Extrap. from 0.139 mol/kg is -0.46 kJ/mol	$\Delta H=$	-44.16	0.40	61RES
323	$NaOH{:}H_2O(cr) = NaOH(ai) + H_2O(l)$	$\Delta H=$	-21.405	0.050	62MUR/GIA
324	$NaOH{:}H_2O(cr) = NaOH(ai) + H_2O(l)$ Soly. = 27.312 mol/kg. Act. coef. (31.8), osmot. coef. (2.715) from Hamer and Wu (1972). Shibata and Murata (1937) obtained soly. = 27.063 mol/kg	$\Delta G=$	-26.906	0.080	69SIE/GIA
325	$NaOH{:}H_2O(cr) = NaOH(cr) + H_2O(l)$ Meas. vap. pres. of H_2O = 1.97 x 10^{-4} atm. See also Shibata (1931) (P/atm = 2.62×10^{-4})	$\Delta G=$	12.55	0.20	16BAX/STA
326	$NaF(cr) = NaF(ai)$ Extrap. from meas. between 0.132-0.016 mol/kg by Debye-Hueckel eqn.	$\Delta H=$	0.937	0.020	78NUT/CHU
327	$NaF(cr) = NaF(ai)$ Extrap. from meas. between 0.09-0.02 mol/kg	$\Delta H=$	0.941	0.040	72COO/DAV
328	$NaF(cr) = NaF(ai)$ Extrap. from 0.111 mol/kg is -0.44 kJ/mol from Parker (1965)	$\Delta H=$	0.96	0.20	69TSV/RAB

Table II-1. Measurements on Aqueous Thermochemistry at 298.15 K and 0.101325 MPa (cont'd).

NO.	REACTION OR SUBSTANCE	PROP. TYPE	OBSVD. VALUE kJ/mol or J/(mol·K)	± UNC.	REFERENCE
329	NaF(cr) = NaF(ai) Extrap. from 0.04 mol/kg is -0.30 kJ/mol from Parker (1965)	ΔH=	0.91	0.20	53HEP/JOL
330	NaF(cr) = NaF(ai) Corr. from 0.25 mol/kg = -0.49 kJ/mol from Parker (1965)	ΔH=	0.99	0.40	65DAV/BEN
330a	NaF(cr) = NaF(ai) Corr. from 0.068 mol/kg = -0.37 kJ/mol from Parker (1965)	ΔH=	0.80	0.15	79GER/PER
331	NaF(cr) = NaF(ai) Soly. = 0.983 mol/kg from Payne (1937); act. coef. = 0.574 from Hamer and Wu (1972)	ΔG=	2.837	0.050	84COD
332	NaCl(cr) = NaCl(ai) Extrap. from 0.126 mol/kg is -0.360 kJ/mol	ΔH=	3.887	0.020	20RAN/BIS
333	NaCl(cr) = NaCl(ai) Extrap. from 0.05 mol/kg is -0.297 kJ/mol	ΔH=	3.883	0.015	55BEN/BEN
334	NaCl(cr) = NaCl(ai) Extrap. from 0.222 mol/kg is -0.360 kJ/mol	ΔH=	3.866	0.080	69TSV/RAB
335	NaCl(cr) = NaCl(ai) Extrap. from 0.018 mol/kg is -0.213 kJ/mol	ΔH=	3.866	0.040	74STE/LOW
336	NaCl(cr) = NaCl(ai) Extrap. from 0.035 mol/kg is -0.268 kJ/mol	ΔH=	3.960	0.080	65BAZ/STA
337	NaCl(cr) = NaCl(ai) Extrap. from 0.052 mol/kg from Parker (1965)	ΔH=	3.76	0.15	77JOL/PER
338	NaCl(cr) = NaCl(ai) Meas. betw. 0.01-0.03 mol/kg. Debye-Hueckel extrap. yields ΔH = 3.800 ± 0.030 kJ/mol	ΔH=	3.828	0.040	71DAD/TAH
339	NaCl(cr) = NaCl(ai) Soly. = 6.150 mol/kg from Akerlof and Turck (1935) (6.162 mol/kg), Pearce and Nelson (1932) (6.138 mol/kg), Raridon et al. (1968) (6.160 mol/kg). Act. coef. (1.005) from Hidalgo and Orr (1968) and Hamer and Wu (1972).	ΔG=	-9.030	0.080	84COD

Table II-1. Measurements on Aqueous Thermochemistry at 298.15 K and 0.101325 MPa (cont'd).

NO.	REACTION OR SUBSTANCE	PROP. TYPE	OBSVD. VALUE kJ/mol or J/(mol·K)	± UNC.	REFERENCE
340	NaCl(cr) = NaCl(ai) Meas. from 0.01-0.5 mol/kg	$\Delta H=$	3.789	0.060	80TAN/PIE
341	NaCl(cr) = NaCl(ai)	$\Delta H=$	3.883	0.045	80VOR/PER
342	Na(cr) + AgCl(cr) = NaCl(ai) + Ag(cr) Includes corr. of -85.492 kJ/mol (0.88606 V) for sodium amalgam electrode used	$\Delta G=$	-283.340	0.040	40SMI/TAY
343	Na(cr) + AgCl(cr) = NaCl(ai) + Ag(cr) Includes corr. of -85.492 kJ/mol (0.88606 V) for sodium amalgam electrode used	$\Delta S=$	10.57	0.40	40SMI/TAY
344	Na(cr) + AgCl(cr) = NaCl(ai) + Ag(cr)	$\Delta G=$	-283.70	0.30	64LEB/ALE
345	NaBr(cr) = NaBr(ai) Extrap. from 0.03-0.06 mol/kg is -0.250 to -0.290 kJ/mol	$\Delta H=$	-0.602	0.040	49WAL
346	NaBr(cr) = NaBr(ai) Extrap. from 0.019 mol/kg is -0.200 kJ/mol	$\Delta H=$	-0.66	0.04	77JOL/PER
347	NaBr(cr) = NaBr(ai)	$\Delta H=$	-0.640	0.040	80VOR/PER
348	NaBr:2H$_2$O(cr) = NaBr(ai) + 2 H$_2$O(l) Review of literature	$\Delta H=$	18.63	0.40	65PAR
349	NaBr:2H$_2$O(cr) = NaBr(ai) + 2 H$_2$O(l) Soly. = 9.17 mol/kg from Pearce and Nelson (1932) (9.131 mol/kg), Makarov et al. (1966) (9.17 mol/kg), Linke (1958) (9.18 mol/kg). Act. coef. (2.00) and act. of H$_2$O (0.580) from Hamer and Wu (1972) and Makarov et al. (1966)	$\Delta G=$	-11.722	0.100	84COD
350	NaBr:2H$_2$O(cr) = NaBr(cr) + 2 H$_2$O(l) log P/atm = 7.4933-(2817)(K/T). See also Bell (1937)	$\Delta G=$	5.134	0.050	38DINb
351	NaI(cr) = NaI(ai) Meas. from 9.8 - 0.63 mol/kg. Diln. data from Parker (1965)	$\Delta H=$	-7.565	0.080	25WUS/LAN
352	NaI(cr) = NaI(ai) Meas. from 0.10-0.02 mol/kg. Diln. data from Parker (1965)	$\Delta H=$	-7.48	0.10	54MIS/SUK

Table II-1. Measurements on Aqueous Thermochemistry at 298.15 K and 0.101325 MPa (cont'd).

NO.	REACTION OR SUBSTANCE	PROP. TYPE	OBSVD. VALUE kJ/mol or J/(mol·K)	± UNC.	REFERENCE
353	NaI(cr) = NaI(ai) Extrap. from 0.111 mol/kg is -0.30 kJ/mol	$\Delta H=$	-7.66	0.10	69TSV/RAB
354	NaI(cr) = NaI(ai) Extrap. from 0.020 mol/kg is -0.205 kJ/mol	$\Delta H=$	-7.575	0.040	78PER/BYVb
355	NaI(cr) = NaI(ai) Details of meas. not given	$\Delta H=$	-7.57	0.20	72KRE/ZVE
356	NaI(cr) = NaI(ai) Extrap. from 0.018 mol/kg is -0.200 kJ/mol	$\Delta H=$	-7.531	0.080	73JOL/THO
357	NaI:2H$_2$O(cr) = NaI(ai) + 2 H$_2$O(l) Review of literature	$\Delta H=$	16.13	0.20	65PAR
358	NaI:2H$_2$O(cr) = NaI(cr) + 2 H$_2$O(l) Log P/atm = 7.3415-(2915.9)(K/T), meas. from 293-303 K	$\Delta H=$	23.60	1.00	38DINa
359	NaI:2H$_2$O(cr) = NaI(cr) + 2 H$_2$O(l) Log P/atm = 7.3415-(2915.9)(K/T), meas. from 293-303 K	$\Delta G=$	10.658	0.080	38DINa
360	NaI:2H$_2$O(cr) = NaI(ai) + 2 H$_2$O(l) Soly. = 12.30 mol/kg from Hill et al. (1933) (12.24 mol/kg), Samuseva et al. (1962) (12.22 mol/kg), and Makarov et al. (1966) (12.34 mol/kg). Act. coef. (6.24) and act. of H$_2$O (0.380) from Makarov et al. (1966) and Hamer and Wu(1972)	$\Delta G=$	-16.732	0.080	84COD
361	NaI(cr) = NaI(ai) Meas. from 0.05-0.01 mol/kg. Diln. corr. from Parker (1965).	$\Delta H=$	-7.536	0.060	80TAN/PIE
362	Na$_2$SO$_4$(cr) = Na$_2$SO$_4$(ai) Meas. from 0.001-0.023 mol/kg. Diln. corr. (-0.285 to -0.916 kJ/mol) from Wallace and Robinson (1941)	$\Delta H=$	-2.502	0.050	69GAR/JEK
363	Na$_2$SO$_4$(cr) = Na$_2$SO$_4$(ai) Corr. from 0.05 mol/kg is -1.021 kJ/mol	$\Delta H=$	-2.360	0.100	38PIT/COU
364	Na$_2$SO$_4$(cr) = Na$_2$SO$_4$(ai) Corr. from 303 K = +1.51 kJ/mol taken from Gardner et al. (1969). Diln. corr. = -1.02 kJ/mol	$\Delta H=$	-2.14	0.20	55COU

Table II-1. Measurements on Aqueous Thermochemistry at 298.15 K and 0.101325 MPa (cont'd).

NO.	REACTION OR SUBSTANCE	PROP. TYPE	OBSVD. VALUE kJ/mol or J/(mol·K)	± UNC.	REFERENCE
365	$Na_2SO_4(cr) = Na_2SO_4(ai)$	$\Delta H=$	-2.372	0.060	69REA/COB
366	$Na_2SO_4(cr) = Na_2SO_4(ai)$ Calc. from meas. on enthalpy of soln. and transit. for $Na_2SO_4(cr, III)$	$\Delta H=$	-2.330	0.080	72BRO/GIA
367	$Na_2SO_4(cr) = Na_2SO_4(ai)$ Soly. = 3.63 mol/kg from Rosenberg (1949) (3.63 mol/kg) and Kangro and Groeneveld (1962) (3.64 mol/kg). Act. coef. (0.137) from Robinson and Stokes (1959)	$\Delta G=$	1.758	0.080	84COD
368	$Na_2SO_4{:}10H_2O(cr) = Na_2SO_4(ai) + 10\ H_2O(l)$	$\Delta H=$	79.454	0.080	58BRO/GIA
369	$Na_2SO_4{:}10H_2O(cr) = Na_2SO_4(cr) + 10\ H_2O(g)$ From 293-305 K; log P/atm = 7.5502 - 2726.75(K/T)	$\Delta H=$	522.0	1.0	37HIG
370	$Na_2SO_4{:}10H_2O(cr) = Na_2SO_4(cr) + 10\ H_2O(g)$ From 293-305 K; log P/atm = 7.5502 - 2726.75(K/T)	$\Delta G=$	91.06	0.10	37HIG
371	$Na_2SO_4{:}10H_2O(cr) = Na_2SO_4(cr) + 10\ H_2O(g)$ From 297-305 K; log P/atm = 7.5366 - 2728.24(K/T)	$\Delta H=$	522.3	2.0	29MAT/OGU
372	$Na_2SO_4{:}10H_2O(cr) = Na_2SO_4(cr) + 10\ H_2O(g)$ From 297-305 K; log P/atm = 7.5366 - 2728.24(K/T)	$\Delta G=$	92.1	1.0	29MAT/OGU
373	$Na_2SO_4{:}10H_2O(cr) = Na_2SO_4(cr) + 10\ H_2O(g)$ Meas. dissoc. pressure from 290-305 K	$\Delta G=$	91.22	0.30	28PER/URR
374	$Na_2SO_4{:}10H_2O(cr) = Na_2SO_4(cr) + 10\ H_2O(g)$ Meas. by air sat. at 298.15 K	$\Delta G=$	91.19	0.40	20BAX/LAN
375	$Na_2SO_4{:}10H_2O(cr) = Na_2SO_4(cr) + 10\ H_2O(g)$	$\Delta G=$	91.16	0.05	21WIL
376	$NaNO_3(cr) = NaNO_3(ai)$ Extrap. from 0.12 mol/kg is -0.090 kJ/mol.	$\Delta H=$	20.470	0.080	37LAN/MAR
377	$NaNO_3(cr) = NaNO_3(ai)$ Extrap. from 0.05 mol/kg is -0.184 kJ/mol	$\Delta H=$	20.36	0.10	67MIS/SHP
378	$NaNO_3(cr) = NaNO_3(ai)$ No details given	$\Delta H=$	20.29	0.10	67KRE/EGO

Table II-1. Measurements on Aqueous Thermochemistry at 298.15 K and 0.101325 MPa (cont'd).

NO.	REACTION OR SUBSTANCE	PROP. TYPE	OBSVD. VALUE kJ/mol or J/(mol·K)	± UNC.	REFERENCE
379	$NaNO_3(cr) = NaNO_3(ai)$ Soly. = 10.83 mol/kg from Akerlof and Turck (1935) (10.825 mol/kg), Pearce and Hopson (1937) (10.83 mol/kg), Kangro and Groeneveld (1962) (10.87 mol/kg). Act. coef. (0.323) from Wu and Hamer (1980) (0.3227); Kangro and Groeneveld (1962) (0.323)	$\Delta G=$	-6.208	0.080	84COD
380	$KF(cr)$ Measurements from (16-323 K). S(16 K) = 0.16 J/(mol·K)	$S=$	66.57	0.20	49WES/PIT
381	$KCl(cr)$ From meas. of Berg and Morrison (1957) (2.8-270 K). See also Southard and Nelson (1933) (17-288 K), Kirkham and Yates (1968) (20-300 K), Clusius et al. (1949) (10-268 K)	$S=$	82.57	0.20	84COD
382	$KClO_3(cr)$ Meas. from 14-294 K. S(14 K) = 0.71 J/(mol·K)	$S=$	142.97	0.20	34LAT/SCH
383	$KBrO_3(cr)$ Meas. by Ahlberg and Latimer (1934) (14-297 K). S(14 K) = 0.96 J/(mol·K)	$S=$	149.20	0.80	84COD
384	$KBr(cr)$ Meas. of Berg and Morrison (1957) (2-270 K). See also Clusius et al. (1949) (10-270 K)	$S=$	95.90	0.20	84COD
385	$KI(cr)$ Meas. of Berg and Morrison (1957) (2-270 K). See also Clusius et al.(1949) (10-270 K)	$S=$	106.06	0.20	84COD
386	$K_2SO_4(cr)$ From meas. by Paukov (1969) (12-300 K). S(12 K) = 0.30 J/(mol·K). See also Moore and Kelley (1942) (53-295 K)	$S=$	175.56	0.30	84COD
387	$KNO_3(cr)$ From meas. by Southard and Nelson (1933) (15-293 K). S(16 K) = 1.00 J/(mol·K)	$S=$	133.00	0.40	84COD
388	$K(cr) + H_2O(l) = KOH(ai) + 0.5 H_2(g)$ Diln. corr. from 0.0555 mol/kg is -0.372 kJ/mol. See Gunn and Green (1958)	$\Delta H=$	-196.472	0.040	67GUN

Table II-1. Measurements on Aqueous Thermochemistry at 298.15 K and 0.101325 MPa (cont'd).

NO.	REACTION OR SUBSTANCE	PROP. TYPE	OBSVD. VALUE kJ/mol or J/(mol·K)	± UNC.	REFERENCE
389	$K(cr) + H_2O(l) = KOH(ai) + 0.5 H_2(g)$	$\Delta H=$	-196.20	0.35	51KET/WAL
390	$K(cr) + H_2O(l) = KOH(ai) + 0.5 H_2(g)$ Includes corr. of +1.09 kJ/mol for Na impurity in sample and -0.75 kJ/mol for diln. Meas. made in bomb at const vol.	$\Delta H=$	-196.65	0.40	55MES/FAS
391	$K(cr) + H^+(ao) = K^+(ao) + 0.5 H_2(g)$ Corr. for sat. calomel electrode = +23.28 kJ/mol; corr. for K(amalgam) = -101.28 kJ/mol from Lewis and Keyes (1912)	$\Delta G=$	-282.24	0.40	30SHI/ODA
392	$K(cr) + AgCl(cr) = Ag(cr) + KCl(ai)$ Corr. for K(amalg) = -101.12 kJ/mol from Lewis and Keyes (1912)	$\Delta G=$	-303.73	0.80	58BRA/STR
393	$K(cr) + H^+(ao) = K^+(ao) + 0.5 H_2(g)$ Meas. K(amalg) vs. Ag|AgCl|Cl⁻(ao) electrode; no details given	$\Delta G=$	-282.02	0.10	68DIL/ITZ
394	$K(cr) + AgCl(cr) = KCl(ai) + Ag(cr)$ Corr. for K(amalg) = +106.293 kJ/mol (calc from Lewis and Keyes (1912)). Corr. for KCl (1.0168 mol/kg) = +2.425 kJ/mol	$\Delta G=$	-303.689	0.050	34ARM/CRE
395	$K(cr) + H^+(ao) = K^+(ao) + 0.5 \dot{H}_2(g)$ Meas. against Hg|Hg$_2$Cl$_2$|Cl⁻(ao) electrode. Corr. for K(amalg) = 106.293 kJ/mol (calc. from Lewis and Keyes (1912)). Corr. for calomel elect. = +25.873 ± 0.030 kJ/mol	$\Delta G=$	-282.239	0.050	34ARM/CRE
396	$K(cr) + H^+(ao) = K^+(ao) + 0.5 H_2(g)$ Meas. against normal calomel electrode, corr. = +27.257 ± 0.050 kJ/mol. Corr. for K(amalg) = +101.155 kJ/mol	$\Delta G=$	-282.11	0.10	12LEW/KEY
397	$K(cr) + AgCl(cr) = KCl(ai) + Ag(cr)$ Corr. for K(amalg) = +104.16 ± 0.10 kJ/mol (amalg. concentration not accurately specified)	$\Delta G=$	-303.95	0.30	64LEB/ALE
398	$KF(cr) = KF(ai)$ Extrap. meas. of enthalpy of soln. from 0.004-0.20 mol/kg by Debye-Hueckel eqn.	$\Delta H=$	-17.757	0.080	27LAN/EIC
399	$KF(cr) = KF(ai)$ Extrap. from 0.111 mol/kg is -0.473 kJ/mol from Parker (1965)	$\Delta H=$	-17.627	0.090	69TSV/RAB

Table II-1. Measurements on Aqueous Thermochemistry at 298.15 K and 0.101325 MPa (cont'd).

NO.	REACTION OR SUBSTANCE	PROP. TYPE	OBSVD. VALUE kJ/mol or	± UNC. J/(mol·K)	REFERENCE
399a	KF(cr) = KF(ai) Corr. from 0.048 mol/kg = -0.36 kJ/mol from Parker (1965)	$\Delta H=$	-17.57	0.20	79GER/PER
400	KF:2H$_2$O(cr) = KF(cr) + 2 H$_2$O(l) From difference in enthalpies of solution	$\Delta H=$	24.736	0.080	27LAN/EIC
401	KF:2H$_2$O(cr) = KF(ai) + 2 H$_2$O(l) From soly. = 17.50 mol/kg, act. coef. = 3.80, act. of H$_2$O = 0.292 from Tamas and Kosza (1964), Hamer and Wu (1972)	$\Delta G=$	-14.711	0.080	84COD
402	KF:2H$_2$O(cr) = KF(cr) + 2 H$_2$O(g) Meas. decomp. pres. from 298-308 K; log P/atm = 7.85 - (3037)(K/T)	$\Delta G=$	26.68	0.10	40BEL
403	KF:2H$_2$O(cr) = KF(cr) + 2 H$_2$O(g) Meas. decomp. pres. from 273-303 K; log P/atm = 7.597 - (2957.9)(K/T)	$\Delta G=$	26.530	0.080	27LAN/EIC
404	KF:2H$_2$O(cr) = KF(cr) + 2 H$_2$O(g) Meas. decomp. pres. from 273-303 K; log P/atm = 7.597 - (2957.9)(K/T)	$\Delta H=$	113.26	0.40	27LAN/EIC
405	KCl(cr) = KCl(ai) Corr. from 0.11 mol/kg is -0.343 kJ/mol	$\Delta H=$	17.241	0.018	80KIL
406	KCl(cr) = KCl(ai)	$\Delta H=$	17.234	0.040	79JOH/GAY
407	KCl(cr) = KCl(ai) Corr. for diln. from 0.11 mol/kg is -0.343 kJ/mol from Parker (1965)	$\Delta H=$	17.23	0.10	68SAM/TSV
408	KCl(cr) = KCl(ai) Meas. at 0.03 and 0.15 mol/kg. Diln. corr. from Parker (1965)	$\Delta H=$	17.175	0.040	77PER/BYV
409	KCl(cr) = KCl(ai) Extrap. from 0.0347 mol/kg is -0.260 kJ/mol	$\Delta H=$	17.242	0.010	67VAS/KOC
410	KCl(cr) = KCl(ai) Soly. = 4.82 mol/kg from Akerlof and Turck (1935) (4.826 mol/kg), Ratner and Makarov (1958) (4.83 mol/kg), Chiang and Hsieh (1949) (4.80 mol/kg), Pearce and Nelson (1932) (4.81 mol/kg), Sherman and Menzies (1937) (4.80	$\Delta G=$	-5.173	0.040	84COD

Table II-1. Measurements on Aqueous Thermochemistry at 298.15 K and 0.101325 MPa (cont'd).

NO.	REACTION OR SUBSTANCE	PROP. TYPE	OBSVD. VALUE kJ/mol or J/(mol·K)	± UNC.	REFERENCE
	mol/kg), Durham et al. (1953) (4.816 mol/kg). Act. coef. (0.589) from Hamer and Wu (1972)				
411	$KClO_3(cr) = KCl(cr) + 1.5\ O_2(g)$ Decomposition by combust. of benzoic acid in bomb	$\Delta H=$	-38.32	0.40	65GIL/WAG
412	$KClO_3(cr) = KCl(cr) + 1.5\ O_2(g)$ Decomposition by combust. of graphite in bomb. Only 2 meas.	$\Delta H=$	-39.20	0.60	70JOH/SMI
413	$KClO_3(cr) = KClO_3(ai)$	$\Delta H=$	41.28	0.04	75MED/EFI
414	$KClO_3(cr) = KClO_3(ai)$	$\Delta H=$	41.38	0.24	60NEL/MOS
415	$KClO_3(cr) = KClO_3(ai)$ Soly. = 0.700 mol/kg from Jones and Froning (1944). Act. coef. = 0.519 from Hamer and Wu (1972)	$\Delta G=$	5.020	0.080	84COD
416	$KBr(cr) = KBr(ai)$ Corr. from 0.111 mol/kg is -0.280 kJ/mol	$\Delta H=$	19.80	0.10	69TSV/RAB
417	$KBr(cr) = KBr(ai)$ Meas. at 0.066 and 0.13 mol/kg. Diln. from Parker (1965)	$\Delta H=$	19.771	0.040	73WAG/KIL
418	$KBr(cr) = KBr(ai)$ Measurements made from 0.1 to 0.008 mol/kg	$\Delta H=$	19.78	0.04	79EFI/KLE
419	$KBr(cr) = KBr(ai)$ Soly. = 5.75 mol/kg from Durham et al. (1953) (5.75 mol/kg), Makarov and Popov (1959) (5.748 mol/kg). Act. coef. (0.642) calc from Hamer and Wu (1972)	$\Delta G=$	-6.475	0.050	84COD
420	$KBr(cr) = KBr(ai)$ Corr. from 0.02 mol/kg is -0.195 kJ/mol from Parker (1965)	$\Delta H=$	19.815	0.020	78PER/BYVa
421	$KBr(cr) = KBr(ai)$ Extrap. from 0.017 mol/kg from Parker (1965)	$\Delta H=$	19.60	0.10	77JOL/PER
422	$KBr(cr) = KBr(ai)$ From EMF meas. vs. calomel electrode	$\Delta G=$	-6.489	0.080	33MUR

Table II-1. Measurements on Aqueous Thermochemistry at 298.15 K and 0.101325 MPa (cont'd).

NO.	REACTION OR SUBSTANCE	PROP. TYPE	OBSVD. VALUE kJ/mol or J/(mol·K)	± UNC.	REFERENCE
423	$KBrO_3(cr) = KBr(cr) + 1.5\ O_2(g)$ Decomp. by combust. of graphite in bomb	$\Delta H=$	-33.89	0.25	70JOH/SMI
424	$KBrO_3(cr) = KBrO_3(ai)$ Soly. = 0.488 mol/kg from Ricci (1934), act. coef. = 0.556 from Hamer and Wu (1972)	$\Delta G=$	6.467	0.050	84COD
425	$KBrO_3(cr) + 9\ I^-(ao) + 6\ H^+(ao) = 3\ I_3^-(ao)$ $+ KBr(ai) + 3\ H_2O(l)$ Diln. corr. = -3.0 kJ/mol	$\Delta H=$	-511.5	3.3	53MEL/JOL
426	$KBrO_3(cr) + 6\ HI(ai) = KBr(cr) + 3\ I_2(cr)$ $+ 3\ H_2O(l)$	$\Delta H=$	-550.67	0.40	70JOH/SMI
427	$KBrO_3(cr) = KBrO_3(ai)$ Diln. corr. from 0.0307 mol/kg is -0.083 kJ/mol	$\Delta H=$	40.77	0.12	70JOH/SMI
428	$KBrO_3(cr) = KBrO_3(ai)$ Extrap. from 0.005 mol/kg by Debye-Hueckel eq.	$\Delta H=$	40.832	0.080	75MED/EFI
429	$KBrO_3(cr) = KBrO_3(ai)$ Diln. corr. from 0.012 mol/kg is -0.08 kJ/mol	$\Delta H=$	40.84	0.20	53MEL/JOL
430	$KBrO_3(cr) + 8\ Br^-(ao) + 6\ H^+(ao) = 3\ Br_3^-(ao)$ $+ K^+(ao) + 3\ H_2O(l)$	$\Delta H=$	-176.43	1.20	53MEL/JOL
431	$KI(cr) = KI(ai)$ Corr. for diln. from 0.020 mol/kg is -0.20 kJ/mol	$\Delta H=$	20.25	0.10	78PER/BYVa
432	$KI(cr) = KI(ai)$ Corr. for diln. from 0.111 mol/kg is -0.238 kJ/mol	$\Delta H=$	20.18	0.10	69TSV/RAB
433	$KI(cr) = KI(ai)$	$\Delta H=$	20.145	0.040	79EFI/KLE
434	$KI(cr) = KI(ai)$ Extrap. from 0.018 mol/kg from Parker (1965)	$\Delta H=$	20.602	0.050	77JOL/PER
435	$KI(cr) = KI(ai)$ Review of literature	$\Delta H=$	20.33	0.20	65PAR
436	$KI(cr) = KI(ai)$ Soly. = 8.96 mol/kg from Pearce et al. (1928) (8.938 mol/kg),Makarov et al. (1958) (8.98 mol/kg). Act. coef. = 0.83 from Makarov et al., loc. cit.	$\Delta G=$	-9.948	0.050	84COD

Table II-1. Measurements on Aqueous Thermochemistry at 298.15 K and 0.101325 MPa (cont'd).

NO.	REACTION OR SUBSTANCE	PROP. TYPE	OBSVD. VALUE	± UNC.	REFERENCE
			kJ/mol or J/(mol·K)		
437	KI(cr) = KI(ai) Diln. corr. from 0.111 mol/kg is -0.238 kJ/mol from Parker (1965)	$\Delta H=$	20.305	0.100	68SAM/TSV
438	KI(cr) = KI(ai)	$\Delta H=$	20.50	0.20	72KRE/ZVE
439	K_2SO_4(cr) = K_2SO_4(ai) Diln. from 0.111 mol/kg is -1.005 kJ/mol from Lange and Streeck (1931)	$\Delta H=$	23.810	0.080	69TSV/RAB
440	K_2SO_4(cr) = K_2SO_4(ai) Diln. from 0.07 mol/kg is -1.050 kJ/mol from Lange and Streeck (1931)	$\Delta H=$	23.75	0.10	69ROM/SAM
441	K_2SO_4(cr) = K_2SO_4(ai) Diln. from 0.138 mol/kg is -0.950 kJ/mol from Lange and Streeck (1931)	$\Delta H=$	23.750	0.080	36MIS/PRO
442	K_2SO_4(cr) = K_2SO_4(ai) Soly. = 0.692 mol/kg from Ishii and Fujita (1978) (0.696 mol/kg), Calvo and Simons (1952) (0.694 mol/kg), Akerlof and Turck (1935) (0.691 mol/kg), Potter and Clynne (1978) (0.683 mol/kg). Act. coef. (0.229) from Robinson and Stokes (1959)	$\Delta G=$	10.264	0.080	84COD
443	KNO_3(cr) = KNO_3(ai) Corr. from 0.0139 mol/kg is -0.105 kJ/mol	$\Delta H=$	34.87	0.40	62MON
444	KNO_3(cr) = KNO_3(ai) Corr. from 0.111 mol/kg is 0.200 kJ/mol	$\Delta H=$	34.928	0.080	69TSV/RAB
445	KNO_3(cr) = KNO_3(ai) Meas. betw. 0.092-0.200 mol/kg	$\Delta H=$	35.040	0.080	67MIS/SHP
446	KNO_3(cr) = KNO_3(ai) Review of literature	$\Delta H=$	34.89	0.12	65PAR
447	KNO_3(cr) = KNO_3(ai) Soly. = 3.75 mol/kg from Akerlof and Turck (1935) (3.774 mol/kg); Chiang and Hsieh (1949) (3.737 mol/kg). Act. coef. (0.237) from Hamer and Wu (1972)	$\Delta G=$	0.585	0.060	84COD
448	RbBr(cr) From meas. by Clusius et al. (1949) (11-273 K). S(11 K) = 0.92 J/(mol·K)	$S=$	109.95	0.80	84COD

Table II-1. Measurements on Aqueous Thermochemistry at 298.15 K and 0.101325 MPa (cont'd).

NO.	REACTION OR SUBSTANCE	PROP. TYPE	OBSVD. VALUE kJ/mol or J/(mol·K)	± UNC.	REFERENCE

449   RbI(cr)
       From meas. by Clusius et al. (1949) (11-273 K).
       S(11 K) = 1.77 J/(mol·K)
      $S=$ 118.95   0.80   84COD

450   Rb_2SO_4(cr)
       From meas. by Paukov and Lavrent'eva (1968)
       (12-303 K). S(12 K) = 0.82 J/(mol·K)
      $S=$ 197.44   0.20   84COD

451    $RbNO_3$(cr)
       From meas. by Berezovskii and Paukov (1982)
       (8-319 K). S(8 K) = 0.14 J/(mol·K)
      $S=$ 144.90   0.20   84COD

452   Rb(cr) + H_2O(l) = RbOH(ai) + 0.5 H_2(g)
       Extrap. from 0.0185 mol/kg is -0.22 kJ/mol.
       Corr. for at. wt. change is negligible
      $\Delta H=$ -195.25   0.10   67GUN

453   Rb(cr) + H_2O(l) = RbOH(ai) + 0.5 H_2(g)
       Meas. in a sealed bomb and corr. to 1 atm
       and const. P
      $\Delta H=$ -195.06   0.40   72VOR/MON

454   RbCl(cr) = RbCl(ai)
       Corr. from 0.037 mol/kg is -0.238 kJ/mol from
       Parker (1965)
      $\Delta H=$ 16.987   0.080   78MON/MEL

455   RbCl(cr) = RbCl(ai)
       Corr. from 0.111 mol/kg is -0.29 kJ/mol from
       Parker (1965)
      $\Delta H=$ 17.03   0.20   69TSV/RAB

456   RbCl(cr) = RbCl(ai)
       Meas. from 0.01-0.03 mol/kg extrap. by Debye-
       Hueckel eq.
      $\Delta H=$ 16.99   0.20   77DAD/TAH

457   RbCl(cr) = RbCl(ai)
       Meas. betw. 0.13-0.03 mol/kg
      $\Delta H=$ 16.917   0.080   82WEI/APE

458   RbCl(cr) = RbCl(ai)
       Corr. from 0.032 mol/kg is -0.230 kJ/mol from
       Fortier et al. (1974)
      $\Delta H=$ 17.004   0.040   79JOH/GAY

459   RbCl(cr) = RbCl(ai)
       Corr. from 0.02 mol/kg is -0.203 kJ/mol from
       Fortier et al. (1974)
      $\Delta H=$ 17.035   0.040   78PER/BYVb

460   RbCl(cr) = RbCl(ai)
       Soly. (7.79 mol/kg) from Ratner and Makarov
       (1958) (7.78 mol/kg), Belyaev and Le Tuyk (1966)
       (7.79 mol/kg), Clynne and Potter (1979) (7.781
      $\Delta G=$ -7.391   0.050   84COD

Table II-1. Measurements on Aqueous Thermochemistry at 298.15 K and 0.101325 MPa (cont'd).

NO.	REACTION OR SUBSTANCE	PROP. TYPE	OBSVD. VALUE kJ/mol or J/(mol·K)	± UNC.	REFERENCE
	mol/kg), Durham et al. (1953) (7.82 mol/kg). Act. coef. (0.570) from Hamer and Wu (1972)				
461	RbCl(cr) = RbCl(ai) Diln. corr. = -0.21 kJ/mol from Parker (1965)	$\Delta H=$	16.74	0.10	66VOR/IBR
462	RbBr(cr) = RbBr(ai) Extrap. from 0.111 mol/kg is -0.19 kJ/mol from Fortier et al. (1974)	$\Delta H=$	21.84	0.40	37LAN/MAR
463	RbBr(cr) = RbBr(ai) Extrap. from 0.01 mol/kg is -0.050 kJ/mol	$\Delta H=$	22.184	0.030	82WEI/APE
464	RbBr(cr) = RbBr(ai) Soly. = 6.75 mol/kg from Durham et al. (1953), Makarov and Popov (1959) (6.737 mol/kg). Act. coef. (0.518) from Hamer and Wu (1972). See also Makarov and Popov (1959)	$\Delta G=$	-6.206	0.080	84COD
465	RbI(cr) = RbI(ai) Diln. corr. from 0.02 mol/kg is -0.178 kJ/mol	$\Delta H=$	25.600	0.040	78PER/BYV2
466	RbI(cr) = RbI(ai) Meas. betw. 0.068-0.041 mol/kg	$\Delta H=$	25.597	0.040	82WEI/APE
467	RbI(cr) = RbI(ai)	$\Delta H=$	25.44	0.20	72KRE/ZVE
468	RbI(cr) = RbI(ai) Soly. = 7.69 mol/kg based on Fajans and Karagunis (1931) (7.70 mol/kg), Briggs et al. (1941) (7.68 mol/kg), Vlasov and Seleznev (1971) (7.63 mol/kg). Act. coef. = 0.534 calc. from Hamer and Wu (1972)	$\Delta G=$	-6.98	0.10	84COD
469	Rb_2SO_4(cr) = Rb_2SO_4(ai) Diln. corr. from -0.75 to -0.88 kJ/mol from Lange and Streeck (1931)	$\Delta H=$	24.05	0.30	66VOR/IBR
470	Rb_2SO_4(cr) = Rb_2SO_4(ai) Meas. from 1.7-0.005 mol/kg	$\Delta H=$	24.42	0.80	75KUZ/LOV
471	Rb_2SO_4(cr) = Rb_2SO_4(ai) Soly. = 1.92 mol/kg from Kuznetsova et al. (1975) (1.91 mol/kg), Lovetskaya et al. (1977) (1.91 mol/kg), and Iskhakova et al. (1979) (1.952 mol/kg). Act. coef. (0.182) extrap. from Robinson and Stokes (1959)	$\Delta G=$	4.38	0.10	84COD

Table II-1. Measurements on Aqueous Thermochemistry at 298.15 K and 0.101325 MPa (cont'd).

NO.	REACTION OR SUBSTANCE	PROP. TYPE	OBSVD. VALUE kJ/mol or J/(mol·K)	± UNC.	REFERENCE
472	$RbNO_3(cr)$ = $RbNO_3(ai)$ Soly. = 5.48 mol/kg from Linke (1958). Act. coef. (0.178) extrap. from Hamer and Wu (1972)	$\Delta G=$	0.12	0.40	84COD
473	$RbNO_3(cr)$ = $RbNO_3(ai)$ Soly. = 4.54 mol/kg from Berkeley (1904) (4.51 mol/kg), Shklovskaya et al. (1976) (4.56 mol/kg),Lovetskaya et al. (1977) (4.52 mol/kg). Act. coef. (0.200) from Hamer and Wu (1972)	$\Delta G=$	0.48	0.05	84COD
474	$CsCl(cr)$ Meas. from 7-298 K. $S(7 K)$ = 0.13 J/(mol·K).	$S=$	101.17	0.20	63TAY/GAR
475	$CsBr(cr)$ Meas. from 12-307 K. $S(12 K)$ = 1.13 J/(mol·K).	$S=$	113.14	0.40	69PAU/LAVb
476	$CsBr(cr)$ Meas. from 3-281 K	$S=$	112.09	0.50	68SOR/SUG
477	$CsBr(cr)$ Meas. from 25-300 K. $S(25 K)$ = 8.24 J/(mol·K)	$S=$	111.88	0.80	68 KIR/YAT
478	$CsI(cr)$ Meas. from 2-281 K	$S=$	121.87	0.40	68SOR/SUG
479	$CsI(cr)$ Meas. from 13-298 K. $S(13 K)$ = 2.26 J/(mol·K)	$S=$	123.05	0.80	63TAY/GAR
480	$CsNO_3(cr)$ Meas. by Flotow et al. (1951) (5-350 K); Frolova and Paukov (1983) (6-302 K). $S(5 K)$ = 0.084 J/(mol·K)	$S=$	153.83	0.30	84COD
481	$Cs(cr)$ + $H_2O(l)$ = $CsOH(ai)$ + 0.5 $H_2(g)$ Corr. from 0.037 mol/kg is -0.34 kJ/mol. Value based on wt. Cs used (corr. for impur.)	$\Delta H=$	-202.23	0.15	72VOR/MON
482	$Cs(cr)$ + $H_2O(l)$ = $CsOH(ai)$ + 0.5 $H_2(g)$ Corr. from 0.0185 mol/kg is -0.250 kJ/mol. See also Vorob'ev et al. (1966) and Friedman and Kahlweit (1956)	$\Delta H=$	-202.325	0.050	67GUN
483	$Cs(cr)$ + $H_2O(l)$ = $CsOH(ai)$ + 0.5 $H_2(g)$ Corr. from 0.055 mol/kg is -0.4 kJ/mol. LISTED FOR INFORMATION ONLY.	$\Delta H=$	-194.6	1.0	66VOR/IBR

Table II-1. Measurements on Aqueous Thermochemistry at 298.15 K and 0.101325 MPa (cont'd).

NO.	REACTION OR SUBSTANCE	PROP. TYPE	OBSVD. VALUE kJ/mol or J/(mol·K)	± UNC.	REFERENCE
484	$Cs(cr) + H^+(ao) = Cs^+(ao) + 0.5 H_2(g)$ Corr. for Cs(amalg.) = -130.5 kJ/mol. LISTED FOR INFORMATION ONLY.	$\Delta H=$	-261.9	1.5	56FRI/KAH
485	$Cs(cr) + AgCl(cr) = CsCl(ai) + Ag(cr)$ Corr. for Cs(amalg) = -114.479 kJ/mol; corr. for CsCl(0.1 mol/kg) = +1.2935 kJ/mol	$\Delta G=$	-312.438	0.080	72MUS/LON
486	$Cs(cr) + AgCl(cr) = CsCl(ai) + Ag(cr)$ Corr. for Cs(amalg) = -118.58 kJ/mol; extrap. to CsCl(ai) by Debye-Hueckel eq.	$\Delta G=$	-312.04	0.50	64LEB/ALE
487	$Cs(cr) + H^+(ao) = Cs^+(ao) + 0.5 H_2(g)$ Meas. E/V = 2.1098 at 298.15 K for the cell Ca(Hg)|CsOH(0.02109 mol/kg)|HgO|Hg. Corr. for Cs(Hg) = -120.19 kJ/mol; corr. for Cs^+(0.0211 mol/kg) = +19.78 kJ/mol, corr. for Hg|HgO(cr) electrode is +9.40 kJ/mol	$\Delta G=$	-294.26	0.40	39BEN/FOR
488	$CsCl(cr) = CsCl(ai)$ Diln. corr. from 0.069 mol/kg is -0.235 kJ/mol	$\Delta H=$	17.200	0.030	78MON/MEL
489	$CsCl(cr) = CsCl(ai)$ Meas. betw. 0.015-0.03 mol/kg. Measurement at 0.01 mol/kg in error (misprint?). Extrap. by Debye-Hueckel eq. yields 17.413 kJ/mol	$\Delta H=$	17.426	0.080	77DAD/TAH
490	$CsCl(cr) = CsCl(ai)$ Corr. from 0.017 mol/kg is -0.17 kJ/mol	$\Delta H=$	17.41	0.20	66VOR/IBR
491	$CsCl(cr) = CsCl(ai)$ Corr. from 0.020 mol/kg is -0.18 kJ/mol	$\Delta H=$	17.35	0.20	78PER/BYVc
492	$CsCl(cr) = CsCl(ai)$ Corr. from 0.092 mol/kg is -0.235 kJ/mol	$\Delta H=$	17.170	0.040	79JOH/GAY
493	$CsCl(cr) = CsCl(ai)$ Soly. = 11.34 mol/kg from Clynne and Potter (1979). Act. coef. (0.513) from Hamer and Wu (1972). See also Makarov et al. (1958) (act. coef. = 0.516)	$\Delta G=$	-8.733	0.050	84COD
494	$CsBr(cr) = CsBr(ai)$ Corr. from 0.062 mol/kg is -0.160 kJ/mol	$\Delta H=$	25.986	0.050	78MON/MEL

Table II-1. Measurements on Aqueous Thermochemistry at 298.15 K and 0.101325 MPa (cont'd).

NO.	REACTION OR SUBSTANCE	PROP. TYPE	OBSVD. VALUE kJ/mol or J/(mol·K)	± UNC.	REFERENCE
495	CsBr(cr) = CsBr(ai) Diln. from 0.020 mol/kg is -0.17 kJ/mol	$\Delta H=$	26.02	0.15	78PER/BYVc
496	CsBr(cr) = CsBr(ai)	$\Delta H=$	26.18	0.10	73DEV/SOM
497	CsBr(cr) = CsBr(ai) Corr. from 0.021 mol/kg is -0.17kJ/mol	$\Delta H=$	26.19	0.40	66VOR/IBR
498	CsBr(cr) = CsBr(ai) Soly. = 5.740 mol/kg and act. coef. (0.421) from Vlasov and Seleznev (1972). Act. coef. values adjusted to Hamer-Wu scale at 1.5 mol/kg	$\Delta G=$	-4.279	0.080	84COD
499	CsBr(cr) = CsBr(ai) Soly. = 5.80 mol/kg from Samuseva et al. (1962b) (5.796 mol/kg), Fajans and Karagunis (1931) (5.80 mol/kg). Act. coef. (0.451) extrap. from Hamer and Wu (1972)	$\Delta G=$	-4.767	0.080	84COD
500	CsI(cr) = CsI(ai) Meas. from 0.021-0.044 mol/kg. Diln. corr. from Parker (1965)	$\Delta H=$	33.28	0.20	65VOR/IBR
501	CsI(cr) = CsI(ai)	$\Delta H=$	33.141	0.060	64MIT/COB
502	CsI(cr) = CsI(ai) Diln. corr. from 0.111 mol/kg is -0.12 kJ/mol	$\Delta H=$	33.10	0.20	69TSV/RAB
503	CsI(cr) = CsI(ai)	$\Delta H=$	33.070	0.060	78MON/MEL
504	CsI(cr) = CsI(ai)	$\Delta H=$	33.01	0.20	72KRE/ZVE
505	CsI(cr) = CsI(ai)	$\Delta H=$	33.18	0.10	78PER/BYVc
506	CsI(cr) = CsI(ai) Soly. = 3.32 mol/kg from Briggs et al. (1930) (3.29 mol/kg), Fajans and Karagunis (1931) (3.40 mol/kg), Samuseva et al. (1962) (3.292 mol/kg), Vlasov and Seleznev (1972) (3.354 mol/kg), Stepin et al. (1976)(3.28 mol/kg). Act. coef. (0.420) from Hamer and Wu (1972) (0.424) and Vlasov and Seleznev (1972) (0.410).	$\Delta G=$	-1.65	0.15	84COD
507	$Cs_2SO_4(cr)$ Meas. from 12 - 308 K. S(12 K) = 1.30 J(mol·K)	$S=$	211.92	0.40	68PAU/KHR

Table II-1. Measurements on Aqueous Thermochemistry at 298.15 K and 0.101325 MPa (cont'd).

NO.	REACTION OR SUBSTANCE	PROP. TYPE	OBSVD. VALUE kJ/mol or J/(mol·K)	± UNC.	REFERENCE
508	$Cs_2SO_4(cr) = Cs_2SO_4(ai)$ Meas from 0.011-0.031 mol/kg, diln. from Lange and Streeck (1931)	$\Delta H=$	17.17	0.30	66VOR/IBR
509	$Cs_2SO_4(cr) = Cs_2SO_4(ai)$ Diln. corr. from 0.111 mol/kg is -0.54 kJ/mol	$\Delta H=$	17.04	0.40	69TSV/RAB
510	$Cs_2SO_4(cr) = Cs_2SO_4(ai)$ Corr. from 0.007 mol/kg = -0.55 kJ/mol	$\Delta H=$	16.98	0.20	74KUZ/LOV
511	$CsNO_3(cr) = CsNO_3(ai)$ Meas. to 0.07 mol/kg	$\Delta H=$	40.25	0.80	64NOS/SAM
512	$CsNO_3(cr) = CsNO_3(ai)$	$\Delta H=$	40.330	0.040	75OHA/BOE
513	$CsNO_3(cr) = CsNO_3(ai)$ No details of meas. given but apparently meas. from 0.05-1 mol/kg	$\Delta H=$	40.37	0.20	75LOV/KUZ
514	$HCl(ai) = H^+(ao) + Cl^-(ao)$ CONSTRAINT - SOLVED EXACTLY.	$\Delta H=$	0.0	0.0	DEF
515	$HCl(ai) = H^+(ao) + Cl^-(ao)$ CONSTRAINT - SOLVED EXACTLY.	$\Delta G=$	0.0	0.0	DEF
516	$HCl(ai) = H^+(ao) + Cl^-(ao)$ CONSTRAINT - SOLVED EXACTLY.	$\Delta S=$	0.0	0.0	DEF
517-666	THESE REACTIONS ARE DEFINITIONS OF AQUEOUS ELECTROLYTES AT THE STANDARD STATE, IN TERMS OF OF THEIR IONS. ALL ARE SIMILAR TO REACTIONS 514-516.				
667	$0.5 H_2(g) + 0.5 O_2(g) = OH^-(ao)$ CONSTRAINT - SOLVED EXACTLY.	$HGS=$	0.0	0.0	
668	$0.5 N_2(g) + 0.5 Cl_2(g) + 2 H_2(g) + 2 O_2(g)$ $= NH_4ClO_4(cr)$ CONSTRAINT - SOLVED EXACTLY.	$HGS=$	0.0	0.0	
667-745	REACTIONS 667-745 WITH PROPERTY TYPE "HGS" REPRESENT THE CONSTRAINT $\Delta G - \Delta H + T*\Delta S = 0$ FOR THE FORMATION OF INDIVIDUAL SUBSTANCES				

Table II-2.   Index to the Catalog of Measurements.

SUBSTANCE	REACTIONS
$OH^-(ao)$	6 8 10 11 12 14 16 17 18 19 20 21 116 117 118 190 See also LiOH(ai), NaOH(ai), KOH(ai), RbOH(ai), CsOH(ai)
$H_2O_2(l)$ $H_2O_2(20H_2O)$	22 23 24 87 23
$F^-(ao)$	See LiF(ai), NaF(ai), KF(ai)
$Cl^-(ao)$	72 73 74 167 168 173 190 197 198 199 200 200a 201 209 256 See also HCl(ai), NH_4Cl(ai), LiCl(ai), NaCl(ai), KCl(ai), RbCl(ai), CsCl(ai)
$ClO_3^-(ao)$	See $KClO_3$(ai)
$ClO_4^-(ao)$	See $HClO_4$(ai), NH_4ClO_4(ai), $AgClO_4$(ai), $LiClO_4$(ai), $NaClO_4$(ai), $KClO_4$(ai), $CsClO_4$(ai)
HCl(ai)	1 2 3 4 7 9 13 15 25 26 27 28 29 30 31 32 33 139 151 152 153 154 155 156 157 158 159 160 161 162 191 192 205 206 221 318
$HClO_4$(ai)	5
$Br^-(ao)$	72 73 74 84 85 88 172 173 209 218 219 220 256 430 See also HBr(ai), NH4Br(ai), LiBr(ai), NaBr(ai), KBr(ai), RbBr(ai), CsBr(ai)
$Br_2(ao)$ $Br_3^-(ao)$ $BrO_3^-(ao)$ HBr(ai)	79 80 82 83 84 85 84 85 430 See $KBrO_3$(ai) 75 76 77 78 81 82 83 169 170 171 210 211 212 213 214 215 216 217 221 222
$I^-(ao)$	88 237 238 425 See also HI(ai), LiI(ai), NaI(ai), KI(ai), RbI(ai), CsI(ai)
$I_3^-(ao)$	425
HI(g)	86 89 90 94 95 96

Table II-2. Index to the Catalog of Measurements (continued)

SUBSTANCE   REACTIONS

HI(ai)
$\quad$ 87 $\quad$ 89 $\quad$ 90 $\quad$ 93 224 225 226 227 228 229
$\quad$ 231 232 233 240 426

HI(50H$_2$O)
$\quad$ 91 $\quad$ 93

SO$_4^{-2}$(ao)
$\quad$ 181 243
$\quad$ See also H$_2$SO$_4$(ai), (NH$_4$)$_2$SO$_4$(ai),
$\quad$ Li$_2$SO$_4$(ai), Na$_2$SO$_4$(ai), K$_2$SO$_4$(ai),
$\quad$ Rb$_2$SO$_4$(ai), Cs$_2$SO$_4$(ai)

H$_2$SO$_4$(l)
$\quad$ 97 $\quad$ 99 100 101

H$_2$SO$_4$(ai)
$\quad$ 99 100 101 175 176 177 178 179 180

H$_2$SO$_4$(115H$_2$O)
$\quad$ 98

NO$_3^-$(ao)
$\quad$ See HNO$_3$(ai), NH$_4$NO$_3$(ai), AgNO$_3$(ai),
$\quad$ NaNO$_3$(ai), KNO$_3$(ai), RbNO$_3$(ai),
$\quad$ CsNO$_3$(ai)

NH$_3$(ao)
$\quad$ 113 114 115 116 117 118 139

NH$_4^+$(ao)
$\quad$ 110 111 112 115 116 117 118
$\quad$ See also NH$_4$NO$_3$(ai), NH$_4$Cl(ai),
$\quad$ NH$_4$ClO$_4$(ai), NH$_4$Br(ai), (NH$_4$)$_2$SO$_4$(ai)

N$_2$H$_4$(300H$_2$O)
$\quad$ 91 $\quad$ 92

HNO$_3$(ai)
$\quad$ 119 125 127 222 240

NH$_4$OH(ai)
$\quad$ 116 117 118 (as NH$_4^+$(ao) + OH$^-$(ao))

NH$_4$NO$_3$(cr)
$\quad$ 109 120 121 122 123 124 128 129 130 131
$\quad$ 132

NH$_4$NO$_3$(ai)
$\quad$ 127 128 129 130 131 132

NH$_4$Cl(cr)
$\quad$ 66 102 103 104 133 134 135 136 137 138

NH$_4$Cl(ai)
$\quad$ 133 134 135 136 137 138 139

NH$_4$ClO$_4$(cr)
$\quad$ 34 $\quad$ 38 $\quad$ 39 $\quad$ 40 $\quad$ 66

NH$_4$ClO$_4$(ai)
$\quad$ 38 $\quad$ 39 $\quad$ 40

NH$_4$Br(cr)
$\quad$ 105 106 107 140 141 142

NH$_4$Br(ai)
$\quad$ 140 141 142

(NH$_4$)$_2$SO$_4$(cr)
$\quad$ 108 143 144 145

(NH$_4$)$_2$SO$_4$(ai)
$\quad$ 143 144 145

Hg^{+2}(ao)
$\quad$ 149 150

Hg$_2^{+2}$(ao)
$\quad$ 148 149 167 168 172 181

Hg$_2$Cl$_2$(cr)
$\quad$ 146 151 152 153 154 155 156 157 158 159
$\quad$ 160 161 162 163 164 167 168 173 249 250
$\quad$ 251 252 256

Hg$_2$Br$_2$(cr)
$\quad$ 169 170 171 172 173 253 254

Hg$_2$SO$_4$(cr)
$\quad$ 147 175 176 177 178 179 180 181

Ag$^+$(ao)
$\quad$ 187 188 189 197 198 199 200 200a 201 218
$\quad$ 219 220 237 238 241 242 243
$\quad$ See also AgClO$_4$(ai), AgNO$_3$(ai)

Table II-2. Index to the Catalog of Measurements (continued)

SUBSTANCE   REACTIONS

AgCl(cr)   41   51   62 182 190 191 192 193 194 195
                         196 197 198 199 200 200a 201 202 203 204
                         205 206 209 221 249 250 251 252 268 342
                         343 344 392 394 397 485 486
AgClO$_4$(cr)   41   42   43   51   62
AgClO$_4$(ai)   42   43
AgBr(cr)   183 207 208 209 210 211 212 213 214 215
                         216 217 218 219 220 221 222 253 254 256
AgI(cr)   184 223 224 225 226 227 228 229 230 231
                         232 233 234 235 236 237 238 239 240

Ag$_2$SO$_4$(cr)   185 241 242 243
AgNO$_3$(cr)   186 244 245 246 247 248
AgNO$_3$(ai)   195 196 222 230 239 240 244 245 246 247
                         248

Li$^+$(ao)   267
                         See also LiOH(ai), LiF(ai), LiCl(ai),

LiOH(cr)   257 269 270 271 273 274 275
LiOH(ai)   264 265 266 269 270 272
LiOH:H$_2$O(cr)   258 271 272 273 274 275
LiF(cr)   259 276 277 278 279 280
LiF(ai)   276 277 278 279 280
LiCl(cr)   64 260 281 282 283 285 286 287
LiCl(ai)   268 281 282 283 284
LiCl:H$_2$O(cr)   284 285 286 287
LiClO$_4$(cr)   44   45   64
LiClO$_4$(ai)   44   45

LiBr(cr)   261 288 289
LiBr(ai)   288 289
LiI(cr)   262 290 291
LiI(ai)   290 291
Li$_2$SO$_4$(cr)   263 292 293 294 295 297 298 299 300 301
                         302 303 304 305 306
Li$_2$SO$_4$(ai)   292 293 294 295 296
Li$_2$SO$_4$:H$_2$O(cr) 296 297 298 299 300 301 302 303 304 305
                         306

Na$^+$(ao)   319
                         See also NaOH(ai), NaF(ai), NaCl(ai),
                         NaClO$_4$(ai), NaBr(ai), NaI(ai),
                         Na$_2$SO$_4$(ai), NaNO$_3$(ai)

Table II-2. Index to the Catalog of Measurements (continued)

Table II-2.   Index to the Catalog of Measurements (continued)

THERMODYNAMIC FUNCTIONS

Thermodynamic functions for 107 substances are given in this Annex. It includes many of the substances in Table 5 for which Key Values are given. Also included are the gaseous dimers of the alkali metals, which are given here because of their use in interpreting vapor pressures. The order of the tables is the same as in Table 5. A table of contents, immediately following this introduction, shows the specific substances.

In many cases these functions cover wider ranges of temperature than those that were used in the Key Value analysis. The wider ranges are given in order to permit the users of the Key Values to obtain values appropriate for their applications.

The functions given are heat capacity, entropy, enthalpy increment and the Gibbs energy function. All values are for substances in their standard states and are molar values. (The subscript "m", meaning molar, is not appended to the symbols for the functions). In this work the standard state pressure is 0.101325 MPa (1 atm). Uncertainties are given for each function. These are estimates of accuracy, at the 95% confidence level.

The table for C(cr, graphite) has been taken from the JANAF Thermochemical Tables (1982); the table for H_2O(cr & l) is documented in the CODATA Thermodynamic Tables (1987). The remaining tables were prepared at the Intitute for High Temperatures, Moscow, U.S.S.R (IVTAN), following critical evaluation of the available pertinent experimental data. Details of the procedures and sources of data used may be found in Gurvich et al. (1978, 1979, 1981, 1982), Hultgren et al. (1973), and the Notes for Table 5 of this Report.

The species treated in Table 5 for which functions are not given here are:

> Br_2(l), I_2(cr), Hg(l), H_2S(aq, undissoc), CO_2(aq, undissoc), H_3BO_3(aq, undissoc), $CdSO_4 \cdot (8/3)H_2O$(cr), Hg_2SO_4(cr), $CuSO_4$(cr), $PbSO_4$(cr), ZnO(cr), CdO(cr), HgO(cr), ThO(cr), AgCl(cr), and the aqueous ions.

CONTENTS

CONTENTS

Table III-1   THERMODYNAMIC FUNCTIONS at 1 atm: O(g)

$\dfrac{T}{K}$	$C_p°$	$-\{G°-H°(0)\}/T$	$S°$	$H°-H°(0)$
	........J·(mol·K)$^{-1}$........			kJ·mol^{-1}
0	0.000	0.000	0.000	0.000
100	23.703	113.769	135.839	2.207
200	22.734	129.350	152.045	4.539
300	21.901	138.533	161.086	6.766
400	21.483	144.990	167.322	8.933
500	21.257	149.952	172.090	11.069
600	21.124	153.974	175.952	13.187
700	21.040	157.352	179.202	15.295
800	20.984	160.262	182.008	17.396
900	20.944	162.818	184.477	19.493
1000	20.915	165.096	186.682	21.585
1100	20.894	167.151	188.674	23.676
1200	20.877	169.021	190.491	25.764
1300	20.864	170.738	192.162	27.851
1400	20.854	172.324	193.708	29.937
1500	20.845	173.798	195.146	32.022
1600	20.839	175.175	196.491	34.106
1700	20.833	176.466	197.755	36.190
1800	20.830	177.682	198.945	38.273
1900	20.827	178.831	200.071	40.356
2000	20.827	179.920	201.140	42.439
2100	20.827	180.955	202.156	44.521
2200	20.830	181.941	203.125	46.604
2300	20.835	182.882	204.051	48.687
2400	20.842	183.783	204.938	50.771
2500	20.851	184.646	205.789	52.856
2600	20.863	185.475	206.607	54.941
2700	20.877	186.273	207.394	57.028
2800	20.894	187.041	208.154	59.117
2900	20.914	187.781	208.887	61.207
3000	20.937	188.497	209.597	63.300
3100	20.963	189.189	210.284	65.395
3200	20.991	189.858	210.950	67.493
3300	21.022	190.507	211.596	69.593
3400	21.056	191.137	212.224	71.697
3500	21.092	191.748	212.835	73.805
3600	21.130	192.342	213.430	75.916
3700	21.171	192.920	214.009	78.031
3800	21.213	193.482	214.574	80.150
3900	21.257	194.030	215.126	82.273
4000	21.302	194.564	215.665	84.401
298.15	21.912	138.393	160.950	6.725

Uncertainties in Functions

300	0.001	0.003	0.003	0.001
1000	0.001	0.003	0.003	0.003
4000	0.003	0.005	0.005	0.010

Table III-2   THERMODYNAMIC FUNCTIONS at 1 atm: $O_2(g)$

$\dfrac{T}{K}$	C_p°	$-\{G^\circ-H^\circ(0)\}/T$	S°	$H^\circ-H^\circ(0)$
	$\cdots\cdots\cdots$ J$\cdot$(mol$\cdot$K)$^{-1}$ $\cdots\cdots\cdots$			kJ$\cdot$mol^{-1}
0	0.000	0.000	0.000	0.000
100	29.112	144.188	173.199	2.901
200	29.128	164.316	193.380	5.813
300	29.387	176.108	205.224	8.735
400	30.108	184.497	213.762	11.706
500	31.093	191.054	220.584	14.765
600	32.093	196.467	226.342	17.925
700	32.986	201.101	231.358	21.180
800	33.739	205.167	235.813	24.517
900	34.363	208.798	239.824	27.924
1000	34.880	212.086	243.473	31.387
1100	35.312	215.094	246.818	34.897
1200	35.681	217.868	249.906	38.446
1300	36.005	220.444	252.776	42.032
1400	36.296	222.850	255.455	45.647
1500	36.566	225.108	257.968	49.290
1600	36.822	227.237	260.336	52.959
1700	37.069	229.250	262.576	56.655
1800	37.310	231.161	264.702	60.374
1900	37.547	232.980	266.725	64.116
2000	37.783	234.716	268.657	67.883
2100	38.016	236.377	270.506	71.671
2200	38.248	237.969	272.280	75.485
2300	38.478	239.498	273.985	79.321
2400	38.705	240.969	275.628	83.182
2500	38.929	242.387	277.212	87.063
2600	39.150	243.756	278.743	90.967
2700	39.366	245.080	280.225	94.892
2800	39.578	246.361	281.661	98.841
2900	39.784	247.602	283.053	102.809
3000	39.986	248.806	284.405	106.798
3100	40.181	249.976	285.719	110.804
3200	40.371	251.113	286.998	114.833
3300	40.555	252.220	288.243	118.877
3400	40.734	253.297	289.457	122.945
3500	40.906	254.347	290.640	127.026
3600	41.074	255.371	291.796	131.128
3700	41.236	256.371	292.922	135.239
3800	41.393	257.347	294.024	139.374
3900	41.546	258.302	295.102	143.517
4000	41.695	259.235	296.156	147.681
298.15	29.378	175.928	205.043	8.680

Uncertainties in Functions

300	0.003	0.005	0.005	0.002
1000	0.005	0.010	0.010	0.010
4000	0.010	0.010	0.020	0.020

Table III-3 THERMODYNAMIC FUNCTIONS at 1 atm: H(g)

$\dfrac{T}{K}$	$C_p°$	$-\{G°-H°(0)\}/T$	$S°$	$H°-H°(0)$
	$\ldots\ldots\ldots J\cdot(mol\cdot K)^{-1}\ldots\ldots\ldots$			$kJ\cdot mol^{-1}$
0	0.000	0.000	0.000	0.000
100	20.786	71.114	91.900	2.079
200	20.786	85.522	106.308	4.157
300	20.786	93.950	114.736	6.236
400	20.786	99.930	120.716	8.314
500	20.786	104.568	125.355	10.393
600	20.786	108.358	129.144	12.472
700	20.786	111.562	132.349	14.550
800	20.786	114.338	135.124	16.629
900	20.786	116.786	137.572	18.708
1000	20.786	118.976	139.762	20.786
1100	20.786	120.957	141.744	22.865
1200	20.786	122.766	143.552	24.943
1300	20.786	124.430	145.216	27.022
1400	20.786	125.970	146.757	29.101
1500	20.786	127.404	148.191	31.179
1600	20.786	128.746	149.532	33.258
1700	20.786	130.006	150.792	35.337
1800	20.786	131.194	151.980	37.415
1900	20.786	132.318	153.104	39.494
2000	20.786	133.384	154.170	41.572
2100	20.786	134.398	155.185	43.651
2200	20.786	135.365	156.151	45.730
2300	20.786	136.289	157.076	47.808
2400	20.786	137.174	157.960	49.887
2500	20.786	138.022	158.809	51.965
2600	20.786	138.838	159.624	54.044
2700	20.786	139.622	160.408	56.123
2800	20.786	140.378	161.164	58.201
2900	20.786	141.108	161.894	60.280
3000	20.786	141.812	162.598	62.359
3100	20.786	142.494	163.280	64.437
3200	20.786	143.154	163.940	66.516
3300	20.786	143.793	164.580	68.594
3400	20.786	144.414	165.200	70.673
3500	20.786	145.016	165.803	72.752
3600	20.786	145.602	166.388	74.830
3700	20.786	146.172	166.958	76.909
3800	20.786	146.726	167.512	78.988
3900	20.786	147.266	168.052	81.066
4000	20.786	147.792	168.578	83.145
298.15	20.786	93.822	114.608	6.197

Uncertainties in Functions

300	0.001	0.002	0.002	0.001
1000	0.001	0.002	0.002	0.002
4000	0.002	0.003	0.003	0.005

Table III-4   THERMODYNAMIC FUNCTIONS at 1 atm: $H_2(g)$

$\dfrac{T}{K}$	$C_p°$	$-\{G°-H°(0)\}/T$	$S°$	$H°-H°(0)$
	........J·(mol·K)$^{-1}$........			kJ·mol^{-1}
0	0.000	0.000	0.000	0.000
100	28.155	70.625	100.617	2.999
200	27.447	90.837	119.302	5.693
300	28.849	102.346	130.748	8.521
400	29.181	110.539	139.105	11.426
500	29.260	116.928	145.627	14.350
600	29.327	122.170	150.967	17.278
700	29.440	126.615	155.495	20.216
800	29.623	130.477	159.438	23.169
900	29.880	133.893	162.941	26.143
1000	30.204	136.958	166.105	29.147
1100	30.580	139.741	169.001	32.186
1200	30.991	142.293	171.679	35.264
1300	31.422	144.650	174.177	38.385
1400	31.860	146.844	176.521	41.548
1500	32.296	148.897	178.736	44.757
1600	32.724	150.828	180.834	48.008
1700	33.138	152.652	182.830	51.301
1800	33.535	154.382	184.735	54.634
1900	33.915	156.028	186.559	58.007
2000	34.277	157.598	188.308	61.419
2100	34.622	159.101	189.989	64.863
2200	34.949	160.542	191.607	68.341
2300	35.259	161.927	193.167	71.850
2400	35.555	163.260	194.674	75.392
2500	35.837	164.546	196.131	78.961
2600	36.106	165.788	197.542	82.559
2700	36.363	166.989	198.910	86.185
2800	36.610	168.153	200.237	89.833
2900	36.847	169.282	201.526	93.505
3000	37.076	170.377	202.778	97.201
3100	37.298	171.442	203.998	100.921
3200	37.513	172.478	205.185	104.660
3300	37.723	173.487	206.343	108.422
3400	37.928	174.470	207.472	112.204
3500	38.129	175.429	208.575	116.009
3600	38.326	176.365	209.651	119.831
3700	38.520	177.279	210.704	123.674
3800	38.711	178.172	211.734	127.537
3900	38.899	179.046	212.742	131.415
4000	39.085	179.901	213.729	135.313
298.15	28.836	102.170	130.571	8.468
Uncertainties in Functions				
300	0.002	0.003	0.003	0.001
1000	0.003	0.005	0.005	0.005
4000	0.005	0.005	0.005	0.010

Table III-5   THERMODYNAMIC FUNCTIONS: H_2O(cr & l)

$\dfrac{T}{K}$	$\dfrac{P}{MPa}$	C_p°	$-\{G^\circ-H^\circ(0)\}/T$	S°	$\dfrac{H^\circ-H^\circ(0)}{kJ\cdot mol^{-1}}$
		........$J\cdot(mol\cdot K)^{-1}$......			
0 (a)	0.101	0.000	0.000	3.392	0.000
5	0.101	0.020	3.392	3.404	0.000
10	0.101	0.270	3.408	3.468	0.001
15	0.101	1.000	3.467	3.700	0.004
20	0.101	1.980	3.570	4.120	0.011
25	0.101	3.110	3.723	4.683	0.024
30	0.101	4.100	3.943	5.343	0.042
35	0.101	5.150	4.186	6.043	0.065
40	0.101	6.120	4.481	6.806	0.093
45	0.101	7.030	4.781	7.581	0.126
50	0.101	7.900	5.105	8.365	0.163
60	0.101	9.610	5.775	9.958	0.251
70	0.101	11.260	6.492	11.563	0.355
80	0.101	12.820	7.219	13.169	0.476
90	0.101	14.380	7.969	14.769	0.612
100	0.101	15.830	8.729	16.359	0.763
110	0.101	17.220	9.497	17.933	0.928
120	0.101	18.490	10.263	19.488	1.107
130	0.101	19.710	11.030	21.015	1.298
140	0.101	20.820	11.798	22.519	1.501
150	0.101	21.950	12.564	23.991	1.714
160	0.101	23.140	13.323	25.448	1.940
170	0.101	24.370	14.079	26.885	2.177
180	0.101	25.610	14.833	28.316	2.427
190	0.101	26.840	15.578	29.731	2.689
200	0.101	28.120	16.322	31.142	2.964
210	0.101	29.490	17.058	32.544	3.252
220	0.101	30.820	17.795	33.950	3.554
230	0.101	32.110	18.528	35.345	3.868
240	0.101	33.410	19.260	36.743	4.196
250	0.101	34.710	19.986	38.130	4.536
260	0.101	36.030	20.708	39.520	4.891
270	0.101	37.360	21.431	40.901	5.257
273.15 cr	0.101	37.779	21.656	41.338	5.376
273.15 l	0.101	76.172	21.656	63.337	11.385
273.16 (b)	0.0006	76.171	21.658	63.340	11.386
280	0.101	75.510	22.697	65.215	11.905
290	0.101	75.358	24.214	67.862	12.658
300	0.101	75.349	25.709	70.416	13.412
320	0.101	75.344	28.657	75.279	14.919
340	0.101	75.393	31.535	79.847	16.426
360	0.101	75.678	34.339	84.164	17.937
373.15 (a,c)	0.101	75.972	36.139	86.884	18.936
380 (a)	0.129	76.151	37.065	88.267	19.457
400 (a)	0.246	76.769	39.718	92.186	20.988

Table III-5   THERMODYNAMIC FUNCTIONS: H_2O(cr & l) (continued)

$\dfrac{T}{K}$	$\dfrac{P}{MPa}$	C_p°	$-\{G^{\circ}-H^{\circ}(0)\}/T$	S°	$\dfrac{H^{\circ}-H^{\circ}(0)}{kJ \cdot mol^{-1}}$
		........J·(mol·K)$^{-1}$.......			
440 (a)	0.733	78.525	44.806	99.569	24.096
460 (a)	1.170	79.808	47.244	103.076	25.683
480 (a)	1.789	81.510	49.615	106.489	27.300
500 (a)	2.637	83.796	51.921	109.833	28.956
520 (a)	3.767	86.908	54.164	113.133	30.664
540 (a)	5.234	91.246	56.346	116.419	32.439
560 (a)	7.103	97.560	58.471	119.732	34.306
580 (a)	9.444	107.492	60.541	123.127	36.300
600 (a)	12.339	125.462	62.557	126.705	38.489
298.15	0.101	75.351	25.432	69.950	13.273

Uncertainties in Functions

300	-	0.080	0.073	0.030	0.020
373	0.00001	0.040	0.061	0.030	0.020
500	0.0009	0.170	0.053	0.032	0.021
600	0.0012	0.380	0.047	0.032	0.021

a.   At 0.101325 MPa, 0-373.15 K; at saturation pressure, 373.15-600 K
b.   Triple point, at 0.0006117 MPa.
c.   Nominal normal boiling point, at 0.101325 MPa.

Table III-6   THERMODYNAMIC FUNCTIONS at 1 atm: $H_2O(g)$

$\dfrac{T}{K}$	$C_p°$	$-\{G°-H°(0)\}/T$	$S°$	$H°-H°(0)$
	........J·(mol·K)$^{-1}$........			kJ·mol^{-1}
0	0.000	0.000	0.000	0.000
100	33.310	119.380	152.272	3.289
200	33.354	142.261	175.371	6.622
300	33.618	155.711	188.934	9.967
400	34.146	165.280	198.717	13.375
500	35.071	172.764	206.427	16.831
600	36.289	178.930	212.926	20.398
700	37.606	184.201	218.619	24.093
800	38.942	188.828	223.728	27.920
900	40.261	192.968	228.391	31.881
1000	41.544	196.729	232.699	35.970
1100	42.782	200.183	236.718	40.189
1200	43.969	203.387	240.491	44.525
1300	45.105	206.379	244.056	48.980
1400	46.186	209.192	247.438	53.545
1500	47.215	211.850	250.660	58.215
1600	48.191	214.373	253.739	62.986
1700	49.117	216.776	256.688	67.851
1800	49.992	219.072	259.521	72.809
1900	50.819	221.273	262.246	77.850
2000	51.600	223.388	264.873	82.971
2100	52.336	225.424	267.408	88.167
2200	53.030	227.389	269.859	93.435
2300	53.682	229.287	272.231	98.772
2400	54.296	231.125	274.529	104.170
2500	54.873	232.906	276.757	109.628
2600	55.415	234.634	278.920	115.145
2700	55.924	236.313	281.021	120.713
2800	56.402	237.947	283.064	126.329
2900	56.851	239.537	285.051	131.992
3000	57.274	241.086	286.985	137.698
3100	57.671	242.598	288.870	143.444
3200	58.046	244.072	290.708	149.233
3300	58.399	245.513	292.498	155.052
3400	58.733	246.920	294.247	160.913
3500	59.049	248.297	295.955	166.801
3600	59.350	249.644	297.622	172.722
3700	59.636	250.963	299.253	178.671
3800	59.910	252.255	300.847	184.647
3900	60.173	253.521	302.406	190.649
4000	60.428	254.762	303.933	196.682
298.15	33.609	155.504	188.726	9.905

Uncertainties in Functions

300	0.030	0.005	0.010	0.005
1000	0.100	0.030	0.050	0.050
4000	1.000	0.500	0.500	2.000

Table III-7   THERMODYNAMIC FUNCTIONS at 1 atm: He(g)

$\dfrac{T}{K}$	$C_p°$	$-\{G°-H°(0)\}/T$	$S°$	$H°-H°(0)$
	........J·(mol·K)$^{-1}$........			kJ·mol^{-1}
0	0.000	0.000	0.000	0.000
100	20.786	82.550	103.336	2.079
200	20.786	96.958	117.744	4.157
300	20.786	105.386	126.172	6.236
400	20.786	111.366	132.152	8.314
500	20.786	116.004	136.790	10.393
600	20.786	119.794	140.580	12.472
700	20.786	122.998	143.784	14.550
800	20.786	125.774	146.560	16.629
900	20.786	128.222	149.008	18.708
1000	20.786	130.412	151.198	20.786
1100	20.786	132.393	153.179	22.865
1200	20.786	134.202	154.988	24.943
1300	20.786	135.866	156.652	27.022
1400	20.786	137.406	158.192	29.101
1500	20.786	138.840	159.626	31.179
1600	20.786	140.182	160.968	33.258
1700	20.786	141.442	162.228	35.337
1800	20.786	142.630	163.416	37.415
1900	20.786	143.754	164.540	39.494
2000	20.786	144.820	165.606	41.572
2100	20.786	145.834	166.620	43.651
2200	20.786	146.801	167.587	45.730
2300	20.786	147.725	168.511	47.808
2400	20.786	148.610	169.396	49.887
2500	20.786	149.458	170.244	51.965
2600	20.786	150.274	171.060	54.044
2700	20.786	151.058	171.844	56.123
2800	20.786	151.814	172.600	58.201
2900	20.786	152.543	173.330	60.280
3000	20.786	153.248	174.034	62.359
3100	20.786	153.930	174.716	64.437
3200	20.786	154.590	175.376	66.516
3300	20.786	155.229	176.015	68.595
3400	20.786	155.850	176.636	70.673
3500	20.786	156.452	177.238	72.752
3600	20.786	157.038	177.824	74.830
3700	20.786	157.607	178.394	76.909
3800	20.786	158.162	178.948	78.988
3900	20.786	158.702	179.488	81.066
4000	20.786	159.228	180.014	83.145
298.15	20.786	105.257	126.044	6.197

Uncertainties in Functions

300	0.001	0.002	0.002	0.001
1000	0.001	0.002	0.003	0.002
4000	0.002	0.003	0.004	0.004

Table III-8   THERMODYNAMIC FUNCTIONS at 1 atm: Ne(g)

$\dfrac{T}{K}$	$C_p°$	$-\{G°-H°(0)\}/T$	$S°$	$H°-H°(0)$
	........J·(mol·K)$^{-1}$.......			kJ·mol^{-1}
0	0.000	0.000	0.000	0.000
100	20.786	102.726	123.512	2.079
200	20.786	117.133	137.920	4.157
300	20.786	125.562	146.348	6.236
400	20.786	131.541	152.328	8.314
500	20.786	136.180	156.966	10.393
600	20.786	139.969	160.756	12.472
700	20.786	143.174	163.960	14.550
800	20.786	145.949	166.735	16.629
900	20.786	148.397	169.184	18.708
1000	20.786	150.588	171.374	20.786
1100	20.786	152.569	173.355	22.865
1200	20.786	154.377	175.164	24.943
1300	20.786	156.041	176.827	27.022
1400	20.786	157.582	178.368	29.101
1500	20.786	159.016	179.802	31.179
1600	20.786	160.357	181.143	33.258
1700	20.786	161.617	182.403	35.337
1800	20.786	162.805	183.592	37.415
1900	20.786	163.929	184.715	39.494
2000	20.786	164.995	185.782	41.572
2100	20.786	166.010	186.796	43.651
2200	20.786	166.977	187.763	45.730
2300	20.786	167.901	188.687	47.808
2400	20.786	168.785	189.571	49.887
2500	20.786	169.634	190.420	51.965
2600	20.786	170.449	191.235	54.044
2700	20.786	171.233	192.020	56.123
2800	20.786	171.989	192.776	58.201
2900	20.786	172.719	193.505	60.280
3000	20.786	173.423	194.210	62.359
3100	20.786	174.105	194.891	64.437
3200	20.786	174.765	195.551	66.516
3300	20.786	175.405	196.191	68.594
3400	20.786	176.025	196.811	70.673
3500	20.786	176.628	197.414	72.752
3600	20.786	177.213	197.999	74.830
3700	20.786	177.783	198.569	76.909
3800	20.786	178.337	199.123	78.988
3900	20.786	178.877	199.663	81.066
4000	20.786	179.403	200.190	83.145
298.15	20.786	125.433	146.219	6.197

Uncertainties in Functions

300	0.001	0.003	0.003	0.001
1000	0.001	0.003	0.003	0.002
4000	0.002	0.004	0.005	0.004

Table III-9 THERMODYNAMIC FUNCTIONS at 1 atm: Ar(g)

$\dfrac{T}{K}$	$C_p°$	$-\{G°-H°(0)\}/T$	$S°$	$H°-H°(0)$
	........J·(mol·K)$^{-1}$........			kJ·mol^{-1}
0	0.000	0.000	0.000	0.000
100	20.786	111.243	132.029	2.079
200	20.786	125.651	146.437	4.157
300	20.786	134.079	154.865	6.236
400	20.786	140.059	160.845	8.314
500	20.786	144.697	165.483	10.393
600	20.786	148.487	169.273	12.472
700	20.786	151.691	172.477	14.550
800	20.786	154.467	175.253	16.629
900	20.786	156.915	177.701	18.708
1000	20.786	159.105	179.891	20.786
1100	20.786	161.086	181.872	22.865
1200	20.786	162.895	183.681	24.943
1300	20.786	164.559	185.345	27.022
1400	20.786	166.099	186.885	29.101
1500	20.786	167.533	188.319	31.179
1600	20.786	168.875	189.661	33.258
1700	20.786	170.135	190.921	35.337
1800	20.786	171.323	192.109	37.415
1900	20.786	172.447	193.233	39.494
2000	20.786	173.513	194.299	41.572
2100	20.786	174.527	195.313	43.651
2200	20.786	175.494	196.280	45.730
2300	20.786	176.418	197.204	47.808
2400	20.786	177.303	198.089	49.887
2500	20.786	178.151	198.937	51.965
2600	20.786	178.966	199.753	54.044
2700	20.786	179.751	200.537	56.123
2800	20.786	180.507	201.293	58.201
2900	20.786	181.236	202.022	60.280
3000	20.786	181.941	202.727	62.359
3100	20.786	182.622	203.409	64.437
3200	20.786	183.282	204.069	66.516
3300	20.786	183.922	204.708	68.594
3400	20.786	184.543	205.329	70.673
3500	20.786	185.145	205.931	72.752
3600	20.786	185.731	206.517	74.830
3700	20.786	186.300	207.086	76.909
3800	20.786	186.855	207.641	78.988
3900	20.786	187.394	208.181	81.066
4000	20.786	187.921	208.707	83.145
298.15	20.786	133.950	154.737	6.197

Uncertainties in Functions

300	0.001	0.003	0.003	0.001
1000	0.001	0.003	0.003	0.002
4000	0.003	0.004	0.005	0.005

Table III-10   THERMODYNAMIC FUNCTIONS at 1 atm: Kr(g)

$\dfrac{T}{K}$	$C_p°$	$-\{G°-H°(0)\}/T$	$S°$	$H°-H°(0)$
	$\ldots\ldots\ldots$J$\cdot$(mol$\cdot$K)$^{-1}\ldots\ldots\ldots$			kJ$\cdot$mol^{-1}
0	0.000	0.000	0.000	0.000
100	20.786	120.483	141.269	2.079
200	20.786	134.891	155.677	4.157
300	20.786	143.319	164.105	6.236
400	20.786	149.298	170.085	8.314
500	20.786	153.937	174.723	10.393
600	20.786	157.727	178.513	12.472
700	20.786	160.931	181.717	14.550
800	20.786	163.706	184.493	16.629
900	20.786	166.155	186.941	18.708
1000	20.786	168.345	189.131	20.786
1100	20.786	170.326	191.112	22.865
1200	20.786	172.134	192.921	24.943
1300	20.786	173.798	194.584	27.022
1400	20.786	175.339	196.125	29.101
1500	20.786	176.773	197.559	31.179
1600	20.786	178.114	198.900	33.258
1700	20.786	179.374	200.161	35.337
1800	20.786	180.563	201.349	37.415
1900	20.786	181.686	202.473	39.494
2000	20.786	182.753	203.539	41.572
2100	20.786	183.767	204.553	43.651
2200	20.786	184.734	205.520	45.730
2300	20.786	185.658	206.444	47.808
2400	20.786	186.542	207.329	49.887
2500	20.786	187.391	208.177	51.965
2600	20.786	188.206	208.992	54.044
2700	20.786	188.991	209.777	56.123
2800	20.786	189.747	210.533	58.201
2900	20.786	190.476	211.262	60.280
3000	20.786	191.181	211.967	62.359
3100	20.786	191.862	212.648	64.437
3200	20.786	192.522	213.308	66.516
3300	20.786	193.162	213.948	68.594
3400	20.786	193.782	214.569	70.673
3500	20.786	194.385	215.171	72.752
3600	20.786	194.970	215.757	74.830
3700	20.786	195.540	216.326	76.909
3800	20.786	196.094	216.880	78.988
3900	20.786	196.634	217.420	81.066
4000	20.786	197.160	217.947	83.145
298.15	20.786	143.190	163.976	6.197

Uncertainties in Functions

300	0.001	0.003	0.003	0.001
1000	0.001	0.003	0.003	0.002
4000	0.003	0.004	0.005	0.005

Table III-11 THERMODYNAMIC FUNCTIONS at 1 atm: Xe(g)

$\dfrac{T}{K}$	$C_p°$	$-\{G°-H°(0)\}/T$	$S°$	$H°-H°(0)$
	$\ldots\ldots\ldots J\cdot(mol\cdot K)^{-1}\ldots\ldots\ldots$			$kJ\cdot mol^{-1}$
0	0.000	0.000	0.000	0.000
100	20.786	126.083	146.869	2.079
200	20.786	140.491	161.277	4.157
300	20.786	148.919	169.705	6.236
400	20.786	154.899	175.685	8.314
500	20.786	159.537	180.323	10.393
600	20.786	163.327	184.113	12.472
700	20.786	166.531	187.317	14.550
800	20.786	169.307	190.093	16.629
900	20.786	171.755	192.541	18.708
1000	20.786	173.945	194.731	20.786
1100	20.786	175.926	196.712	22.865
1200	20.786	177.735	198.521	24.943
1300	20.786	179.399	200.185	27.022
1400	20.786	180.939	201.725	29.101
1500	20.786	182.373	203.159	31.179
1600	20.786	183.715	204.501	33.258
1700	20.786	184.975	205.761	35.337
1800	20.786	186.163	206.949	37.415
1900	20.786	187.287	208.073	39.494
2000	20.786	188.353	209.139	41.572
2100	20.786	189.367	210.153	43.651
2200	20.786	190.334	211.120	45.730
2300	20.786	191.258	212.044	47.808
2400	20.786	192.143	212.929	49.887
2500	20.786	192.991	213.778	51.966
2600	20.786	193.807	214.593	54.044
2700	20.786	194.591	215.377	56.123
2800	20.786	195.347	216.133	58.201
2900	20.786	196.076	216.863	60.280
3000	20.786	196.781	217.567	62.359
3100	20.786	197.463	218.249	64.437
3200	20.786	198.123	218.909	66.516
3300	20.786	198.762	219.548	68.594
3400	20.786	199.383	220.169	70.673
3500	20.786	199.985	220.772	72.752
3600	20.786	200.571	221.357	74.830
3700	20.786	201.140	221.927	76.909
3800	20.786	201.695	222.481	78.988
3900	20.786	202.235	223.021	81.066
4000	20.786	202.761	223.547	83.145
298.15	20.786	148.790	169.576	6.197
Uncertainties in Functions				
300	0.001	0.003	0.003	0.001
1000	0.001	0.003	0.004	0.003
4000	0.003	0.004	0.005	0.005

Table III-12   THERMODYNAMIC FUNCTIONS at 1 atm: F(g)

$\dfrac{T}{K}$	$C_p°$	$-\{G°-H°(0)\}/T$	$S°$	$H°-H°(0)$
	........J·(mol·K)$^{-1}$........			kJ·mol^{-1}
0	0.000	0.000	0.000	0.000
100	21.204	113.512	134.371	2.086
200	22.605	128.132	149.561	4.286
300	22.743	136.914	158.782	6.561
400	22.432	143.235	165.286	8.820
500	22.100	148.162	170.255	11.047
600	21.833	152.188	174.259	13.243
700	21.629	155.587	177.609	15.415
800	21.475	158.524	180.486	17.570
900	21.357	161.107	183.009	19.711
1000	21.266	163.412	185.254	21.842
1100	21.195	165.491	187.277	23.965
1200	21.138	167.384	189.119	26.082
1300	21.091	169.122	190.809	28.193
1400	21.054	170.728	192.371	30.300
1500	21.023	172.219	193.822	32.404
1600	20.996	173.612	195.178	34.505
1700	20.974	174.919	196.450	36.603
1800	20.955	176.149	197.648	38.700
1900	20.939	177.310	198.781	40.795
2000	20.925	178.411	199.855	42.888
2100	20.913	179.456	200.875	44.980
2200	20.903	180.452	201.848	47.071
2300	20.893	181.403	202.777	49.160
2400	20.885	182.312	203.666	51.249
2500	20.878	183.183	204.518	53.337
2600	20.871	184.020	205.337	55.425
2700	20.865	184.824	206.125	57.512
2800	20.860	185.598	206.883	59.598
2900	20.855	186.345	207.615	61.684
3000	20.851	187.066	208.322	63.769
3100	20.847	187.763	209.006	65.854
3200	20.843	188.437	209.668	67.938
3300	20.840	189.090	210.309	70.022
3400	20.837	189.723	210.931	72.106
3500	20.834	190.338	211.535	74.190
3600	20.832	190.935	212.122	76.273
3700	20.829	191.515	212.693	78.356
3800	20.827	192.080	213.248	80.439
3900	20.825	192.630	213.789	82.522
4000	20.823	193.165	214.316	84.604
298.15	22.746	136.779	158.642	6.518

Uncertainties in Functions

300	0.002	0.002	0.004	0.001
1000	0.002	0.002	0.003	0.001
4000	0.003	0.003	0.004	0.004

Table III-13   THERMODYNAMIC FUNCTIONS at 1 atm: $F_2(g)$

$\dfrac{T}{K}$	C_p°	$-\{G^\circ-H^\circ(0)\}/T$	S°	$H^\circ-H^\circ(0)$
	........J·(mol·K)$^{-1}$........			kJ·mol^{-1}
0	0.000	0.000	0.000	0.000
100	29.114	141.192	170.263	2.907
200	29.686	161.364	190.545	5.836
300	31.337	173.266	202.876	8.883
400	32.995	181.870	212.126	12.102
500	34.259	188.695	219.632	15.468
600	35.172	194.392	225.963	18.942
700	35.839	199.302	231.438	22.495
800	36.344	203.626	236.258	26.105
900	36.740	207.496	240.562	29.760
1000	37.065	211.000	244.450	33.451
1100	37.343	214.204	247.996	37.171
1200	37.588	217.158	251.256	40.918
1300	37.812	219.898	254.274	44.688
1400	38.020	222.455	257.084	48.480
1500	38.215	224.852	259.713	52.292
1600	38.396	227.109	262.186	56.122
1700	38.563	229.242	264.518	59.970
1800	38.711	231.263	266.727	63.834
1900	38.837	233.186	268.823	67.712
2000	38.935	235.018	270.818	71.601
2100	39.003	236.768	272.720	75.498
2200	39.036	238.444	274.535	79.400
2300	39.033	240.051	276.270	83.304
2400	38.993	241.595	277.931	87.206
2500	38.914	243.081	279.521	91.101
2600	38.799	244.512	281.045	94.987
2700	38.648	245.892	282.507	98.860
2800	38.464	247.225	283.909	102.716
2900	38.249	248.513	285.255	106.551
3000	38.006	249.760	286.548	110.365
3100	37.739	250.966	287.790	114.152
3200	37.450	252.136	288.983	117.911
3300	37.143	253.270	290.131	121.641
3400	36.821	254.371	291.235	125.340
3500	36.488	255.439	292.298	129.005
3600	36.145	256.477	293.321	132.637
3700	35.795	257.486	294.306	136.234
3800	35.441	258.468	295.256	139.796
3900	35.085	259.423	296.172	143.322
4000	34.728	260.353	297.056	146.813
298.15	31.304	173.083	202.682	8.825

Uncertainties in Functions

300	0.002	0.003	0.005	0.001
1000	0.050	0.010	0.020	0.030
4000	0.500	0.100	0.100	0.500

Table III-14   THERMODYNAMIC FUNCTIONS at 1 atm: HF(g)

$\dfrac{T}{K}$	$C_p°$	$-\{G°-H°(0)\}/T$	$S°$	$H°-H°(0)$
	$\ldots\ldots\ldots J\cdot(mol\cdot K)^{-1}\ldots\ldots\ldots$			$kJ\cdot mol^{-1}$
0	0.000	0.000	0.000	0.000
100	29.133	113.577	141.845	2.827
200	29.128	133.337	162.035	5.740
300	29.137	145.004	173.848	8.653
400	29.149	153.313	182.232	11.567
500	29.172	159.771	188.738	14.483
600	29.230	165.056	194.062	17.403
700	29.350	169.530	198.575	20.331
800	29.549	173.411	202.506	23.275
900	29.827	176.843	206.002	26.243
1000	30.169	179.919	209.162	29.243
1100	30.558	182.711	212.055	32.279
1200	30.974	185.269	214.732	35.356
1300	31.403	187.633	217.228	38.474
1400	31.831	189.831	219.571	41.636
1500	32.250	191.888	221.782	44.841
1600	32.654	193.822	223.876	48.087
1700	33.039	195.649	225.867	51.371
1800	33.404	197.381	227.766	54.694
1900	33.747	199.029	229.581	58.049
2000	34.070	200.600	231.321	61.443
2100	34.373	202.103	232.990	64.863
2200	34.657	203.544	234.596	68.315
2300	34.923	204.928	236.142	71.793
2400	35.173	206.259	237.634	75.301
2500	35.407	207.543	239.075	78.831
2600	35.628	208.783	240.468	82.382
2700	35.836	209.982	241.816	85.952
2800	36.032	211.142	243.123	89.548
2900	36.218	212.267	244.391	93.160
3000	36.394	213.358	245.622	96.793
3100	36.562	214.418	246.818	100.441
3200	36.722	215.449	247.981	104.103
3300	36.876	216.452	249.114	107.785
3400	37.023	217.429	250.217	111.480
3500	37.164	218.381	251.292	115.189
3600	37.300	219.310	252.341	118.913
3700	37.432	220.217	253.364	122.645
3800	37.560	221.102	254.364	126.397
3900	37.684	221.968	255.342	130.160
4000	37.805	222.814	256.297	133.933
298.15	29.137	144.826	173.670	8.599

Uncertainties in Functions

300	0.002	0.003	0.003	0.001
1000	0.003	0.005	0.005	0.005
4000	0.005	0.012	0.012	0.020

Table III-15   THERMODYNAMIC FUNCTIONS at 1 atm: Cl(g)

$\dfrac{T}{K}$	$C_p°$	$-\{G°-H°(0)\}/T$	$S°$	$H°-H°(0)$
	........J·(mol·K)$^{-1}$........			kJ·mol^{-1}
0	0.000	0.000	0.000	0.000
100	20.788	121.281	142.067	2.079
200	21.079	135.696	156.528	4.166
300	21.852	144.177	165.217	6.312
400	22.467	150.268	171.595	8.531
500	22.744	155.056	176.644	10.794
600	22.782	159.011	180.797	13.072
700	22.692	162.380	184.303	15.346
800	22.549	165.314	187.324	17.608
900	22.389	167.910	189.971	19.855
1000	22.234	170.235	192.322	22.086
1100	22.089	172.341	194.434	24.302
1200	21.959	174.263	196.350	26.505
1300	21.844	176.031	198.103	28.695
1400	21.742	177.666	199.718	30.874
1500	21.652	179.186	201.215	33.043
1600	21.574	180.607	202.610	35.205
1700	21.504	181.940	203.916	37.358
1800	21.443	183.196	205.143	39.506
1900	21.389	184.381	206.301	41.647
2000	21.341	185.505	207.397	43.784
2100	21.298	186.573	208.437	45.916
2200	21.260	187.589	209.427	48.044
2300	21.226	188.559	210.371	50.168
2400	21.195	189.487	211.274	52.289
2500	21.167	190.376	212.139	54.407
2600	21.142	191.229	212.968	56.523
2700	21.120	192.049	213.766	58.636
2800	21.099	192.838	214.534	60.747
2900	21.080	193.599	215.274	62.855
3000	21.063	194.334	215.988	64.963
3100	21.047	195.044	216.678	67.068
3200	21.033	195.730	217.346	69.172
3300	21.019	196.395	217.993	71.275
3400	21.007	197.040	218.621	73.376
3500	20.996	197.665	219.230	75.476
3600	20.985	198.272	219.821	77.575
3700	20.975	198.862	220.396	79.673
3800	20.966	199.436	220.955	81.770
3900	20.958	199.995	221.499	83.866
4000	20.950	200.539	222.030	85.962
298.15	21.838	144.047	165.081	6.272
	Uncertainties in Functions			
300	0.001	0.004	0.004	0.001
1000	0.002	0.004	0.005	0.001
4000	0.003	0.005	0.006	0.004

Table III-16   THERMODYNAMIC FUNCTIONS at 1 atm: $Cl_2(g)$

$\dfrac{T}{K}$	$C_p°$	$-\{G°-H°(0)\}/T$	$S°$	$H°-H°(0)$
	$\ldots\ldots\ldots J\cdot(mol\cdot K)^{-1}\ldots\ldots$			$kJ\cdot mol^{-1}$
0	0.000	0.000	0.000	0.000
100	29.299	159.781	188.900	2.912
200	31.721	180.109	209.856	5.950
300	33.981	192.370	223.182	9.244
400	35.296	201.371	233.156	12.714
500	36.064	208.551	241.121	16.285
600	36.547	214.547	247.742	19.917
700	36.874	219.703	253.402	23.589
800	37.111	224.231	258.342	27.289
900	37.294	228.269	262.724	31.010
1000	37.442	231.915	266.661	34.747
1100	37.568	235.239	270.236	38.497
1200	37.678	238.293	273.510	42.260
1300	37.778	241.120	276.530	46.033
1400	37.872	243.751	279.333	49.815
1500	37.961	246.211	281.949	53.607
1600	38.048	248.522	284.401	57.407
1700	38.133	250.701	286.711	61.216
1800	38.243	252.763	288.895	65.037
1900	38.349	254.719	290.965	68.867
2000	38.468	256.581	292.935	72.707
2100	38.604	258.358	294.815	76.561
2200	38.760	260.056	296.615	80.429
2300	38.940	261.683	298.341	84.313
2400	39.146	263.246	300.003	88.218
2500	39.379	264.748	301.605	92.144
2600	39.640	266.196	303.155	96.094
2700	39.927	267.592	304.656	100.073
2800	40.239	268.942	306.114	104.081
2900	40.571	270.249	307.532	108.121
3000	40.921	271.514	308.913	112.195
3100	41.281	272.743	310.261	116.306
3200	41.647	273.936	311.577	120.452
3300	42.013	275.096	312.864	124.635
3400	42.373	276.225	314.124	128.854
3500	42.721	277.326	315.357	133.109
3600	43.050	278.399	316.565	137.398
3700	43.356	279.447	317.749	141.718
3800	43.635	280.470	318.909	146.068
3900	43.882	281.470	320.046	150.444
4000	44.094	282.449	321.159	154.844
298.15	33.949	192.179	222.972	9.181

Uncertainties in Functions

300	0.002	0.010	0.010	0.001
1000	0.300	0.030	0.030	0.020
4000	1.000	0.100	0.100	1.000

Table III-17   THERMODYNAMIC FUNCTIONS at 1 atm: HCl(g)

$\dfrac{T}{K}$	$C_p°$	$-\{G°-H°(0)\}/T$	$S°$	$\dfrac{H°-H°(0)}{kJ \cdot mol^{-1}}$
	........J·(mol·K)$^{-1}$........			
0	0.000	0.000	0.000	0.000
100	29.117	126.293	154.979	2.869
200	29.124	146.260	175.163	5.781
300	29.137	157.995	186.974	8.694
400	29.175	166.338	195.360	11.609
500	29.305	172.818	201.882	14.532
600	29.576	178.122	207.246	17.474
700	29.988	182.618	211.835	20.452
800	30.500	186.527	215.872	23.475
900	31.063	189.993	219.496	26.553
1000	31.640	193.111	222.799	29.689
1100	32.202	195.950	225.841	32.881
1200	32.734	198.560	228.666	36.128
1300	33.229	200.978	231.306	39.426
1400	33.685	203.234	233.786	42.772
1500	34.101	205.349	236.124	46.162
1600	34.479	207.343	238.337	49.591
1700	34.825	209.228	240.438	53.057
1800	35.139	211.018	242.438	56.555
1900	35.426	212.722	244.345	60.084
2000	35.690	214.349	246.169	63.640
2100	35.932	215.906	247.916	67.221
2200	36.156	217.400	249.593	70.825
2300	36.363	218.835	251.205	74.451
2400	36.556	220.216	252.757	78.098
2500	36.737	221.548	254.253	81.762
2600	36.908	222.834	255.697	85.445
2700	37.068	224.077	257.093	89.143
2800	37.221	225.280	258.444	92.858
2900	37.367	226.446	259.752	96.588
3000	37.506	227.578	261.022	100.331
3100	37.641	228.677	262.254	104.089
3200	37.770	229.745	263.451	107.859
3300	37.895	230.784	264.615	111.642
3400	38.017	231.796	265.748	115.438
3500	38.136	232.782	266.852	119.246
3600	38.252	233.743	267.928	123.065
3700	38.365	234.681	268.977	126.896
3800	38.475	235.597	270.002	130.738
3900	38.584	236.492	271.003	134.591
4000	38.689	237.367	271.981	138.455
298.15	29.136	157.816	186.793	8.640

Uncertainties in Functions

300	0.002	0.005	0.005	0.001
1000	0.003	0.005	0.005	0.005
4000	0.005	0.010	0.015	0.020

Table III-18 THERMODYNAMIC FUNCTIONS at 1 atm: Br(g)

$\dfrac{T}{K}$	$C_p°$	$-\{G°-H°(0)\}/T$	$S°$	$H°-H°(0)$
	........J·(mol·K)$^{-1}$........			kJ·mol^{-1}
0	0.000	0.000	0.000	0.000
100	20.786	131.415	152.201	2.079
200	20.786	145.823	166.609	4.157
300	20.786	154.251	175.037	6.236
400	20.787	160.231	181.017	8.315
500	20.798	164.869	185.657	10.394
600	20.833	168.660	189.451	12.475
700	20.909	171.865	192.668	14.562
800	21.027	174.644	195.467	16.658
900	21.184	177.099	197.952	18.768
1000	21.365	179.298	200.194	20.896
1100	21.559	181.292	202.239	23.042
1200	21.753	183.117	204.123	25.208
1300	21.937	184.801	205.872	27.392
1400	22.107	186.365	207.504	29.594
1500	22.258	187.826	209.034	31.813
1600	22.388	189.197	210.475	34.045
1700	22.497	190.489	211.836	36.290
1800	22.587	191.711	213.124	38.544
1900	22.657	192.870	214.347	40.807
2000	22.710	193.973	215.511	43.075
2100	22.748	195.026	216.620	45.348
2200	22.772	196.031	217.679	47.624
2300	22.785	196.995	218.691	49.902
2400	22.788	197.919	219.661	52.181
2500	22.782	198.808	220.591	54.459
2600	22.769	199.663	221.485	56.737
2700	22.751	200.487	222.344	59.013
2800	22.727	201.282	223.171	61.287
2900	22.699	202.051	223.968	63.558
3000	22.669	202.794	224.737	65.827
3100	22.636	203.514	225.480	68.092
3200	22.600	204.212	226.198	70.354
3300	22.564	204.889	226.893	72.612
3400	22.527	205.546	227.566	74.867
3500	22.489	206.184	228.218	77.117
3600	22.450	206.805	228.851	79.364
3700	22.412	207.410	229.466	81.607
3800	22.374	207.998	230.063	83.847
3900	22.336	208.571	230.643	86.082
4000	22.298	209.130	231.208	88.314
298.15	20.786	154.123	174.909	6.197

Uncertainties in Functions

300	0.001	0.003	0.004	0.001
1000	0.002	0.003	0.004	0.001
4000	0.003	0.004	0.005	0.004

Table III-19   THERMODYNAMIC FUNCTIONS at 1 atm: $Br_2(g)$

$\dfrac{T}{K}$	$C_p°$	$-\{G°-H°(0)\}/T$	$S°$	$H°-H°(0)$
	........J·(mol·K)$^{-1}$........			kJ·mol^{-1}
0	0.000	0.000	0.000	0.000
100	30.899	179.028	208.513	2.948
200	34.596	199.992	231.230	6.248
300	36.075	212.942	245.582	9.792
400	36.729	222.471	256.061	13.436
500	37.082	230.042	264.298	17.128
600	37.305	236.333	271.080	20.848
700	37.465	241.719	276.843	24.587
800	37.590	246.429	281.854	28.340
900	37.697	250.616	286.288	32.105
1000	37.793	254.386	290.265	35.879
1100	37.883	257.814	293.871	39.663
1200	37.970	260.958	297.172	43.456
1300	38.060	263.862	300.214	47.258
1400	38.159	266.561	303.039	51.069
1500	38.264	269.082	305.675	54.890
1600	38.387	271.447	308.148	58.722
1700	38.527	273.675	310.480	62.568
1800	38.690	275.782	312.686	66.429
1900	38.877	277.780	314.783	70.307
2000	39.089	279.680	316.783	74.205
2100	39.325	281.493	318.695	78.125
2200	39.582	283.226	320.531	82.070
2300	39.853	284.886	322.296	86.042
2400	40.134	286.481	323.998	90.041
2500	40.417	288.015	325.642	94.069
2600	40.695	289.493	327.233	98.125
2700	40.960	290.919	328.774	102.207
2800	41.207	292.298	330.268	106.316
2900	41.429	293.632	331.718	110.448
3000	41.619	294.925	333.126	114.601
3100	41.775	296.180	334.493	118.771
3200	41.892	297.398	335.821	122.954
3300	41.969	298.582	337.112	127.148
3400	42.005	299.734	338.365	131.347
3500	41.998	300.855	339.583	135.548
3600	41.950	301.947	340.765	139.745
3700	41.861	303.012	341.914	143.936
3800	41.734	304.050	343.028	148.116
3900	41.571	305.064	344.110	152.282
4000	41.375	306.053	345.160	156.429
298.15	36.057	212.740	245.359	9.725
Uncertainties in Functions				
300	0.002	0.004	0.005	0.001
1000	0.400	0.030	0.050	0.020
4000	1.000	0.100	0.100	1.000

Table III-20   THERMODYNAMIC FUNCTIONS at 1 atm: HBr(g)

$\dfrac{T}{K}$	$C_p°$	$-\{G°-H°(0)\}/T$	$S°$	$H°-H°(0)$
	$\dots\dots\dots$J$\cdot$(mol$\cdot$K)$^{-1}\dots\dots\dots$			kJ$\cdot$mol^{-1}
0	0.000	0.000	0.000	0.000
100	29.115	138.004	166.775	2.877
200	29.124	158.013	186.958	5.789
300	29.142	169.763	198.770	8.702
400	29.220	178.113	207.162	11.619
500	29.454	184.601	213.704	14.551
600	29.873	189.914	219.108	17.516
700	30.432	194.424	223.754	20.531
800	31.063	198.352	227.858	23.605
900	31.710	201.839	231.554	26.744
1000	32.335	204.981	234.928	29.946
1100	32.919	207.847	238.037	33.209
1200	33.454	210.485	240.925	36.529
1300	33.938	212.931	243.622	39.899
1400	34.374	215.215	246.154	43.315
1500	34.766	217.357	248.539	46.772
1600	35.119	219.377	250.794	50.266
1700	35.437	221.289	252.933	53.795
1800	35.726	223.104	254.967	57.353
1900	35.989	224.832	256.905	60.939
2000	36.230	226.483	258.758	64.550
2100	36.451	228.062	260.531	68.184
2200	36.656	229.577	262.231	71.840
2300	36.848	231.032	263.865	75.515
2400	37.027	232.433	265.437	79.209
2500	37.197	233.784	266.952	82.920
2600	37.357	235.088	268.414	86.648
2700	37.511	236.348	269.827	90.391
2800	37.658	237.569	271.194	94.150
2900	37.799	238.751	272.518	97.923
3000	37.936	239.898	273.801	101.710
3100	38.069	241.012	275.047	105.510
3200	38.198	242.095	276.258	109.323
3300	38.324	243.148	277.435	113.149
3400	38.447	244.173	278.581	116.988
3500	38.566	245.172	279.698	120.838
3600	38.682	246.146	280.786	124.701
3700	38.794	247.097	281.847	128.575
3800	38.902	248.025	282.883	132.460
3900	39.007	248.932	283.895	136.355
4000	39.107	249.818	284.884	140.261
298.15	29.141	169.583	198.591	8.648

Uncertainties in Functions

300	0.003	0.004	0.004	0.001
1000	0.005	0.010	0.015	0.001
4000	0.010	0.050	0.080	0.100

Table III-21 THERMODYNAMIC FUNCTIONS at 1 atm: I(g)

$\dfrac{T}{K}$	$C_p°$	$-\{G°-H°(0)\}/T$	$S°$	$H°-H°(0)$
	.........J·(mol·K)$^{-1}$........			kJ·mol^{-1}
0	0.000	0.000	0.000	0.000
100	20.786	137.185	157.971	2.079
200	20.786	151.593	172.379	4.157
300	20.786	160.021	180.807	6.236
400	20.786	166.001	186.787	8.314
500	20.786	170.639	191.425	10.393
600	20.786	174.429	195.215	12.472
700	20.786	177.633	198.419	14.550
800	20.787	180.408	201.195	16.629
900	20.789	182.857	203.643	18.708
1000	20.795	185.047	205.834	20.787
1100	20.806	187.028	207.816	22.867
1200	20.824	188.837	209.627	24.948
1300	20.851	190.501	211.295	27.032
1400	20.889	192.042	212.842	29.119
1500	20.937	193.478	214.285	31.210
1600	20.995	194.821	215.637	33.307
1700	21.062	196.083	216.912	35.409
1800	21.138	197.274	218.118	37.519
1900	21.220	198.402	219.263	39.637
2000	21.308	199.472	220.354	41.764
2100	21.399	200.492	221.396	43.899
2200	21.493	201.465	222.393	46.044
2300	21.588	202.395	223.351	48.198
2400	21.682	203.288	224.272	50.361
2500	21.775	204.145	225.159	52.534
2600	21.865	204.970	226.014	54.716
2700	21.953	205.765	226.841	56.907
2800	22.037	206.532	227.641	59.107
2900	22.116	207.273	228.416	61.314
3000	22.191	207.990	229.167	63.530
3100	22.261	208.685	229.896	65.752
3200	22.327	209.359	230.604	67.982
3300	22.387	210.014	231.292	70.217
3400	22.443	210.649	231.961	72.459
3500	22.494	211.268	232.612	74.706
3600	22.540	211.869	233.246	76.958
3700	22.581	212.455	233.865	79.214
3800	22.618	213.027	234.467	81.474
3900	22.651	213.584	235.055	83.737
4000	22.679	214.128	235.629	86.004
298.15	20.786	159.892	180.678	6.197

Uncertainties in Functions

300	0.001	0.003	0.004	0.001
1000	0.002	0.003	0.004	0.001
4000	0.003	0.004	0.005	0.004

Table III-22   THERMODYNAMIC FUNCTIONS at 1 atm: $I_2(g)$

$\dfrac{T}{K}$	$C_p°$	$-\{G°-H°(0)\}/T$	$S°$	$H°-H°(0)$
	.........J·(mol·K)$^{-1}$.......			kJ·mol^{-1}
0	0.000	0.000	0.000	0.000
100	33.135	191.544	221.896	3.035
200	36.062	213.352	245.997	6.529
300	36.897	226.859	260.805	10.184
400	37.256	236.741	271.474	13.893
500	37.464	244.551	279.811	17.630
600	37.613	251.015	286.655	21.384
700	37.735	256.532	292.464	25.152
800	37.847	261.345	297.510	28.931
900	37.956	265.616	301.974	32.722
1000	38.070	269.456	305.979	36.522
1100	38.196	272.944	309.613	40.335
1200	38.341	276.140	312.943	44.163
1300	38.514	279.091	316.018	48.004
1400	38.719	281.832	318.880	51.866
1500	38.959	284.392	321.559	55.750
1600	39.231	286.794	324.082	59.660
1700	39.528	289.060	326.469	63.596
1800	39.842	291.200	328.737	67.565
1900	40.161	293.234	330.900	71.566
2000	40.474	295.169	332.968	75.597
2100	40.768	297.017	334.950	79.660
2200	41.035	298.784	336.853	83.750
2300	41.266	300.480	338.682	87.865
2400	41.456	302.109	340.443	92.002
2500	41.600	303.676	342.138	96.156
2600	41.697	305.187	343.771	100.319
2700	41.747	306.646	345.346	104.491
2800	41.751	308.055	346.865	108.669
2900	41.711	309.419	348.329	112.840
3000	41.630	310.739	349.742	117.010
3100	41.511	312.019	351.105	121.168
3200	41.359	313.262	352.421	125.310
3300	41.176	314.467	353.691	129.440
3400	40.966	315.639	354.917	133.546
3500	40.733	316.779	356.101	137.628
3600	40.480	317.887	357.245	141.690
3700	40.210	318.966	358.351	145.726
3800	39.925	320.016	359.419	149.733
3900	39.629	321.040	360.453	153.712
4000	39.323	322.038	361.452	157.657
298.15	36.888	226.650	260.578	10.116

Uncertainties in Functions

300	0.002	0.004	0.005	0.001
1000	0.400	0.030	0.050	0.020
4000	1.000	0.100	0.100	1.000

Table III-23   THERMODYNAMIC FUNCTIONS at 1 atm: HI(g)

$\frac{T}{K}$	$C_p°$	$-(G°-H°(0))/T$	$S°$	$H°-H°(0)$
	........J·(mol·K)$^{-1}$........			kJ·mol^{-1}
0	0.000	0.000	0.000	0.000
100	29.114	145.816	174.665	2.885
200	29.125	165.864	194.848	5.797
300	29.158	177.627	206.661	8.710
400	29.329	185.986	215.068	11.633
500	29.738	192.483	221.652	14.584
600	30.351	197.813	227.126	17.587
700	31.070	202.346	231.858	20.658
800	31.808	206.302	236.055	23.802
900	32.511	209.822	239.842	27.018
1000	33.157	212.999	243.302	30.302
1100	33.735	215.901	246.489	33.647
1200	34.249	218.575	249.447	37.047
1300	34.704	221.057	252.207	40.495
1400	35.106	223.375	254.794	43.986
1500	35.464	225.552	257.228	47.515
1600	35.783	227.604	259.528	51.078
1700	36.070	229.546	261.706	54.671
1800	36.331	231.391	263.775	58.291
1900	36.568	233.148	265.746	61.936
2000	36.787	234.825	267.627	65.604
2100	36.990	236.430	269.427	69.293
2200	37.179	237.970	271.152	73.001
2300	37.358	239.448	272.809	76.728
2400	37.526	240.872	274.402	80.473
2500	37.687	242.244	275.937	84.233
2600	37.840	243.569	277.418	88.010
2700	37.987	244.849	278.849	91.801
2800	38.128	246.088	280.233	95.607
2900	38.262	247.289	281.574	99.427
3000	38.390	248.453	282.873	103.259
3100	38.512	249.584	284.134	107.104
3200	38.627	250.683	285.358	110.961
3300	38.734	251.752	286.549	114.830
3400	38.834	252.792	287.706	118.708
3500	38.924	253.806	288.834	122.596
3600	39.005	254.794	289.931	126.493
3700	39.077	255.759	291.001	130.397
3800	39.137	256.700	292.044	134.308
3900	39.186	257.619	293.061	138.224
4000	39.224	258.518	294.054	142.144
298.15	29.157	177.447	206.481	8.657

Uncertainties in Functions

300	0.003	0.004	0.004	0.001
1000	0.010	0.080	0.100	0.020
4000	0.100	0.100	0.150	0.200

Table III-24   THERMODYNAMIC FUNCTIONS: S(cr & l)

$\dfrac{T}{K}$	$C_p°$	$-\{G°-H°(0)\}/T$	$S°$	$H°-H°(0)$
	.........J·(mol·K)$^{-1}$.......			kJ·mol^{-1}
0	0.000	0.000	0.000	0.000
100	12.770	5.640	12.489	0.685
200	19.389	11.975	23.623	2.330
300	22.791	17.348	32.195	4.454
368.54 α	24.252	20.570	37.033	6.067
368.54 β	24.757	20.570	38.123	6.469
388.36 β	25.326	21.499	39.435	6.966
388.36 l	31.718	21.499	43.859	8.684
400	32.377	22.164	44.808	9.058
428.15 *	36.603	23.727	47.088	10.002
432.25 λ tr	48.841	23.950	47.486	10.174
432.25	48.841	23.950	47.486	10.174
453.15 *	42.480	25.087	49.623	11.119
500	38.034	27.577	53.564	12.994
600	34.379	32.478	60.103	16.575
700	32.459	36.806	65.267	19.922
800	32.000	40.638	69.545	23.126
900	32.000	44.063	73.314	26.326
1000	32.000	47.160	76.686	29.526
298.15	22.750	17.256	32.054	4.412

Uncertainties in Functions

300	0.050	0.040	0.050	0.006
1000	2.000	1.000	1.000	1.000

* Rapid change in functions within this region.

Table III-25   THERMODYNAMIC FUNCTIONS at 1 atm: S(g)

$\dfrac{T}{K}$	$C_p°$	$-\{G°-H°(0)\}/T$	$S°$	$H°-H°(0)$
	$\cdots\cdots\cdots$ J·(mol·K)$^{-1}$ $\cdots\cdots\cdots$			kJ·mol^{-1}
0	0.000	0.000	0.000	0.000
100	21.356	121.898	142.783	2.088
200	23.388	136.599	158.284	4.337
300	23.669	145.529	167.866	6.701
400	23.233	152.001	174.621	9.048
500	22.741	157.059	179.751	11.346
600	22.338	161.195	183.860	13.600
700	22.032	164.684	187.280	15.817
800	21.800	167.695	190.206	18.008
900	21.624	170.342	192.763	20.179
1000	21.490	172.699	195.034	22.335
1100	21.386	174.824	197.077	24.478
1200	21.307	176.757	198.934	26.613
1300	21.249	178.530	200.637	28.740
1400	21.209	180.166	202.211	30.863
1500	21.186	181.685	203.673	32.983
1600	21.178	183.102	205.040	35.101
1700	21.184	184.431	206.324	37.219
1800	21.203	185.681	207.535	39.338
1900	21.234	186.862	208.683	41.460
2000	21.276	187.980	209.773	43.585
2100	21.327	189.043	210.812	45.715
2200	21.386	190.055	211.805	47.851
2300	21.452	191.022	212.758	49.993
2400	21.523	191.946	213.672	52.141
2500	21.598	192.833	214.552	54.297
2600	21.677	193.685	215.401	56.461
2700	21.757	194.504	216.220	58.633
2800	21.838	195.294	217.013	60.813
2900	21.919	196.056	217.781	63.000
3000	21.999	196.793	218.525	65.196
3100	22.078	197.506	219.248	67.400
3200	22.156	198.196	219.950	69.612
3300	22.231	198.866	220.633	71.831
3400	22.303	199.516	221.298	74.058
3500	22.372	200.148	221.945	76.292
3600	22.439	200.762	222.576	78.532
3700	22.502	201.360	223.192	80.779
3800	22.562	201.942	223.793	83.032
3900	22.618	202.510	224.380	85.292
4000	22.671	203.064	224.953	87.556
298.15	23.674	145.391	167.720	6.657

Uncertainties in Functions

300	0.001	0.005	0.006	0.001
1000	0.002	0.005	0.006	0.001
4000	0.003	0.006	0.007	0.004

Table III-26   THERMODYNAMIC FUNCTIONS at 1 atm: $S_2(g)$

$\dfrac{T}{K}$	$C_p°$	$-\{G°-H°(0)\}/T$	$S°$	$\dfrac{H°-H°(0)}{kJ \cdot mol^{-1}}$
	.... J·(mol·K)$^{-1}$			
0	0.000	0.000	0.000	0.000
100	29.369	164.262	194.950	3.069
200	30.462	185.304	215.508	6.041
300	32.540	197.617	228.258	9.192
400	34.108	206.524	237.849	12.530
500	35.133	213.587	245.578	15.996
600	35.815	219.473	252.048	19.545
700	36.305	224.533	257.607	23.152
800	36.697	228.978	262.482	26.803
900	37.045	232.946	266.824	30.491
1000	37.377	236.533	270.745	34.212
1100	37.704	239.808	274.322	37.966
1200	38.030	242.823	277.617	41.753
1300	38.353	245.619	280.674	45.572
1400	38.669	248.226	283.528	49.423
1500	38.976	250.670	286.206	53.304
1600	39.268	252.970	288.731	57.218
1700	39.546	255.145	291.120	61.158
1800	39.805	257.207	293.388	65.126
1900	40.047	259.168	295.547	69.121
2000	40.270	261.039	297.607	73.135
2100	40.475	262.828	299.577	77.171
2200	40.663	264.541	301.465	81.231
2300	40.834	266.186	303.276	85.305
2400	40.990	267.768	305.017	89.396
2500	41.132	269.292	306.694	93.503
2600	41.262	270.762	308.309	97.620
2700	41.380	272.181	309.869	101.756
2800	41.490	273.554	311.376	105.900
2900	41.592	274.883	312.834	110.056
3000	41.689	276.172	314.245	114.217
3100	41.780	277.423	315.614	118.390
3200	41.869	278.637	316.941	122.571
3300	41.956	279.817	318.231	126.764
3400	42.043	280.966	319.485	130.962
3500	42.129	282.083	320.705	135.175
3600	42.218	283.173	321.893	139.390
3700	42.309	284.235	323.051	143.617
3800	42.402	285.271	324.180	147.852
3900	42.500	286.283	325.283	152.097
4000	42.601	287.272	326.360	156.349
298.15	32.505	197.427	228.058	9.132

Uncertainties in Functions

300	0.010	0.010	0.010	0.002
1000	0.500	0.100	0.100	0.100
4000	1.500	0.400	0.500	1.000

Table III-27   THERMODYNAMIC FUNCTIONS at 1 atm: $SO_2(g)$

$\dfrac{T}{K}$	$C_p°$	$-\{G°-H°(0)\}/T$	$S°$	$H°-H°(0)$
	.........J·(mol·K)$^{-1}$........			kJ·mol^{-1}
0	0.000	0.000	0.000	0.000
100	33.543	175.642	208.934	3.329
200	36.380	198.898	232.949	6.810
300	39.909	212.950	248.360	10.623
400	43.427	223.349	260.329	14.792
500	46.490	231.774	270.360	19.293
600	48.938	238.946	279.062	24.070
700	50.829	245.237	286.754	29.062
800	52.282	250.865	293.640	34.220
900	53.407	255.969	299.867	39.508
1000	54.290	260.646	305.541	44.894
1100	54.993	264.967	310.750	50.361
1200	55.564	268.986	315.560	55.888
1300	56.033	272.742	320.027	61.470
1400	56.426	276.270	324.194	67.093
1500	56.759	279.596	328.098	72.752
1600	57.045	282.744	331.771	78.442
1700	57.295	285.730	335.237	84.161
1800	57.516	288.572	338.518	89.902
1900	57.713	291.284	341.633	95.662
2000	57.891	293.877	344.598	101.443
2100	58.054	296.359	347.427	107.242
2200	58.204	298.742	350.131	113.055
2300	58.343	301.034	352.721	118.881
2400	58.474	303.240	355.207	124.722
2500	58.598	305.367	357.597	130.576
2600	58.717	307.420	359.897	136.441
2700	58.831	309.405	362.115	142.318
2800	58.941	311.326	364.257	148.208
2900	59.047	313.187	366.327	154.107
3000	59.152	314.992	368.331	160.018
3100	59.254	316.744	370.272	165.938
3200	59.354	318.446	372.155	171.870
3300	59.454	320.102	373.982	177.806
3400	59.552	321.713	375.759	183.758
3500	59.651	323.282	377.486	189.715
3600	59.749	324.811	379.168	195.687
3700	59.847	326.302	380.806	201.667
3800	59.946	327.758	382.404	207.656
3900	60.045	329.179	383.962	213.655
4000	60.144	330.568	385.484	219.666
298.15	39.842	212.731	248.114	10.549

Uncertainties in Functions

300	0.020	0.050	0.050	0.010
1000	0.050	0.100	0.200	0.050
4000	0.300	0.400	0.500	0.200

Table III-28   THERMODYNAMIC FUNCTIONS at 1 atm: $H_2S(g)$

$\dfrac{T}{K}$	$C_p°$	$-\{G°-H°(0)\}/T$	$S°$	$H°-H°(0)$
	........J·(mol·K)$^{-1}$........			kJ·mol^{-1}
0	0.000	0.000	0.000	0.000
100	33.293	136.035	169.116	3.308
200	33.441	159.009	192.219	6.642
300	34.270	172.505	205.908	10.021
400	35.661	182.163	215.949	13.515
500	37.291	189.758	224.079	17.161
600	39.052	196.070	231.032	20.977
700	40.873	201.513	237.188	24.973
800	42.673	206.326	242.764	29.151
900	44.385	210.663	247.890	33.505
1000	45.968	214.627	252.650	38.023
1100	47.405	218.288	257.100	42.694
1200	48.694	221.698	261.281	47.500
1300	49.845	224.896	265.225	52.428
1400	50.868	227.911	268.957	57.465
1500	51.779	230.766	272.498	62.598
1600	52.590	233.481	275.866	67.817
1700	53.316	236.069	279.076	73.113
1800	53.967	238.544	282.143	78.479
1900	54.553	240.917	285.076	83.903
2000	55.084	243.196	287.888	89.385
2100	55.566	245.389	290.588	94.919
2200	56.007	247.502	293.184	100.499
2300	56.412	249.543	295.682	106.121
2400	56.785	251.516	298.091	111.778
2500	57.131	253.426	300.417	117.476
2600	57.453	255.277	302.664	123.205
2700	57.754	257.072	304.838	128.967
2800	58.036	258.816	306.943	134.754
2900	58.302	260.511	308.985	140.573
3000	58.553	262.160	310.966	146.416
3100	58.792	263.765	312.889	152.282
3200	59.019	265.330	314.759	158.171
3300	59.237	266.855	316.579	164.087
3400	59.445	268.344	318.351	170.022
3500	59.645	269.797	320.076	175.974
3600	59.838	271.217	321.759	181.949
3700	60.025	272.605	323.402	187.947
3800	60.205	273.963	325.005	193.957
3900	60.380	275.292	326.571	199.986
4000	60.551	276.594	328.102	206.030
298.15	34.248	172.298	205.696	9.957

Uncertainties in Functions

300	0.010	0.050	0.050	0.010
1000	0.050	0.100	0.200	0.200
4000	0.300	0.500	0.800	1.000

Table III-29 THERMODYNAMIC FUNCTIONS at 1 atm: N(g)

$\dfrac{T}{K}$	$C_p°$	$-\{G°-H°(0)\}/T$	$S°$	$H°-H°(0)$
	$\cdots\cdots\cdots$ J·(mol·K)$^{-1}$ $\cdots\cdots\cdots$			kJ·mol^{-1}
0	0.000	0.000	0.000	0.000
100	20.786	109.698	130.485	2.079
200	20.786	124.106	144.892	4.157
300	20.786	132.534	153.321	6.236
400	20.786	138.514	159.300	8.314
500	20.786	143.152	163.939	10.393
600	20.786	146.942	167.728	12.472
700	20.786	150.146	170.933	14.550
800	20.786	152.922	173.708	16.629
900	20.786	155.370	176.156	18.708
1000	20.786	157.560	178.347	20.786
1100	20.786	159.541	180.328	22.865
1200	20.786	161.350	182.136	24.943
1300	20.786	163.014	183.800	27.022
1400	20.786	164.554	185.341	29.101
1500	20.786	165.988	186.775	31.179
1600	20.786	167.330	188.116	33.258
1700	20.786	168.590	189.376	35.337
1800	20.786	169.778	190.564	37.415
1900	20.788	170.902	191.688	39.494
2000	20.790	171.968	192.755	41.573
2100	20.793	172.982	193.769	43.652
2200	20.798	173.949	194.737	45.732
2300	20.804	174.873	195.661	47.812
2400	20.813	175.758	196.547	49.893
2500	20.826	176.607	197.397	51.974
2600	20.843	177.422	198.214	54.058
2700	20.864	178.207	199.001	56.143
2800	20.891	178.963	199.760	58.231
2900	20.924	179.693	200.494	60.322
3000	20.963	180.398	201.204	62.416
3100	21.010	181.081	201.892	64.515
3200	21.064	181.742	202.560	66.618
3300	21.127	182.382	203.209	68.728
3400	21.198	183.004	203.841	70.844
3500	21.277	183.608	204.456	72.968
3600	21.366	184.196	205.057	75.100
3700	21.463	184.768	205.643	77.241
3800	21.569	185.324	206.217	79.392
3900	21.685	185.867	206.779	81.555
4000	21.809	186.397	207.330	83.730
298.15	20.786	132.406	153.192	6.197

Uncertainties in Functions

300	0.001	0.003	0.003	0.001
1000	0.001	0.003	0.003	0.001
4000	0.003	0.005	0.005	0.005

Table III-30   THERMODYNAMIC FUNCTIONS at 1 atm: $N_2(g)$

$\dfrac{T}{K}$	$C_p°$	$-\{G°-H°(0)\}/T$	$S°$	$H°-H°(0)$
	........$J\cdot(mol\cdot K)^{-1}$.......			$kJ\cdot mol^{-1}$
0	0.000	0.000	0.000	0.000
100	29.104	130.679	159.701	2.902
200	29.107	150.812	179.876	5.813
300	29.125	162.599	191.680	8.724
400	29.249	170.968	200.072	11.641
500	29.580	177.467	206.630	14.581
600	30.109	182.794	212.067	17.564
700	30.754	187.318	216.756	20.607
800	31.433	191.262	220.907	23.716
900	32.090	194.767	224.648	26.893
1000	32.696	197.929	228.061	30.132
1100	33.241	200.813	231.203	33.429
1200	33.723	203.468	234.117	36.779
1300	34.147	205.931	236.833	40.173
1400	34.517	208.230	239.377	43.606
1500	34.842	210.387	241.770	47.075
1600	35.127	212.420	244.028	50.573
1700	35.377	214.343	246.165	54.098
1800	35.598	216.167	248.194	57.649
1900	35.795	217.904	250.124	61.218
2000	35.969	219.562	251.964	64.805
2100	36.126	221.147	253.723	68.410
2200	36.267	222.666	255.407	72.031
2300	36.394	224.125	257.022	75.664
2400	36.509	225.528	258.573	79.309
2500	36.614	226.880	260.066	82.966
2600	36.710	228.184	261.504	86.633
2700	36.799	229.444	262.891	90.308
2800	36.881	230.663	264.231	93.991
2900	36.956	231.843	265.526	97.681
3000	37.027	232.986	266.780	101.383
3100	37.093	234.096	267.996	105.091
3200	37.155	235.174	269.174	108.801
3300	37.213	236.222	270.318	112.518
3400	37.268	237.241	271.430	116.244
3500	37.320	238.234	272.511	119.970
3600	37.369	239.200	273.563	123.708
3700	37.417	240.143	274.588	127.448
3800	37.462	241.063	275.586	131.188
3900	37.505	241.960	276.560	134.941
4000	37.547	242.837	277.510	138.693
298.15	29.124	162.419	191.500	8.670

Uncertainties in Functions

300	0.001	0.003	0.004	0.001
1000	0.001	0.003	0.005	0.001
4000	0.005	0.005	0.010	0.020

Table III-31   THERMODYNAMIC FUNCTIONS at 1 atm: $NH_3(g)$

$\dfrac{T}{K}$	$C_p°$	$-\{G°-H°(0)\}/T$	$S°$	$\dfrac{H°-H°(0)}{kJ \cdot mol^{-1}}$
	........$J \cdot (mol \cdot K)^{-1}$........			
0	0.000	0.000	0.000	0.000
100	33.284	122.653	155.731	3.308
200	33.750	145.631	178.881	6.650
300	35.678	159.183	192.882	10.109
400	38.674	168.986	203.540	13.821
500	41.994	176.816	212.526	17.855
600	45.229	183.442	220.471	22.218
700	48.269	189.254	227.674	26.894
800	51.112	194.476	234.307	31.865
900	53.769	199.249	240.482	37.110
1000	56.244	203.665	246.277	42.612
1100	58.535	207.790	251.746	48.352
1200	60.645	211.671	256.931	54.313
1300	62.577	215.343	261.862	60.475
1400	64.340	218.835	266.566	66.824
1500	65.946	222.168	271.060	73.339
1600	67.408	225.360	275.363	80.005
1700	68.739	228.423	279.491	86.816
1800	69.950	231.371	283.454	93.750
1900	71.056	234.213	287.266	100.802
2000	72.067	236.958	290.937	107.959
2100	72.993	239.614	294.476	115.211
2200	73.844	242.185	297.893	122.556
2300	74.627	244.679	301.192	129.979
2400	75.351	247.101	304.384	137.478
2500	76.022	249.454	307.474	145.049
2600	76.646	251.744	310.467	152.678
2700	77.227	253.973	313.371	160.373
2800	77.770	256.145	316.190	168.125
2900	78.279	258.262	318.928	175.930
3000	78.758	260.329	321.590	183.781
3100	79.209	262.347	324.179	191.678
3200	79.636	264.319	326.701	199.621
3300	80.040	266.247	329.158	207.605
3400	80.424	268.132	331.553	215.630
3500	80.790	269.978	333.889	223.687
3600	81.139	271.785	336.170	231.784
3700	81.472	273.555	338.398	239.918
3800	81.793	275.290	340.575	248.081
3900	82.100	276.992	342.703	256.271
4000	82.396	278.661	344.786	264.498
298.15	35.630	158.975	192.660	10.043

Uncertainties in Functions

300	0.005	0.004	0.050	0.010
1000	0.500	0.200	0.400	0.200
4000	3.000	1.000	2.000	4.000

Table III-32 THERMODYNAMIC FUNCTIONS: P(cr & l)

$\dfrac{T}{K}$	$C_p°$	$-\{G°-H°(0)\}/T$	$S°$	$H°-H°(0)$
	.........J·(mol·K)$^{-1}$........			kJ·mol^{-1}
0	0.000	0.000	0.000	0.000
100	13.728	8.823	17.702	0.888
200	21.092	16.385	32.141	3.151
300	23.849	23.224	41.237	5.404
317.3 cr	24.267	24.243	42.585	5.820
317.3 l	26.120	24.243	44.661	6.479
400	26.120	29.114	50.711	8.639
500	26.120	34.038	56.540	11.251
600	26.120	38.197	61.302	13.863
700	26.120	41.793	65.328	16.475
800	26.120	44.958	68.816	19.087
900	26.120	47.783	71.893	21.699
1000	26.120	50.334	74.645	24.311
298.15	23.824	23.112	41.090	5.360

Uncertainties in Functions

300	0.200	0.250	0.250	0.015
1000	5.000	3.000	4.000	2.000

Table III-33   THERMODYNAMIC FUNCTIONS at 1 atm: P(g)

$\dfrac{T}{K}$	$C_p°$	$-\{G°-H°(0)\}/T$	$S°$	$H°-H°(0)$
	........J·(mol·K)$^{-1}$........			kJ·mol^{-1}
0	0.000	0.000	0.000	0.000
100	20.786	119.596	140.382	2.079
200	20.786	134.004	154.790	4.157
300	20.786	142.432	163.218	6.236
400	20.786	148.412	169.198	8.314
500	20.786	153.050	173.836	10.393
600	20.786	156.840	177.626	12.472
700	20.786	160.044	180.830	14.550
800	20.786	162.820	183.606	16.629
900	20.786	165.268	186.054	18.708
1000	20.786	167.458	188.244	20.786
1100	20.788	169.439	190.225	22.865
1200	20.791	171.248	192.034	24.944
1300	20.797	172.912	193.699	27.023
1400	20.810	174.452	195.240	29.104
1500	20.832	175.886	196.677	31.186
1600	20.865	177.228	198.022	33.270
1700	20.914	178.489	199.289	35.359
1800	20.981	179.678	200.486	37.454
1900	21.069	180.803	201.622	39.556
2000	21.179	181.872	202.706	41.668
2100	21.313	182.889	203.742	43.793
2200	21.471	183.859	204.737	45.931
2300	21.653	184.788	205.696	48.087
2400	21.859	185.679	206.621	50.263
2500	22.088	186.534	207.518	52.460
2600	22.337	187.358	208.389	54.681
2700	22.605	188.153	209.237	56.928
2800	22.890	188.921	210.065	59.203
2900	23.189	189.664	210.873	61.507
3000	23.500	190.384	211.664	63.841
3100	23.821	191.083	212.440	66.207
3200	24.148	191.762	213.202	68.605
3300	24.480	192.423	213.950	71.037
3400	24.814	193.067	214.685	73.501
3500	25.148	193.695	215.410	75.999
3600	25.480	194.309	216.123	78.531
3700	25.808	194.908	216.825	81.095
3800	26.130	195.494	217.518	83.692
3900	26.444	196.067	218.201	86.321
4000	26.750	196.629	218.874	88.981
298.15	20.786	142.303	163.090	6.197

Uncertainties in Functions

300	0.001	0.002	0.003	0.001
1000	0.002	0.002	0.003	0.001
4000	0.003	0.003	0.004	0.004

Table III-34   THERMODYNAMIC FUNCTIONS at 1 atm: $P_2(g)$

$\dfrac{T}{K}$	$C_p°$	$-\{G°-H°(0)\}/T$	$S°$	$H°-H°(0)$
	........J·(mol·K)$^{-1}$........			kJ·mol^{-1}
0	0.000	0.000	0.000	0.000
100	29.120	156.162	185.254	2.909
200	30.101	176.361	205.637	5.855
300	32.067	188.335	218.212	8.963
400	33.678	197.033	227.671	12.255
500	34.772	203.949	235.312	15.681
600	35.503	209.725	241.721	19.198
700	36.006	214.699	247.234	22.775
800	36.364	219.074	252.066	26.394
900	36.630	222.983	256.365	30.044
1000	36.833	226.518	260.236	33.718
1100	36.994	229.746	263.754	37.410
1200	37.125	232.716	266.979	41.116
1300	37.234	235.467	269.955	44.834
1400	37.328	238.031	272.718	48.562
1500	37.409	240.430	275.296	52.299
1600	37.481	242.685	277.713	56.044
1700	37.546	244.813	279.987	59.795
1800	37.606	246.828	282.135	63.553
1900	37.665	248.740	284.170	67.317
2000	37.720	250.560	286.103	71.086
2100	37.774	252.297	287.945	74.861
2200	37.829	253.958	289.703	78.641
2300	37.884	255.549	291.386	82.426
2400	37.943	257.076	293.000	86.218
2500	38.005	258.544	294.550	90.015
2600	38.071	259.958	296.042	93.819
2700	38.144	261.321	297.480	97.630
2800	38.230	262.637	298.869	101.449
2900	38.321	263.910	300.212	105.277
3000	38.422	265.142	301.513	109.114
3100	38.538	266.335	302.775	112.962
3200	38.664	267.493	304.000	116.822
3300	38.804	268.618	305.192	120.696
3400	38.961	269.711	306.353	124.584
3500	39.133	270.774	307.485	128.489
3600	39.322	271.809	308.590	132.411
3700	39.529	272.818	309.670	136.354
3800	39.755	273.801	310.727	140.318
3900	40.000	274.762	311.763	144.305
4000	40.266	275.699	312.779	148.318
298.15	32.032	188.150	218.014	8.904
	Uncertainties in Functions			
300	0.002	0.003	0.004	0.001
1000	0.050	0.010	0.010	0.010
4000	0.200	0.040	0.050	0.300

Table III-35   THERMODYNAMIC FUNCTIONS at 1 atm: $P_4(g)$

$\dfrac{T}{K}$	$C_p°$	$-\{G°-H°(0)\}/T$	$S°$	$H°-H°(0)$
	.... $J\cdot(mol\cdot K)^{-1}$			$kJ\cdot mol^{-1}$
0	0.000	0.000	0.000	0.000
100	37.139	189.876	223.803	3.393
200	55.502	215.046	255.199	8.031
300	67.233	232.743	280.186	14.233
400	73.210	247.221	300.439	21.287
500	76.452	259.586	317.157	28.785
600	78.362	270.389	331.279	36.534
700	79.568	279.978	343.456	44.435
800	80.374	288.594	354.137	52.434
900	80.938	296.414	363.638	60.502
1000	81.347	303.571	372.188	68.617
1100	81.653	310.167	379.956	76.768
1200	81.887	316.284	387.071	84.945
1300	82.070	321.985	393.633	93.143
1400	82.216	327.323	399.721	101.358
1500	82.335	332.340	405.398	109.586
1600	82.432	337.074	410.715	117.824
1700	82.513	341.555	415.714	126.072
1800	82.580	345.807	420.433	134.326
1900	82.638	349.853	424.899	142.587
2000	82.687	353.712	429.139	150.854
2100	82.729	357.401	433.174	159.124
2200	82.766	360.933	437.024	167.399
2300	82.798	364.322	440.704	175.678
2400	82.826	367.579	444.228	183.959
2500	82.851	370.713	447.610	192.243
2600	82.873	373.733	450.860	200.529
2700	82.893	376.648	453.988	208.817
2800	82.910	379.464	457.003	217.108
2900	82.926	382.189	459.913	225.399
3000	82.941	384.827	462.724	233.693
3100	82.954	387.383	465.444	241.987
3200	82.965	389.864	468.078	250.283
3300	82.976	392.273	470.631	258.581
3400	82.986	394.615	473.108	266.879
3500	82.995	396.892	475.514	275.177
3600	83.003	399.108	477.852	283.477
3700	83.010	401.267	480.126	291.778
3800	83.017	403.372	482.340	300.080
3900	83.024	405.424	484.497	308.382
4000	83.030	407.428	486.599	316.684
298.15	67.081	232.610	279.902	14.100

Uncertainties in Functions

300	1.500	0.400	0.500	0.200
1000	2.000	2.000	4.000	2.000
4000	2.500	5.000	7.000	8.000

Table III-36   THERMODYNAMIC FUNCTIONS: C(graphite)

$\dfrac{T}{K}$	C_p°	$-\{(G^{\circ}-H^{\circ}(0)\}/T$	S°	$H^{\circ}-H^{\circ}(0)$
	........J·(mol·K)$^{-1}$........			kJ·mol^{-1}
0	0.000	0.000	0.000	0.000
25	0.126	0.044	0.073	0.001
50	0.507	0.130	0.266	0.007
75	1.038	0.227	0.571	0.026
100	1.674	0.356	0.952	0.060
150	3.229	0.708	1.912	0.181
200	5.006	1.152	3.082	0.386
250	6.816	1.667	4.394	0.681
300	8.581	2.240	5.793	1.066
350	10.241	2.851	7.242	1.537
400	11.817	3.491	8.713	2.089
450	13.289	4.154	10.191	2.717
500	14.623	4.831	11.662	3.415
600	16.844	6.211	14.533	4.994
700	18.537	7.596	17.263	6.766
800	19.827	8.967	19.826	8.687
900	20.824	10.308	22.221	10.722
1000	21.610	11.612	24.457	12.845
1100	22.244	12.876	26.548	15.039
1200	22.766	14.098	28.506	17.290
1300	23.204	15.278	30.346	19.589
1400	23.578	16.416	32.080	21.929
1500	23.904	17.516	33.718	24.303
1600	24.191	18.577	35.270	26.708
1700	24.448	19.603	36.744	29.140
1800	24.681	20.595	38.149	31.597
1900	24.895	21.554	39.489	34.076
2000	25.094	22.483	40.771	36.576
2200	25.453	24.257	43.180	41.631
2400	25.775	25.928	45.408	46.754
2600	26.071	27.507	47.483	51.939
2800	26.348	29.004	49.426	57.181
3000	26.611	30.427	51.253	62.478
3200	26.863	31.783	52.978	67.825
3400	27.106	33.078	54.614	73.222
3600	27.342	34.318	56.170	78.667
3800	27.574	35.508	57.655	84.159
4000	27.801	36.651	59.075	89.696
298.15	8.517	2.218	5.740	1.050

Uncertainties in Functions

300	0.080	0.100	0.100	0.020
1000	0.080	0.380	0.340	0.050
3000	0.650	1.800	0.550	0.500
4000	3.300	7.000	1.000	2.100

0.008 87(OD)

Table III-37   THERMODYNAMIC FUNCTIONS at 1 atm: C(g)

$\dfrac{T}{K}$	$C_p°$	$-\{G°-H°(0)\}/T$	$S°$	$H°-H°(0)$
	$\cdots\cdots\cdots$J·(mol·K)$^{-1}\cdots\cdots\cdots$			kJ·mol^{-1}
0	0.000	0.000	0.000	0.000
100	21.271	111.218	135.072	2.385
200	20.904	127.220	149.660	4.488
300	20.838	136.206	158.120	6.574
400	20.815	142.469	164.111	8.657
500	20.805	147.279	168.755	10.738
600	20.799	151.184	172.548	12.818
700	20.796	154.471	175.753	14.898
800	20.793	157.309	178.530	16.977
900	20.792	159.805	180.979	19.056
1000	20.791	162.034	183.170	21.136
1100	20.791	164.047	185.151	23.215
1200	20.793	165.882	186.960	25.294
1300	20.797	167.569	188.625	27.373
1400	20.803	169.128	190.166	29.453
1500	20.814	170.579	191.602	31.534
1600	20.829	171.936	192.946	33.616
1700	20.851	173.209	194.209	35.700
1800	20.878	174.409	195.402	37.786
1900	20.912	175.544	196.531	39.876
2000	20.952	176.620	197.605	41.969
2100	20.999	177.644	198.628	44.067
2200	21.052	178.620	199.606	46.169
2300	21.110	179.553	200.543	48.277
2400	21.174	180.447	201.443	50.391
2500	21.242	181.304	202.309	52.512
2600	21.313	182.128	203.143	54.640
2700	21.387	182.921	203.949	56.775
2800	21.464	183.686	204.728	58.917
2900	21.542	184.425	205.483	61.068
3000	21.621	185.139	206.215	63.226
3100	21.701	185.831	206.925	65.392
3200	21.780	186.501	207.615	67.566
3300	21.859	187.151	208.286	69.748
3400	21.936	187.782	208.940	71.938
3500	22.013	188.396	209.577	74.135
3600	22.087	188.993	210.198	76.340
3700	22.160	189.574	210.804	78.553
3800	22.230	190.141	211.396	80.772
3900	22.298	190.693	211.975	82.998
4000	22.363	191.232	212.540	85.231
298.15	20.839	136.070	157.991	6.536

Uncertainties in Functions

300	0.001	0.002	0.003	0.001
1000	0.002	0.002	0.003	0.001
4000	0.003	0.003	0.004	0.004

Table III-38   THERMODYNAMIC FUNCTIONS at 1 atm: CO(g)

$\frac{T}{K}$	$C_p°$	$-\{G°-H°(0)\}/T$	$S°$	$H°-H°(0)$
	........J·(mol·K)$^{-1}$........			kJ·mol^{-1}
0	0.000	0.000	0.000	0.000
100	29.104	136.723	165.749	2.903
200	29.109	156.858	185.925	5.813
300	29.142	168.646	197.731	8.725
400	29.340	177.018	206.136	11.647
500	29.792	183.523	212.727	14.602
600	30.440	188.860	218.214	17.613
700	31.170	193.400	222.960	20.692
800	31.898	197.363	227.171	23.847
900	32.573	200.889	230.967	27.070
1000	33.178	204.073	234.431	30.358
1100	33.709	206.979	237.619	33.704
1200	34.169	209.657	240.572	37.098
1300	34.568	212.142	243.323	40.536
1400	34.914	214.463	245.898	44.009
1500	35.213	216.640	248.317	47.516
1600	35.475	218.691	250.598	51.052
1700	35.703	220.632	252.756	54.611
1800	35.905	222.473	254.801	58.191
1900	36.083	224.227	256.749	61.792
2000	36.241	225.900	258.604	65.409
2100	36.383	227.500	260.375	69.038
2200	36.510	229.033	262.071	72.684
2300	36.625	230.505	263.696	76.340
2400	36.730	231.921	265.257	80.007
2500	36.826	233.285	266.759	83.686
2600	36.914	234.600	268.205	87.374
2700	36.995	235.871	269.599	91.066
2800	37.070	237.099	270.946	94.772
2900	37.140	238.289	272.248	98.482
3000	37.205	239.442	273.508	102.199
3100	37.266	240.561	274.729	105.922
3200	37.324	241.647	275.913	109.652
3300	37.378	242.703	277.063	113.389
3400	37.430	243.730	278.179	117.128
3500	37.479	244.730	279.265	120.874
3600	37.526	245.704	280.322	124.626
3700	37.571	246.654	281.350	128.376
3800	37.615	247.580	282.353	132.138
3900	37.657	248.484	283.331	135.904
4000	37.697	249.367	284.285	139.673
298.15	29.141	168.466	197.551	8.671

Uncertainties in Functions

300	0.002	0.003	0.004	0.001
1000	0.002	0.003	0.004	0.001
4000	0.003	0.004	0.005	0.005

Table III-39   THERMODYNAMIC FUNCTIONS at 1 atm: $CO_2(g)$

$\dfrac{T}{K}$	$C_p°$	$-\{(G°-H°(0)\}/T$	$S°$	$H°-H°(0)$
	........J·(mol·K)$^{-1}$........			kJ·mol^{-1}
0	0.000	0.000	0.000	0.000
100	29.206	149.792	178.888	2.910
200	32.360	170.093	199.856	5.952
300	37.220	182.459	213.906	9.434
400	41.328	191.775	225.198	13.369
500	44.627	199.442	234.787	17.673
600	47.327	206.046	243.170	22.275
700	49.569	211.892	250.639	27.123
800	51.442	217.164	257.384	32.176
900	53.008	221.980	263.536	37.401
1000	54.320	226.422	269.191	42.769
1100	55.423	230.551	274.422	48.258
1200	56.354	234.412	279.285	53.848
1300	57.144	238.040	283.828	59.525
1400	57.818	241.464	288.088	65.274
1500	58.397	244.708	292.097	71.084
1600	58.899	247.789	295.883	76.951
1700	59.335	250.724	299.468	82.864
1800	59.718	253.527	302.870	88.816
1900	60.055	256.210	306.108	94.805
2000	60.355	258.783	309.196	100.825
2100	60.623	261.254	312.148	106.876
2200	60.863	263.632	314.973	112.949
2300	61.081	265.924	317.684	119.047
2400	61.279	268.135	320.287	125.164
2500	61.459	270.271	322.793	131.304
2600	61.626	272.338	325.206	137.455
2700	61.779	274.340	327.535	143.625
2800	61.922	276.280	329.784	149.810
2900	62.055	278.162	331.960	156.013
3000	62.180	279.991	334.065	162.220
3100	62.299	281.768	336.106	168.446
3200	62.411	283.497	338.086	174.683
3300	62.519	285.181	340.008	180.927
3400	62.624	286.821	341.876	187.185
3500	62.725	288.421	343.693	193.453
3600	62.825	289.980	345.461	199.730
3700	62.924	291.503	347.184	206.018
3800	63.023	292.990	348.863	212.315
3900	63.123	294.445	350.501	218.620
4000	63.224	295.865	352.101	224.942
298.15	37.135	182.264	213.676	9.365

Uncertainties in Functions

300	0.002	0.005	0.010	0.003
1000	0.005	0.008	0.020	0.005
4000	0.050	0.010	0.030	0.050

Table III-40 THERMODYNAMIC FUNCTIONS: Si(cr & l)

$\frac{T}{K}$	$C_p°$	$-\{G°-H°(0)\}/T$	$S°$	$H°-H°(0)$
	........J·(mol·K)$^{-1}$........			kJ·mol^{-1}
0	0.000	0.000	0.000	0.000
100	7.280	1.158	3.828	0.267
200	15.650	4.431	11.657	1.443
300	19.855	8.087	18.933	3.254
400	22.301	11.582	25.023	5.377
500	23.610	14.797	30.152	7.678
600	24.472	17.731	34.537	10.084
700	25.124	20.410	38.361	12.565
800	25.662	22.870	41.752	15.105
900	26.135	25.140	44.802	17.695
1000	26.568	27.247	47.578	20.331
1100	26.974	29.213	50.130	23.008
1200	27.362	31.056	52.493	25.725
1300	27.737	32.791	54.698	28.480
1400	28.103	34.430	56.767	31.272
1500	28.462	35.985	58.719	34.100
1600	28.816	37.464	60.567	36.964
1690 cr	29.131	38.737	62.152	39.572
1690 l	27.200	38.737	91.862	89.782
1700	27.200	39.050	92.023	90.054
1800	27.200	42.036	93.578	92.774
1900	27.200	44.788	95.048	95.494
2000	27.200	47.336	96.443	98.214
2100	27.200	49.707	97.770	100.934
2200	27.200	51.920	99.036	103.654
2300	27.200	53.995	100.245	106.374
2400	27.200	55.947	101.402	109.094
2500	27.200	57.787	102.513	111.814
298.15	19.789	8.020	18.810	3.217

Uncertainties in Functions

300	0.030	0.060	0.080	0.008
1000	0.100	0.200	0.250	0.100
2500	1.000	0.700	0.650	0.700

Table III-41   THERMODYNAMIC FUNCTIONS at 1 atm: Si(g)

$\frac{T}{K}$	$C_p°$	$-\{G°-H°(0)\}/T$	$S°$	$H°-H°(0)$
	........J·(mol·K)$^{-1}$........			kJ·mol^{-1}
0	0.000	0.000	0.000	0.000
100	28.022	113.370	140.778	2.741
200	23.796	132.195	158.709	5.303
300	22.234	142.705	168.009	7.591
400	21.613	149.860	174.309	9.780
500	21.316	155.247	179.096	11.924
600	21.153	159.555	182.967	14.047
700	21.057	163.138	186.220	16.157
800	21.000	166.202	189.027	18.260
900	20.971	168.879	191.499	20.358
1000	20.968	171.253	193.708	22.455
1100	20.989	173.387	195.707	24.553
1200	21.033	175.324	197.535	26.654
1300	21.099	177.098	199.221	28.760
1400	21.183	178.735	200.788	30.874
1500	21.283	180.255	202.253	32.997
1600	21.394	181.673	203.630	35.131
1700	21.513	183.003	204.930	37.276
1800	21.638	184.256	206.163	39.434
1900	21.764	185.440	207.337	41.604
2000	21.889	186.563	208.456	43.786
2100	22.012	187.631	209.527	45.982
2200	22.129	188.650	210.554	48.189
2300	22.241	189.624	211.540	50.407
2400	22.347	190.557	212.489	52.637
2500	22.444	191.453	213.403	54.876
2600	22.535	192.314	214.285	57.125
2700	22.617	193.144	215.137	59.383
2800	22.692	193.944	215.961	61.649
2900	22.759	194.717	216.759	63.921
3000	22.819	195.465	217.531	66.200
3100	22.872	196.189	218.280	68.485
3200	22.918	196.890	219.007	70.774
3300	22.958	197.571	219.713	73.068
3400	22.993	198.233	220.399	75.366
3500	23.022	198.876	221.066	77.667
3600	23.046	199.501	221.715	79.970
3700	23.065	200.110	222.347	82.276
3800	23.080	200.703	222.962	84.583
3900	23.092	201.282	223.562	86.892
4000	23.100	201.846	224.146	89.201
298.15	22.251	142.548	167.872	7.550
Uncertainties in Functions				
300	0.001	0.003	0.004	0.001
1000	0.002	0.003	0.004	0.001
4000	0.003	0.004	0.005	0.004

Table III-42   THERMODYNAMIC FUNCTIONS: SiO_2(cr, quartz)

$\dfrac{T}{K}$	$C_p°$	$-\{G°-H°(0)\}/T$	$S°$	$H°-H°(0)$
	........J·(mol·K)$^{-1}$........			kJ·mol^{-1}
0	0.000	0.000	0.000	0.000
100	15.690	3.334	9.694	0.636
200	32.640	10.607	26.087	3.096
300	44.712	18.408	41.736	6.999
400	53.477	26.021	55.744	11.889
500	60.533	33.262	68.505	17.621
600	64.452	40.106	79.919	23.887
700	68.234	46.534	90.114	30.506
800	76.224	52.581	99.674	37.674
848 α	82.967	55.377	104.298	41.485
848 β	67.446	55.377	104.782	41.895
900	67.953	58.349	108.811	45.415
1000	68.941	63.761	116.021	52.260
1100	69.940	68.817	122.639	59.204
1200	70.947	73.561	128.768	66.248
1300	71.960	78.030	134.486	73.394
1400	72.977	82.256	139.856	80.640
1500	73.997	86.267	144.926	87.989
1600	75.019	90.084	149.734	95.440
1696 (a)	76.002	93.587	154.134	102.689
298.15	44.602	18.264	41.460	6.916

Uncertainties in Functions

300	0.300	0.100	0.200	0.020
1000	0.700	0.400	0.800	0.500
1696 (a)	1.000	0.600	1.500	1.000

(a). Metastable fusion temperature.

Table III-43   THERMODYNAMIC FUNCTIONS at 1 atm: $SiF_4(g)$

$\dfrac{T}{K}$	$C_p°$	$-\{G°-H°(0)\}/T$	$S°$	$H°-H°(0)$
	.........J·(mol·K)$^{-1}$........			kJ·mol^{-1}
0	0.000	0.000	0.000	0.000
100	41.816	185.336	220.561	3.523
200	60.879	212.224	255.834	8.722
300	73.826	231.467	283.108	15.492
400	83.143	247.281	305.699	23.367
500	89.627	260.941	324.994	32.026
600	94.091	273.045	341.754	41.226
700	97.208	283.935	356.506	50.800
800	99.434	293.843	369.640	60.638
900	101.064	302.932	381.451	70.667
1000	102.286	311.328	392.165	80.837
1100	103.222	319.128	401.960	91.115
1200	103.953	326.411	410.974	101.475
1300	104.533	333.241	419.318	111.900
1400	105.001	339.670	427.083	122.378
1500	105.383	345.742	434.340	132.897
1600	105.700	351.495	441.152	143.452
1700	105.964	356.959	447.568	154.036
1800	106.187	362.163	453.631	164.644
1900	106.377	367.129	459.378	175.272
2000	106.540	371.880	464.839	185.918
2100	106.681	376.431	470.040	196.579
2200	106.804	380.800	475.006	207.254
2300	106.912	385.000	479.756	217.940
2400	107.006	389.043	484.308	228.636
2500	107.090	392.942	488.678	239.341
2600	107.164	396.705	492.880	250.053
2700	107.230	400.343	496.925	260.773
2800	107.290	403.862	500.826	271.499
2900	107.343	407.271	504.592	282.231
3000	107.392	410.576	508.232	292.968
3100	107.436	413.783	511.754	303.709
3200	107.475	416.899	515.166	314.455
3300	107.512	419.927	518.474	325.204
3400	107.545	422.873	521.683	335.957
3500	107.575	425.741	524.801	346.713
3600	107.603	428.534	527.832	357.472
3700	107.629	431.258	530.781	368.233
3800	107.652	433.915	533.651	378.997
3900	107.675	436.509	536.448	389.764
4000	107.695	439.041	539.174	400.532
298.15	73.622	231.148	282.652	15.356

Uncertainties in Functions

300	0.500	0.300	0.500	0.050
1000	0.700	1.000	1.500	0.500
4000	1.500	2.500	4.000	7.000

Table III-44   THERMODYNAMIC FUNCTIONS: Ge(cr & l)

$\dfrac{T}{K}$	$C_p°$	$-\{G°-H°(0)\}/T$	$S°$	$H°-H°(0)$
	........J·(mol·K)$^{-1}$........			kJ·mol^{-1}
0	0.000	0.000	0.000	0.000
100	13.820	3.666	9.866	0.620
200	21.130	10.000	22.175	2.435
300	23.249	15.637	31.234	4.679
400	24.310	20.427	38.083	7.062
500	24.962	24.526	43.582	9.528
600	25.452	28.096	48.178	12.050
700	25.867	31.254	52.133	14.616
800	26.240	34.085	55.612	17.222
900	26.591	36.653	58.723	19.863
1000	26.926	39.003	61.542	22.539
1100	27.252	41.171	64.124	25.248
1200	27.571	43.185	66.509	27.989
1211.4 cr	27.608	43.405	66.770	28.304
1211.4 l	27.600	43.405	97.338	65.334
1300	27.600	47.148	99.286	67.779
1400	27.600	50.946	101.331	70.539
1500	27.600	54.369	103.236	73.299
1600	27.600	57.480	105.017	76.059
1700	27.600	60.326	106.690	78.819
1800	27.600	62.946	108.268	81.579
1900	27.600	65.371	109.760	84.339
2000	27.600	67.626	111.176	87.099
2100	27.600	69.732	112.522	89.859
2200	27.600	71.707	113.806	92.619
2300	27.600	73.564	115.033	95.379
2400	27.600	75.316	116.208	98.139
2500	27.600	76.975	117.334	100.899
298.15	23.222	15.541	31.090	4.636

Uncertainties in Functions

300	0.100	0.100	0.150	0.020
1000	0.150	0.300	0.400	0.200
2500	1.400	1.100	1.000	1.300

Table III-45 THERMODYNAMIC FUNCTIONS at 1 atm: Ge(g)

$\dfrac{T}{K}$	$C_p°$	$-\{G°-H°(0)\}/T$	$S°$	$H°-H°(0)$
	........J·(mol·K)$^{-1}$........			kJ·mol^{-1}
0	0.000	0.000	0.000	0.000
100	21.314	118.700	139.552	2.085
200	27.485	133.542	156.067	4.505
300	30.757	143.134	167.985	7.455
400	31.072	150.515	176.915	10.560
500	30.361	156.510	183.783	13.637
600	29.265	161.525	189.225	16.620
700	28.102	165.809	193.648	19.488
800	27.029	169.525	197.329	22.243
900	26.108	172.793	200.458	24.899
1000	25.349	175.698	203.168	27.470
1100	24.741	178.306	205.555	29.974
1200	24.264	180.667	207.686	32.423
1300	23.898	182.821	209.613	34.830
1400	23.624	184.798	211.374	37.205
1500	23.426	186.625	212.996	39.557
1600	23.288	188.321	214.503	41.893
1700	23.197	189.903	215.912	44.217
1800	23.143	191.385	217.237	46.533
1900	23.117	192.779	218.487	48.846
2000	23.113	194.094	219.673	51.157
2100	23.123	195.339	220.801	53.469
2200	23.144	196.521	221.877	55.782
2300	23.171	197.646	222.906	58.098
2400	23.202	198.719	223.893	60.417
2500	23.234	199.745	224.841	62.738
2600	23.266	200.728	225.752	65.063
2700	23.297	201.671	226.631	67.392
2800	23.326	202.578	227.479	69.723
2900	23.351	203.451	228.298	72.057
3000	23.374	204.292	229.090	74.393
3100	23.393	205.105	229.857	76.731
3200	23.408	205.890	230.600	79.071
3300	23.420	206.650	231.320	81.413
3400	23.428	207.385	232.019	83.755
3500	23.433	208.099	232.699	86.098
3600	23.435	208.792	233.359	88.442
3700	23.434	209.464	234.001	90.785
3800	23.430	210.118	234.626	93.129
3900	23.424	210.754	235.234	95.471
4000	23.415	211.374	235.827	97.813
298.15	30.733	142.980	167.795	7.398

Uncertainties in Functions

300	0.001	0.004	0.005	0.001
1000	0.002	0.004	0.005	0.001
4000	0.003	0.005	0.006	0.004

Table III-46   THERMODYNAMIC FUNCTIONS: GeO_2(cr & l)

$\dfrac{T}{K}$	C_p°	$-\{G^\circ-H^\circ(0)\}/T$	S°	$H^\circ-H^\circ(0)$
	........J·(mol·K)$^{-1}$........			kJ·mol^{-1}
0	0.000	0.000	0.000	0.000
100	13.710	1.598	5.874	0.428
200	35.740	7.732	22.502	2.954
300	50.475	15.611	40.021	7.323
400	61.281	23.775	56.248	12.989
500	66.273	31.731	70.519	19.394
600	69.089	39.249	82.872	26.173
700	70.974	46.269	93.671	33.182
800	72.449	52.804	103.247	40.355
900	73.764	58.895	111.857	47.666
1000	75.049	64.589	119.696	55.107
1100	76.378	69.931	126.910	62.677
1200	77.796	74.962	133.616	70.385
1300	79.332	79.718	139.903	78.240
1308 tet	79.460	80.087	140.390	78.876
1308 hex	80.075	80.087	156.827	100.376
1388 hex	81.297	84.650	161.617	106.831
1388 l	78.500	84.650	174.009	124.031
1400	78.500	85.418	174.685	124.973
1500	78.500	91.552	180.100	132.823
1600	78.500	97.246	185.167	140.673
1700	78.500	102.560	189.926	148.523
1800	78.500	107.539	194.413	156.373
1900	78.500	112.224	198.657	164.223
2000	78.500	116.647	202.684	172.073
2100	78.500	120.836	206.514	179.923
2200	78.500	124.814	210.165	187.773
2300	78.500	128.602	213.655	195.623
2400	78.500	132.215	216.996	203.473
2500	78.500	135.671	220.200	211.323
2600	78.500	138.982	223.279	219.173
2700	78.500	142.159	226.242	227.023
2800	78.500	145.214	229.097	234.873
2900	78.500	148.154	231.851	242.723
3000	78.500	150.988	234.513	250.573
298.15	50.166	15.460	39.710	7.230

Uncertainties in Functions

300	0.300	0.100	0.150	0.020
1000	0.700	0.400	0.800	0.500
3000	10.000	7.000	15.000	10.000

Table III-47   THERMODYNAMIC FUNCTIONS at 1 atm: $GeF_4(g)$

$\dfrac{T}{K}$	$C_p°$	$-\{G°-H°(0)\}/T$	$S°$	$H°-H°(0)$
	$\ldots\ldots\ldots$ J·(mol·K)$^{-1}$ $\ldots\ldots\ldots$			kJ·mol^{-1}
0	0.000	0.000	0.000	0.000
100	49.708	192.545	230.532	3.799
200	69.286	222.454	271.709	9.851
300	81.792	244.178	302.324	17.444
400	90.034	261.906	327.072	26.066
500	95.252	277.067	347.767	35.350
600	98.612	290.360	365.452	45.055
700	100.854	302.210	380.833	55.036
800	102.406	312.903	394.407	65.203
900	103.518	322.645	406.536	75.502
1000	104.338	331.591	417.488	85.897
1100	104.959	339.859	427.462	96.363
1200	105.439	347.546	436.617	106.884
1300	105.818	354.727	445.072	117.448
1400	106.122	361.464	452.925	128.045
1500	106.369	367.809	460.256	138.670
1600	106.573	373.804	467.127	149.318
1700	106.742	379.485	473.593	159.984
1800	106.885	384.885	479.699	170.665
1900	107.006	390.028	485.481	181.360
2000	107.110	394.940	490.973	192.066
2100	107.200	399.638	496.201	202.782
2200	107.278	404.141	501.189	213.506
2300	107.346	408.465	505.960	224.237
2400	107.406	412.624	510.530	234.974
2500	107.459	416.628	514.915	245.718
2600	107.506	420.490	519.131	256.467
2700	107.548	424.219	523.189	267.219
2800	107.586	427.824	527.101	277.976
2900	107.620	431.313	530.877	288.736
3000	107.650	434.693	534.526	299.500
3100	107.678	437.970	538.056	310.266
3200	107.703	441.152	541.475	321.035
3300	107.726	444.242	544.790	331.807
3400	107.747	447.247	548.006	342.580
3500	107.766	450.171	551.130	353.356
3600	107.783	453.017	554.166	364.133
3700	107.800	455.791	557.119	374.913
3800	107.814	458.496	559.994	385.693
3900	107.829	461.135	562.795	396.475
4000	107.841	463.710	565.525	407.259
298.15	81.602	243.818	301.819	17.293
Uncertainties in Functions				
300	1.000	0.500	1.000	0.100
1000	2.000	1.500	2.500	0.800
4000	3.000	3.000	5.000	10.000

Table III-48   THERMODYNAMIC FUNCTIONS: Sn(cr & l)

$\frac{T}{K}$	$C_p°$	$-\{G°-H°(0)\}/T$	$S°$	$H°-H°(0)$
	........J·(mol·K)$^{-1}$........			kJ·mol^{-1}
0	0.000	0.000	0.000	0.000
100	22.210	10.840	24.020	1.318
200	25.500	21.970	40.670	3.740
300	27.147	30.104	51.348	6.373
400	28.903	36.459	59.398	9.176
500	31.033	41.730	66.065	12.167
505.118 cr	31.160	41.979	66.381	12.326
505.118 l	29.415	41.979	80.626	19.521
600	28.663	48.498	85.619	22.273
700	28.249	54.123	90.003	25.116
800	28.043	58.848	93.760	27.929
900	27.957	62.914	97.057	30.729
1000	27.945	66.478	100.002	33.523
1100	27.979	69.649	102.666	36.319
1200	28.044	72.503	105.103	39.120
1300	28.130	75.099	107.351	41.929
1400	28.229	77.478	109.440	44.746
1500	28.339	79.674	111.391	47.575
1600	28.455	81.714	113.223	50.414
1700	28.575	83.619	114.952	53.266
1800	28.698	85.406	116.589	56.129
1900	28.822	87.088	118.144	59.005
2000	28.947	88.678	119.625	61.894
2100	29.071	90.186	121.041	64.795
2200	29.195	91.619	122.396	67.708
2300	29.318	92.986	123.696	70.634
2400	29.439	94.292	124.947	73.572
2500	29.559	95.542	126.151	76.522
2600	29.676	96.742	127.313	79.483
2700	29.792	97.895	128.435	82.457
2800	29.905	99.005	129.520	85.442
2900	30.015	100.076	130.572	88.438
3000	30.123	101.109	131.591	91.444
3100	30.228	102.109	132.580	94.462
3200	30.331	103.076	133.542	97.490
3300	30.431	104.013	134.477	100.528
3400	30.528	104.923	135.386	103.576
3500	30.622	105.806	136.273	106.634
3600	30.713	106.664	137.137	109.700
3700	30.801	107.499	137.979	112.776
3800	30.886	108.312	138.802	115.860
3900	30.968	109.104	139.605	118.953
4000	31.047	109.877	140.390	122.054
298.15	27.112	29.973	51.180	6.323

Uncertainties in Functions

300	0.030	0.050	0.080	0.008
1000	0.200	0.300	0.500	0.300
4000	5.000	6.000	7.000	10.000

Table III-49   THERMODYNAMIC FUNCTIONS at 1 atm: Sn(g)

$\dfrac{T}{K}$	C_p°	$-\{G^\circ-H^\circ(0)\}/T$	S°	$H^\circ-H^\circ(0)$
	........J·(mol·K)$^{-1}$........			kJ·mol^{-1}
0	0.000	0.000	0.000	0.000
100	20.786	124.824	145.610	2.079
200	20.805	139.232	160.020	4.158
300	21.277	147.667	168.514	6.254
400	22.888	153.696	174.828	8.453
500	25.328	158.470	180.189	10.860
600	27.880	162.499	185.035	13.522
700	30.057	166.042	189.503	16.423
800	31.651	169.236	193.628	19.513
900	32.641	172.160	197.419	22.733
1000	33.096	174.862	200.885	26.024
1100	33.135	177.373	204.044	29.338
1200	32.880	179.717	206.918	32.641
1300	32.439	181.912	209.534	35.908
1400	31.896	183.971	211.918	39.125
1500	31.312	185.908	214.099	42.286
1600	30.728	187.734	216.101	45.388
1700	30.169	189.457	217.947	48.432
1800	29.648	191.088	219.656	51.423
1900	29.172	192.634	221.246	54.363
2000	28.740	194.102	222.732	57.259
2100	28.353	195.499	224.124	60.113
2200	28.006	196.830	225.435	62.931
2300	27.695	198.101	226.673	65.716
2400	27.417	199.316	227.846	68.471
2500	27.166	200.480	228.960	71.200
2600	26.940	201.596	230.021	73.905
2700	26.735	202.668	231.034	76.589
2800	26.548	203.698	232.003	79.253
2900	26.376	204.690	232.931	81.899
3000	26.218	205.646	233.823	84.528
3100	26.071	206.569	234.680	87.143
3200	25.934	207.461	235.505	89.743
3300	25.805	208.323	236.301	92.330
3400	25.684	209.157	237.070	94.904
3500	25.569	209.965	237.813	97.467
3600	25.460	210.749	238.532	100.018
3700	25.356	211.509	239.228	102.559
3800	25.257	212.247	239.903	105.090
3900	25.163	212.965	240.557	107.611
4000	25.071	213.663	241.193	110.122
298.15	21.259	147.538	168.383	6.215

Uncertainties in Functions

300	0.001	0.003	0.004	0.001
1000	0.002	0.004	0.005	0.001
4000	0.003	0.005	0.006	0.004

Table III-50   THERMODYNAMIC FUNCTIONS: SnO(cr & l)

$\dfrac{T}{K}$	C_p°	$-\{G^\circ-H^\circ(0)\}/T$	S°	$H^\circ-H^\circ(0)$
	........J·(mol·K)$^{-1}$.......			kJ·mol^{-1}
0	0.000	0.000	0.000	0.000
100	24.160	6.510	17.430	1.092
200	39.860	17.680	39.540	4.373
300	47.808	28.051	57.466	8.824
400	49.164	37.222	71.405	13.673
500	50.520	45.207	82.521	18.657
600	51.876	52.224	91.852	23.777
700	53.232	58.476	99.950	29.032
800	54.588	64.118	107.147	34.423
900	55.944	69.266	113.655	39.950
1000	57.300	74.007	119.619	45.612
1100	58.656	78.408	125.144	51.410
1200	60.012	82.520	130.306	57.343
1250 cr	60.690	84.481	132.770	60.361
1250 l	63.000	84.481	154.930	88.061
1300	63.000	87.238	157.401	91.211
1400	63.000	92.419	162.069	97.511
1500	63.000	97.209	166.416	103.811
1600	63.000	101.663	170.482	110.111
1700	63.000	105.824	174.301	116.411
1800	63.000	109.729	177.902	122.711
1900	63.000	113.408	181.308	129.011
2000	63.000	116.884	184.540	135.311
298.15	47.783	27.869	57.170	8.736
Uncertainties in Functions				
300	0.300	0.150	0.300	0.020
1000	6.000	4.000	6.000	3.500
2000	8.000	10.000	13.000	10.000

Table III-51   THERMODYNAMIC FUNCTIONS: SnO_2(cr & l)

$\dfrac{T}{K}$	$C_p°$	$-\{G°-H°(0)\}/T$	$S°$	$H°-H°(0)$
	........J·(mol·K)$^{-1}$........			$kJ·mol^{-1}$
0	0.000	0.000	0.000	0.000
100	19.250	3.102	10.042	0.694
200	40.380	11.619	30.314	3.739
300	53.540	21.094	49.370	8.483
400	65.087	30.349	66.577	14.491
500	70.827	39.150	81.779	21.315
600	74.281	47.383	95.020	28.583
700	76.656	55.037	106.659	36.136
800	78.457	62.149	117.017	43.895
900	79.922	68.772	126.345	51.816
1000	81.180	74.960	134.833	59.872
1100	82.303	80.762	142.623	68.048
1200	83.332	86.221	149.829	76.330
1300	84.297	91.375	156.538	84.712
1400	85.215	96.256	162.819	93.188
1500	86.098	100.893	168.728	101.754
1600	86.954	105.308	174.312	110.406
1700	87.789	109.525	179.609	119.144
1800	88.609	113.559	184.650	127.964
1900	89.416	117.429	189.463	136.865
1903 cr	89.440	117.542	189.604	137.133
1903 l	92.000	117.542	201.900	160.533
2000	92.000	121.745	206.474	169.457
2100	92.000	125.888	210.963	178.657
2200	92.000	129.853	215.243	187.857
2300	92.000	133.655	219.332	197.057
2400	92.000	137.307	223.248	206.257
2500	92.000	140.820	227.003	215.457
298.15	53.219	20.920	49.040	8.384

Uncertainties in Functions

300	0.200	0.100	0.100	0.020
1000	1.000	4.000	5.000	1.000
2500	10.000	8.000	10.000	20.000

Table III-52   THERMODYNAMIC FUNCTIONS: Pb(cr & l)

$\dfrac{T}{K}$	$C_p°$	$-\{G°-H°(0)\}/T$	$S°$	$H°-H°(0)$
	.........J·(mol·K)$^{-1}$........			kJ·mol^{-1}
0	0.000	0.000	0.000	0.000
100	24.430	19.239	36.899	1.766
200	25.770	32.901	54.346	4.289
300	26.673	41.901	64.965	6.919
400	27.788	48.686	72.796	9.644
500	28.785	54.158	79.105	12.473
600	29.736	58.771	84.437	15.399
600.65 cr	29.742	58.799	84.469	15.419
600.65 l	30.627	58.799	92.481	20.231
700	30.313	63.920	97.146	23.258
800	29.979	68.331	101.172	26.273
900	29.660	72.179	104.684	29.255
1000	29.369	75.588	107.794	32.206
1100	29.116	78.645	110.581	35.130
1200	28.903	81.413	113.105	38.030
1300	28.731	83.941	115.411	40.912
1400	28.602	86.265	117.535	43.778
1500	28.513	88.416	119.505	46.633
1600	28.463	90.418	121.344	49.482
1700	28.451	92.288	123.069	52.327
1800	28.475	94.044	124.695	55.173
1900	28.532	95.698	126.236	58.023
2000	28.620	97.262	127.702	60.881
2100	28.737	98.745	129.101	63.748
2200	28.881	100.155	130.441	66.629
2300	29.048	101.500	131.729	69.525
2400	29.238	102.786	132.969	72.439
2500	29.446	104.017	134.167	75.373
2600	29.671	105.199	135.326	78.329
2700	29.909	106.336	136.450	81.308
2800	30.160	107.431	137.542	84.311
2900	30.419	108.488	138.605	87.340
3000	30.684	109.509	139.641	90.395
3100	30.953	110.497	140.651	93.477
3200	31.223	111.455	141.638	96.586
3300	31.492	112.384	142.603	99.722
3400	31.757	113.287	143.547	102.884
3500	32.015	114.165	144.471	106.073
3600	32.264	115.019	145.377	109.287
298.15	26.650	41.758	64.800	6.870

Uncertainties in Functions

300	0.100	0.250	0.300	0.030
1000	0.150	0.500	0.500	0.150
3600	5.000	5.000	5.000	6.000

Table III-53   THERMODYNAMIC FUNCTIONS at 1 atm: Pb(g)

$\dfrac{T}{K}$	$C_p°$	$-\{G°-H°(0)\}/T$	$S°$	$H°-H°(0)$
	........J·(mol·K)$^{-1}$........			kJ·mol^{-1}
0	0.000	0.000	0.000	0.000
100	20.786	131.773	152.559	2.079
200	20.786	146.181	166.967	4.157
300	20.786	154.609	175.395	6.236
400	20.786	160.589	181.375	8.314
500	20.786	165.227	186.013	10.393
600	20.786	169.017	189.803	12.472
700	20.786	172.221	193.007	14.550
800	20.790	174.996	195.783	16.629
900	20.801	177.445	198.232	18.709
1000	20.827	179.635	200.425	20.790
1100	20.888	181.617	202.413	22.876
1200	20.991	183.427	204.234	24.969
1300	21.156	185.093	205.921	27.076
1400	21.394	186.637	207.497	29.203
1500	21.717	188.078	208.983	31.357
1600	22.132	189.429	210.397	33.549
1700	22.640	190.703	211.754	35.787
1800	23.239	191.909	213.065	38.080
1900	23.922	193.056	214.339	40.438
2000	24.678	194.151	215.585	42.867
2100	25.499	195.201	216.809	45.376
2200	26.365	196.211	218.015	47.969
2300	27.262	197.185	219.206	50.650
2400	28.173	198.127	220.386	53.422
2500	29.084	199.041	221.555	56.284
2600	29.979	199.929	222.713	59.238
2700	30.845	200.794	223.861	62.279
2800	31.669	201.638	224.997	65.405
2900	32.444	202.463	226.122	68.612
3000	33.160	203.270	227.234	71.892
3100	33.812	204.061	228.333	75.242
3200	34.397	204.837	229.416	78.653
3300	34.912	205.598	230.482	82.119
3400	35.358	206.345	231.531	85.633
3500	35.734	207.079	232.562	89.188
3600	36.044	207.801	233.573	92.777
3700	36.290	208.511	234.564	96.394
3800	36.475	209.210	235.534	100.033
3900	36.605	209.897	236.483	103.688
4000	36.683	210.573	237.411	107.352
298.15	20.786	154.480	175.266	6.197
Uncertainties in Functions				
300	0.001	0.004	0.005	0.001
1000	0.002	0.005	0.006	0.001
4000	0.003	0.006	0.007	0.004

Table III-54   THERMODYNAMIC FUNCTIONS: B(cr & l)

$\dfrac{T}{K}$	$C_p°$	$-\{G°-H°(0)\}/T$	$S°$	$H°-H°(0)$
	........J·(mol·K)$^{-1}$........			kJ·mol^{-1}
0	0.000	0.000	0.000	0.000
100	1.070	0.089	0.339	0.025
200	6.050	0.660	2.460	0.360
300	11.210	1.827	5.969	1.243
400	15.915	3.351	9.913	2.625
500	18.634	5.054	13.779	4.362
600	20.513	6.810	17.351	6.324
700	21.951	8.553	20.625	8.450
800	23.116	10.253	23.635	10.705
900	24.094	11.896	26.416	13.067
1000	24.931	13.479	28.999	15.520
1100	25.658	15.001	31.410	18.050
1200	26.294	16.463	33.670	20.648
1300	26.855	17.869	35.797	23.306
1400	27.353	19.222	37.806	26.017
1500	27.798	20.525	39.709	28.775
1600	28.199	21.781	41.516	31.575
1700	28.565	22.993	43.236	34.414
1800	28.903	24.164	44.879	37.287
1900	29.221	25.296	46.450	40.194
2000	29.526	26.391	47.957	43.131
2100	29.824	27.453	49.405	46.099
2200	30.123	28.483	50.799	49.096
2300	30.429	29.482	52.145	52.123
2348 cr	30.580	29.952	52.775	53.588
2348 l	31.400	29.952	74.155	103.788
2400	31.400	30.917	74.842	105.421
2500	31.400	32.700	76.124	108.561
2600	31.400	34.394	77.356	111.701
2700	31.400	36.007	78.541	114.840
2800	31.400	37.547	79.683	117.980
2900	31.400	39.019	80.785	121.120
3000	31.400	40.429	81.849	124.260
3100	31.400	41.782	82.879	127.400
3200	31.400	43.082	83.876	130.540
3300	31.400	44.333	84.842	133.681
3400	31.400	45.538	85.779	136.820
3500	31.400	46.701	86.689	139.961
3600	31.400	47.824	87.574	143.100
3700	31.400	48.910	88.434	146.241
3800	31.400	49.961	89.272	149.380
3900	31.400	50.979	90.087	152.521
4000	31.400	51.967	90.882	155.660
298.15	11.087	1.801	5.900	1.222
Uncertainties in Functions				
300	0.100	0.050	0.080	0.008
1000	0.200	0.200	0.300	0.200
4000	3.000	3.000	4.000	10.000

Table III-55   THERMODYNAMIC FUNCTIONS at 1 atm: B(g)

$\dfrac{T}{K}$	$C_p°$	$-\{G°-H°(0)\}/T$	$S°$	$H°-H°(0)$
	........J·(mol·K)$^{-1}$.......			kJ·mol^{-1}
0	0.000	0.000	0.000	0.000
100	20.881	108.668	130.578	2.191
200	20.809	123.650	145.021	4.274
300	20.796	132.274	153.456	6.354
400	20.792	138.353	159.438	8.434
500	20.790	143.051	164.077	10.513
600	20.789	146.881	167.867	12.592
700	20.788	150.114	171.072	14.671
800	20.788	152.911	173.848	16.749
900	20.787	155.376	176.296	18.828
1000	20.787	157.579	178.486	20.907
1100	20.787	159.572	180.468	22.986
1200	20.787	161.389	182.276	25.064
1300	20.787	163.061	183.940	27.143
1400	20.787	164.608	185.481	29.222
1500	20.787	166.048	186.915	31.300
1600	20.787	167.394	188.256	33.379
1700	20.787	168.659	189.516	35.458
1800	20.786	169.851	190.705	37.536
1900	20.786	170.978	191.828	39.615
2000	20.786	172.048	192.895	41.694
2100	20.786	173.065	193.909	43.772
2200	20.786	174.034	194.876	45.851
2300	20.786	174.961	195.800	47.929
2400	20.786	175.848	196.684	50.008
2500	20.786	176.698	197.533	52.087
2600	20.786	177.515	198.348	54.165
2700	20.786	178.302	199.133	56.244
2800	20.788	179.059	199.889	58.323
2900	20.788	179.790	200.618	60.402
3000	20.789	180.496	201.323	62.481
3100	20.791	181.179	202.005	64.560
3200	20.793	181.840	202.665	66.639
3300	20.795	182.481	203.305	68.718
3400	20.799	183.103	203.925	70.798
3500	20.803	183.706	204.528	72.878
3600	20.808	184.293	205.115	74.959
3700	20.814	184.863	205.685	77.039
3800	20.822	185.419	206.240	79.121
3900	20.831	185.959	206.781	81.204
4000	20.842	186.487	207.308	83.288
298.15	20.796	132.143	153.327	6.316

Uncertainties in Functions

300	0.005	0.015	0.015	0.002
1000	0.005	0.015	0.015	0.005
4000	0.005	0.020	0.020	0.020

Table III-56   THERMODYNAMIC FUNCTIONS: B_2O_3(cr & l)

$\dfrac{T}{K}$	$C_p°$	$-\{(G°-H°(0)\}/T$	$S°$	$H°-H°(0)$
	........J·(mol·K)$^{-1}$........			kJ·mol^{-1}
0	0.000	0.000	0.000	0.000
100	20.870	3.409	10.979	0.757
200	43.930	12.614	32.844	4.046
300	63.135	22.968	54.359	9.417
400	78.524	33.403	74.814	16.564
500	89.119	43.587	93.525	24.969
600	97.827	53.351	110.562	34.327
700	105.644	62.658	126.237	44.506
723 cr	107.366	64.735	129.680	46.955
723 l	133.050	64.735	163.650	71.515
800	131.950	74.913	177.057	81.716
900	130.921	87.140	192.536	94.856
1000	130.185	98.380	206.289	107.909
1100	129.640	108.762	218.671	120.899
1200	129.226	118.397	229.932	133.842
1300	128.904	127.380	240.263	146.748
1400	128.648	135.788	249.806	159.625
1500	128.442	143.688	258.674	172.479
1600	128.273	151.137	266.958	185.314
1700	128.133	158.181	274.730	198.134
1800	128.016	164.861	282.051	210.942
1900	127.916	171.212	288.969	223.738
2000	127.831	177.266	295.528	236.525
2100	127.759	183.047	301.764	249.305
2200	127.695	188.579	307.705	262.077
2300	127.640	193.883	313.380	274.844
2400	127.592	198.976	318.812	287.606
2500	127.549	203.874	324.019	300.363
2600	127.511	208.592	329.021	313.116
2700	127.477	213.142	333.833	325.865
2800	127.447	217.536	338.468	338.611
2900	127.420	221.783	342.940	351.355
3000	127.396	225.894	347.259	364.095
3100	127.374	229.877	351.436	376.834
3200	127.353	233.739	355.480	389.570
298.15	62.761	22.774	53.970	9.301

Uncertainties in Functions

300	0.300	0.300	0.300	0.040
1000	2.000	0.800	1.500	0.700
3200	20.000	4.000	5.000	10.000

Table III-57 THERMODYNAMIC FUNCTIONS: H_3BO_3(cr & l)

$\dfrac{T}{K}$	$C_p°$	$-\{G°-H°(0)\}/T$	$S°$	$H°-H°(0)$
	........J·(mol·K)$^{-1}$........			kJ·mol^{-1}
0	0.000	0.000	0.000	0.000
100	36.170	12.020	29.620	1.760
200	59.300	29.160	61.890	6.546
300	86.564	44.885	90.484	13.680
400	113.817	59.872	119.119	23.699
444.1 cr	125.836	66.378	131.640	28.983
444.1 l	180.000	66.378	181.854	51.283
500	180.000	80.505	203.195	61.345
600	180.000	103.771	236.013	79.345
700	180.000	124.695	263.760	97.345
800	180.000	143.614	287.795	115.345
298.15	86.060	44.604	89.950	13.520

Uncertainties in Functions

300	0.400	0.300	0.600	0.040
800	30.000	15.000	25.000	8.000

Table III-58   THERMODYNAMIC FUNCTIONS at 1 atm: BF$_3$(g)

$\dfrac{T}{K}$	$C_p°$	$-\{G°-H°(0)\}/T$	$S°$	$H°-H°(0)$
		.........J·(mol·K)$^{-1}$........		kJ·mol^{-1}
0	0.000	0.000	0.000	0.000
100	34.097	176.772	210.150	3.338
200	41.987	200.414	235.925	7.102
300	50.607	215.481	254.630	11.745
400	57.549	227.264	270.180	17.166
500	62.950	237.221	283.628	23.203
600	67.073	245.964	295.486	29.714
700	70.196	253.808	306.072	36.584
800	72.569	260.947	315.607	43.728
900	74.389	267.509	324.264	51.080
1000	75.802	273.586	332.178	58.592
1100	76.913	279.248	339.457	66.230
1200	77.800	284.549	346.188	73.967
1300	78.516	289.534	352.445	81.784
1400	79.101	294.239	358.286	89.666
1500	79.584	298.693	363.760	97.601
1600	79.988	302.922	368.910	105.580
1700	80.328	306.948	373.769	113.597
1800	80.616	310.789	378.369	121.644
1900	80.863	314.462	382.735	129.719
2000	81.076	317.980	386.888	137.816
2100	81.261	321.356	390.848	145.933
2200	81.423	324.602	394.632	154.067
2300	81.565	327.726	398.255	162.217
2400	81.690	330.738	401.729	170.380
2500	81.801	333.644	405.066	178.554
2600	81.900	336.453	408.276	186.739
2700	81.988	339.171	411.369	194.934
2800	82.068	341.803	414.352	203.137
2900	82.139	344.355	417.233	211.347
3000	82.204	346.831	420.019	219.564
3100	82.262	349.235	422.715	227.788
3200	82.316	351.573	425.328	236.016
3300	82.365	353.846	427.862	244.251
3400	82.409	356.060	430.321	252.490
3500	82.450	358.216	432.711	260.732
3600	82.487	360.317	435.034	268.979
3700	82.522	362.367	437.294	277.230
3800	82.554	364.368	439.495	285.484
3900	82.584	366.322	441.640	293.740
4000	82.611	368.231	443.732	302.000
298.15	50.463	215.239	254.314	11.650

Uncertainties in Functions

300	0.100	0.100	0.200	0.020
1000	0.300	1.000	1.500	0.300
4000	1.500	2.000	3.500	5.000

Table III-59 THERMODYNAMIC FUNCTIONS: Al(cr & l)

$\dfrac{T}{K}$	C_p°	$-\{G^\circ - H^\circ(0)\}/T$	S°	$H^\circ - H^\circ(0)$
	........J·(mol·K)$^{-1}$........			kJ·mol^{-1}
0	0.000	0.000	0.000	0.000
100	12.996	2.133	6.983	0.485
200	21.340	7.720	19.140	2.284
300	24.234	13.167	28.450	4.585
400	25.735	17.921	35.639	7.088
500	26.911	22.069	41.511	9.721
600	28.043	25.736	46.517	12.468
700	29.345	29.026	50.934	15.336
800	31.006	32.019	54.956	18.349
900	33.210	34.779	58.729	21.555
933.61 cr	34.105	35.664	59.963	22.686
933.61 l	31.750	35.664	71.424	33.386
1000	31.750	38.111	73.605	35.494
1100	31.750	41.477	76.631	38.669
1200	31.750	44.524	79.393	41.844
1300	31.750	47.305	81.935	45.019
1400	31.750	49.864	84.288	48.194
1500	31.750	52.232	86.478	51.369
1600	31.750	54.438	88.527	54.544
1700	31.750	56.500	90.452	57.719
1800	31.750	58.437	92.267	60.894
1900	31.750	60.263	93.984	64.069
2000	31.750	61.990	95.612	67.244
2100	31.750	63.629	97.161	70.419
2200	31.750	65.187	98.638	73.594
2300	31.750	66.672	100.050	76.769
2400	31.750	68.091	101.401	79.944
2500	31.750	69.449	102.697	83.119
2600	31.750	70.752	103.942	86.294
2700	31.750	72.004	105.140	89.469
2800	31.750	73.208	106.295	92.644
2900	31.750	74.368	107.409	95.819
3000	31.750	75.488	108.486	98.994
298.15	24.200	13.073	28.300	4.540

Uncertainties in Functions

300	0.070	0.080	0.100	0.020
1000	0.200	0.200	0.300	0.300
3000	4.000	2.000	2.000	8.000

Table III-60   THERMODYNAMIC FUNCTIONS at 1 atm: Al(g)

$\dfrac{T}{K}$	$C_p°$	$-\{G°-H°(0)\}/T$	$S°$	$H°-H°(0)$
	........J·(mol·K)$^{-1}$........			kJ·mol^{-1}
0	0.000	0.000	0.000	0.000
100	25.192	114.903	139.511	2.461
200	22.133	131.826	155.775	4.790
300	21.383	141.383	164.578	6.958
400	21.117	147.984	170.687	9.081
500	20.995	153.012	175.385	11.186
600	20.930	157.069	179.206	13.282
700	20.891	160.468	182.429	15.373
800	20.866	163.391	185.217	17.461
900	20.849	165.955	187.674	19.546
1000	20.837	168.239	189.870	21.631
1100	20.828	170.297	191.855	23.714
1200	20.821	172.170	193.667	25.796
1300	20.816	173.889	195.333	27.878
1400	20.812	175.476	196.876	29.959
1500	20.808	176.951	198.312	32.040
1600	20.806	178.329	199.654	34.121
1700	20.803	179.621	200.916	36.202
1800	20.801	180.837	202.105	38.282
1900	20.800	181.986	203.229	40.362
2000	20.799	183.075	204.296	42.442
2100	20.797	184.110	205.311	44.522
2200	20.796	185.096	206.278	46.601
2300	20.795	186.037	207.203	48.681
2400	20.795	186.938	208.088	50.760
2500	20.794	187.801	208.937	52.840
2600	20.793	188.630	209.752	54.919
2700	20.793	189.426	210.537	56.998
2800	20.794	190.194	211.293	59.078
2900	20.794	190.934	212.023	61.157
3000	20.795	191.649	212.728	63.237
3100	20.796	192.340	213.410	65.316
3200	20.798	193.009	214.070	67.396
3300	20.801	193.657	214.710	69.476
3400	20.805	194.285	215.331	71.556
3500	20.810	194.895	215.934	73.637
3600	20.816	195.488	216.521	75.719
3700	20.824	196.064	217.091	77.800
3800	20.833	196.625	217.647	79.883
3900	20.845	197.171	218.188	81.967
4000	20.859	197.703	218.716	84.052
298.15	21.391	141.240	164.445	6.919

Uncertainties in Functions

300	0.001	0.004	0.004	0.001
1000	0.002	0.005	0.008	0.001
4000	0.005	0.010	0.020	0.050

Table III-61   THERMODYNAMIC FUNCTIONS: $Al_2O_3(cr \& l)$

$\dfrac{T}{K}$	$C_p°$	$-\{G°-H°(0)\}/T$	$S°$	$H°-H°(0)$
	........J·(mol·K)$^{-1}$........			kJ·mol^{-1}
0	0.000	0.000	0.000	0.000
100	12.841	1.024	4.284	0.326
200	51.141	7.273	24.878	3.519
300	79.470	17.535	51.410	10.163
400	96.187	29.233	76.826	19.037
500	106.021	41.059	99.419	29.180
600	112.851	52.484	119.404	40.152
700	117.241	63.336	137.152	51.671
800	120.332	73.573	153.020	63.557
900	122.667	83.209	167.333	75.712
1000	124.533	92.282	180.357	88.075
1100	126.093	100.839	192.302	100.609
1200	127.444	108.927	203.333	113.287
1300	129.111	116.588	213.600	126.115
1400	130.768	123.865	223.228	139.109
1500	132.424	130.794	232.307	152.269
1600	134.081	137.410	240.906	165.594
1700	135.738	143.741	249.085	179.085
1800	137.395	149.811	256.890	192.742
1900	139.052	155.645	264.363	206.564
2000	140.709	161.262	271.538	220.552
2100	142.366	166.678	278.443	234.706
2200	144.023	171.911	285.104	249.025
2300	145.680	176.973	291.543	263.510
2327 cr	146.127	178.312	293.246	267.450
2327 l	162.900	178.312	341.118	378.850
2400	162.900	183.341	346.150	390.742
2500	162.900	189.987	352.800	407.032
2600	162.900	196.373	359.189	423.322
2700	162.900	202.518	365.337	439.612
2800	162.900	208.439	371.261	455.902
2900	162.900	214.153	376.978	472.192
3000	162.900	219.673	382.500	488.482
3100	162.900	225.012	387.842	504.772
3200	162.900	230.182	393.014	521.062
3300	162.900	235.193	398.026	537.351
3400	162.900	240.054	402.889	553.641
3500	162.900	244.774	407.611	569.932
3600	162.900	249.361	412.200	586.222
3700	162.900	253.823	416.664	602.511
3800	162.900	258.165	421.008	618.802
3900	162.900	262.395	425.239	635.092
4000	162.900	266.518	429.364	651.381
298.15	79.033	17.326	50.920	10.016
Uncertainties in Functions				
300	0.200	0.070	0.100	0.020
1000	0.400	0.300	0.400	0.300
4000	20.000	7.000	8.000	25.000

Table III-62   THERMODYNAMIC FUNCTIONS: $AlF_3(cr)$

$\dfrac{T}{K}$	$C_p°$	$-\{G°-H°(0)\}/T$	$S°$	$H°-H°(0)$
	........$J \cdot (mol \cdot K)^{-1}$........			$kJ \cdot mol^{-1}$
0	0.000	0.000	0.000	0.000
100	24.930	3.608	12.058	0.845
200	56.880	14.701	39.991	5.058
300	75.395	27.768	66.966	11.759
400	86.285	40.560	90.312	19.901
500	92.272	52.558	110.264	28.853
600	97.311	63.644	127.517	38.323
700	105.359	73.893	143.058	48.415
728 cr	108.691	76.634	147.253	51.411
728 l	97.578	76.634	148.029	51.976
800	98.518	83.481	157.276	59.036
900	99.716	92.341	168.950	68.948
1000	100.831	100.538	179.514	78.976
1100	101.893	108.163	189.175	89.113
1200	102.918	115.290	198.085	99.353
1300	103.917	121.981	206.362	109.695
1400	104.897	128.288	214.099	120.136
1500	105.864	134.253	221.369	130.674
1600	106.820	139.914	228.232	141.309
1700	107.767	145.302	234.737	152.038
1800	108.709	150.444	240.923	162.862
1900	109.645	155.363	246.826	173.780
2000	110.578	160.078	252.473	184.791
2100	111.507	164.608	257.891	195.895
298.15	75.122	27.526	66.500	11.620

Uncertainties in Functions

300	0.400	0.200	0.500	0.040
1000	1.000	1.000	1.500	0.500
2100	10.000	4.000	5.000	7.000

Table III-63   THERMODYNAMIC FUNCTIONS: Zn(cr & l)

$\dfrac{T}{K}$	$C_p°$	$-\{G°-H°(0)\}/T$	$S°$	$H°-H°(0)$
	........J·(mol·K)$^{-1}$........			kJ·mol^{-1}
0	0.000	0.000	0.000	0.000
100	19.460	6.560	16.520	0.996
200	24.050	15.660	31.810	3.230
300	25.410	22.774	41.787	5.704
400	26.226	28.495	49.215	8.288
500	27.220	33.253	55.165	10.956
600	28.820	37.338	60.258	13.752
692.73 cr	30.975	40.696	64.543	16.519
692.73 l	31.400	40.696	75.081	23.819
700	31.400	41.055	75.409	24.048
800	31.400	45.617	79.602	27.188
900	31.400	49.603	83.300	30.328
1000	31.400	53.141	86.608	33.468
1100	31.400	56.321	89.601	36.608
1200	31.400	59.210	92.333	39.748
1300	31.400	61.856	94.847	42.888
1400	31.400	64.297	97.174	46.028
1500	31.400	66.562	99.340	49.168
1600	31.400	68.674	101.366	52.308
1700	31.400	70.654	103.270	55.448
1800	31.400	72.516	105.065	58.588
1900	31.400	74.274	106.763	61.728
2000	31.400	75.939	108.373	64.868
298.15	25.390	22.656	41.630	5.657
Uncertainties in Functions				
300	0.040	0.080	0.150	0.020
1000	0.300	0.400	0.600	0.300
2000	5.000	3.000	3.000	4.000

Table III-64   THERMODYNAMIC FUNCTIONS at 1 atm: Zn(g)

$\dfrac{T}{K}$	$C_p°$	$-\{G°-H°(0)\}/T$	$S°$	$H°-H°(0)$
	........J·(mol·K)$^{-1}$........			kJ·mol^{-1}
0	0.000	0.000	0.000	0.000
100	20.786	117.387	138.173	2.079
200	20.786	131.795	152.581	4.157
300	20.786	140.223	161.009	6.236
400	20.786	146.203	166.989	8.314
500	20.786	150.841	171.627	10.393
600	20.786	154.631	175.417	12.472
700	20.786	157.835	178.621	14.550
800	20.786	160.611	181.397	16.629
900	20.786	163.059	183.845	18.708
1000	20.786	165.249	186.035	20.786
1100	20.786	167.230	188.016	22.865
1200	20.786	169.039	189.825	24.943
1300	20.786	170.703	191.489	27.022
1400	20.786	172.243	193.029	29.101
1500	20.786	173.677	194.463	31.179
1600	20.786	175.019	195.805	33.258
1700	20.786	176.279	197.065	35.337
1800	20.786	177.467	198.253	37.415
1900	20.786	178.591	199.377	39.494
2000	20.786	179.657	200.443	41.572
2100	20.786	180.671	201.457	43.651
2200	20.786	181.638	202.424	45.730
2300	20.786	182.562	203.348	47.808
2400	20.786	183.447	204.233	49.887
2500	20.786	184.295	205.081	51.966
2600	20.787	185.110	205.897	54.044
2700	20.787	185.895	206.681	56.123
2800	20.787	186.651	207.437	58.202
2900	20.788	187.380	208.167	60.280
3000	20.789	188.085	208.871	62.359
3100	20.791	188.767	209.553	64.438
3200	20.793	189.426	210.213	66.517
3300	20.796	190.066	210.853	68.597
3400	20.800	190.687	211.474	70.676
3500	20.806	191.289	212.077	72.757
3600	20.813	191.875	212.663	74.838
3700	20.823	192.444	213.234	76.920
3800	20.835	192.999	213.789	79.002
3900	20.849	193.539	214.330	81.087
4000	20.868	194.065	214.858	83.172
298.15	20.786	140.094	160.881	6.197

Uncertainties in Functions

300	0.001	0.003	0.004	0.001
1000	0.001	0.003	0.004	0.001
4000	0.004	0.004	0.005	0.005

Table III-65 THERMODYNAMIC FUNCTIONS: Cd(cr & l)

$\frac{T}{K}$	$C_p°$	$-(G°-H°(0))/T$	$S°$	$H°-H°(0)$
	........J·(mol·K)$^{-1}$........			kJ·mol^{-1}
0	0.000	0.000	0.000	0.000
100	25.110	11.600	25.200	1.360
200	24.920	22.910	41.635	3.745
300	26.036	30.977	51.961	6.295
400	27.080	37.216	59.586	8.948
500	28.337	42.325	65.761	11.718
594.258 cr	29.599	46.448	70.760	14.448
594.258 l	29.900	46.448	80.958	20.508
600	29.900	46.779	81.246	20.680
700	29.900	52.041	85.855	23.670
800	29.900	56.523	89.847	26.660
900	29.900	60.425	93.369	29.650
1000	29.900	63.880	96.519	32.640
1100	29.900	66.978	99.369	35.630
1200	29.900	69.788	101.971	38.620
1300	29.900	72.356	104.364	41.610
1400	29.900	74.723	106.580	44.600
1500	29.900	76.916	108.643	47.590
1600	29.900	78.960	110.572	50.580
1700	29.900	80.873	112.385	53.570
1800	29.900	82.672	114.094	56.560
298.15	26.020	30.847	51.800	6.247

Uncertainties in Functions

300	0.040	0.100	0.150	0.015
1000	0.300	0.400	0.500	0.200
1800	3.000	2.000	2.000	2.000

Table III-66 THERMODYNAMIC FUNCTIONS at 1 atm: Cd(g)

$\dfrac{T}{K}$	$C_p°$	$-\{G°-H°(0)\}/T$	$S°$	$H°-H°(0)$
	........J·(mol·K)$^{-1}$........			kJ·mol^{-1}
0	0.000	0.000	0.000	0.000
100	20.786	124.146	144.932	2.079
200	20.786	138.554	159.340	4.157
300	20.786	146.982	167.768	6.236
400	20.786	152.962	173.748	8.314
500	20.786	157.600	178.386	10.393
600	20.786	161.390	182.176	12.472
700	20.786	164.594	185.380	14.550
800	20.786	167.370	188.156	16.629
900	20.786	169.818	190.604	18.708
1000	20.786	172.008	192.794	20.786
1100	20.786	173.989	194.775	22.865
1200	20.786	175.798	196.584	24.943
1300	20.786	177.461	198.248	27.022
1400	20.786	179.002	199.788	29.101
1500	20.786	180.436	201.222	31.179
1600	20.786	181.777	202.564	33.258
1700	20.786	183.038	203.824	35.337
1800	20.786	184.226	205.012	37.415
1900	20.786	185.350	206.136	39.494
2000	20.786	186.416	207.202	41.572
2100	20.786	187.430	208.216	43.651
2200	20.786	188.397	209.183	45.730
2300	20.786	189.321	210.107	47.808
2400	20.786	190.206	210.992	49.887
2500	20.787	191.054	211.840	51.966
2600	20.787	191.869	212.656	54.044
2700	20.787	192.654	213.440	56.123
2800	20.788	193.410	214.196	58.202
2900	20.790	194.139	214.926	60.281
3000	20.792	194.844	215.630	62.360
3100	20.794	195.525	216.312	64.439
3200	20.798	196.185	216.972	66.519
3300	20.804	196.825	217.613	68.599
3400	20.811	197.446	218.234	70.679
3500	20.820	198.048	218.837	72.761
3600	20.831	198.634	219.424	74.843
3700	20.846	199.204	219.995	76.927
3800	20.864	199.758	220.551	79.013
3900	20.886	200.298	221.093	81.100
4000	20.913	200.825	221.622	83.190
298.15	20.786	146.853	167.640	6.197
	Uncertainties in Functions			
300	0.001	0.003	0.004	0.001
1000	0.001	0.003	0.004	0.001
4000	0.004	0.004	0.005	0.005

Table III-67   THERMODYNAMIC FUNCTIONS at 1 atm: Hg(g)

$\dfrac{T}{K}$	$C_p°$	$-\{G°-H°(0)\}/T$	$S°$	$H°-H°(0)$
	........J·(mol·K)$^{-1}$........			kJ·mol^{-1}
0	0.000	0.000	0.000	0.000
100	20.786	131.368	152.155	2.079
200	20.786	145.776	166.562	4.157
300	20.786	154.204	174.991	6.236
400	20.786	160.184	180.970	8.314
500	20.786	164.822	185.609	10.393
600	20.786	168.612	189.398	12.472
700	20.786	171.816	192.603	14.550
800	20.786	174.592	195.378	16.629
900	20.786	177.040	197.827	18.708
1000	20.786	179.230	200.017	20.786
1100	20.786	181.211	201.998	22.865
1200	20.786	183.020	203.806	24.943
1300	20.786	184.684	205.470	27.022
1400	20.786	186.224	207.011	29.101
1500	20.786	187.658	208.445	31.179
1600	20.786	189.000	209.786	33.258
1700	20.786	190.260	211.046	35.337
1800	20.786	191.448	212.234	37.415
1900	20.786	192.572	213.358	39.494
2000	20.786	193.638	214.424	41.572
2100	20.786	194.652	215.439	43.651
2200	20.786	195.619	216.406	45.730
2300	20.786	196.543	217.330	47.808
2400	20.786	197.428	218.214	49.887
2500	20.786	198.277	219.063	51.965
2600	20.786	199.092	219.878	54.044
2700	20.786	199.876	220.662	56.123
2800	20.786	200.632	221.418	58.201
2900	20.786	201.362	222.148	60.280
3000	20.786	202.066	222.853	62.359
3100	20.786	202.748	223.534	64.437
3200	20.787	203.408	224.194	66.516
3300	20.787	204.047	224.834	68.595
3400	20.787	204.668	225.454	70.673
3500	20.787	205.271	226.057	72.752
3600	20.788	205.856	226.642	74.831
3700	20.789	206.426	227.212	76.909
3800	20.790	206.980	227.766	78.988
3900	20.791	207.520	228.306	81.067
4000	20.793	208.046	228.833	83.147
298.15	20.786	154.076	174.862	6.197

Uncertainties in Functions

300	0.001	0.004	0.005	0.001
1000	0.001	0.004	0.005	0.001
4000	0.004	0.005	0.006	0.005

Table III-68 THERMODYNAMIC FUNCTIONS: Cu(cr & l)

T	$C_p°$	$-\{G°-H°(0)\}/T$	$S°$	$H°-H°(0)$
K	........J·(mol·K)$^{-1}$........			kJ·mol^{-1}
0	0.000	0.000	0.000	0.000
100	16.010	3.340	10.030	0.669
200	22.630	10.320	23.730	2.682
300	24.460	16.470	33.301	5.049
400	25.339	21.612	40.467	7.542
500	25.966	25.974	46.192	10.109
600	26.479	29.753	50.973	12.732
700	26.953	33.085	55.090	15.403
800	27.448	36.067	58.721	18.123
900	28.014	38.768	61.986	20.895
1000	28.700	41.241	64.971	23.730
1100	29.553	43.526	67.745	26.641
1200	30.617	45.654	70.361	29.648
1300	31.940	47.652	72.862	32.773
1358 cr	32.844	48.759	74.275	34.651
1358 l	32.800	48.759	83.951	47.791
1400	32.800	49.829	84.950	49.169
1500	32.800	52.247	87.213	52.449
1600	32.800	54.499	89.330	55.729
1700	32.800	56.607	91.318	59.009
1800	32.800	58.588	93.193	62.289
1900	32.800	60.457	94.967	65.569
2000	32.800	62.225	96.649	68.849
2100	32.800	63.902	98.249	72.129
2200	32.800	65.498	99.775	75.409
2300	32.800	67.021	101.233	78.689
2400	32.800	68.476	102.629	81.969
2500	32.800	69.869	103.968	85.249
2600	32.800	71.205	105.255	88.529
2700	32.800	72.489	106.493	91.809
2800	32.800	73.725	107.685	95.089
2900	32.800	74.916	108.836	98.369
3000	32.800	76.065	109.948	101.649
298.15	24.440	16.367	33.150	5.004

Uncertainties in Functions

300	0.050	0.050	0.080	0.008
1000	0.100	0.200	0.200	0.100
3000	3.000	3.000	3.500	8.000

Table III-69 THERMODYNAMIC FUNCTIONS at 1 atm: Cu(g)

T	$C_p°$	$-\{G°-H°(0)\}/T$	$S°$	$H°-H°(0)$
K	........J·(mol·K)$^{-1}$........			kJ·mol^{-1}
0	0.000	0.000	0.000	0.000
100	20.786	122.795	143.581	2.079
200	20.786	137.203	157.989	4.157
300	20.786	145.631	166.417	6.236
400	20.786	151.611	172.397	8.314
500	20.786	156.249	177.036	10.393
600	20.786	160.039	180.825	12.472
700	20.786	163.243	184.030	14.550
800	20.786	166.019	186.805	16.629
900	20.786	168.467	189.253	18.708
1000	20.786	170.657	191.443	20.786
1100	20.789	172.638	193.425	22.865
1200	20.793	174.447	195.234	24.944
1300	20.804	176.111	196.899	27.024
1400	20.823	177.652	198.441	29.105
1500	20.856	179.086	199.879	31.189
1600	20.909	180.428	201.226	33.277
1700	20.985	181.689	202.496	35.371
1800	21.091	182.879	203.698	37.475
1900	21.231	184.005	204.842	39.591
2000	21.407	185.074	205.935	41.722
2100	21.622	186.093	206.985	43.874
2200	21.878	187.066	207.997	46.048
2300	22.173	187.997	208.975	48.251
2400	22.507	188.891	209.926	50.484
2500	22.878	189.751	210.852	52.753
2600	23.282	190.580	211.757	55.061
2700	23.715	191.381	212.644	57.411
2800	24.173	192.156	213.515	59.805
2900	24.651	192.907	214.371	62.246
3000	25.147	193.637	215.215	64.736
3100	25.651	194.346	216.048	67.276
3200	26.161	195.037	216.871	69.866
3300	26.672	195.711	217.684	72.508
3400	27.178	196.369	218.487	75.201
3500	27.677	197.013	219.282	77.943
3600	28.163	197.642	220.069	80.736
3700	28.634	198.259	220.847	83.576
3800	29.089	198.863	221.617	86.462
3900	29.523	199.457	222.378	89.394
4000	29.941	200.039	223.131	92.369
298.15	20.786	145.503	166.289	6.197

Uncertainties in Functions

300	0.001	0.003	0.004	0.001
1000	0.001	0.003	0.004	0.001
4000	0.004	0.004	0.005	0.005

Table III-70   THERMODYNAMIC FUNCTIONS: Ag(cr & l)

$\dfrac{T}{K}$	$C_p°$	$-\{G°-H°(0)\}/T$	$S°$	$H°-H°(0)$
	.........J·(mol·K)$^{-1}$........			kJ·mol^{-1}
0	0.000	0.000	0.000	0.000
100	20.100	6.700	17.120	1.042
200	24.160	16.120	32.660	3.308
300	25.356	23.400	42.707	5.792
400	25.791	29.187	50.056	8.347
500	26.365	33.962	55.871	10.955
600	26.992	38.029	60.733	13.622
700	27.645	41.580	64.942	16.354
800	28.312	44.738	68.677	19.152
900	28.987	47.588	72.051	22.017
1000	29.667	50.191	75.140	24.949
1100	30.350	52.591	78.000	27.950
1200	31.035	54.821	80.670	31.019
1235.08 cr	31.276	55.567	81.568	32.112
1235.08 l	33.400	55.567	90.474	43.112
1300	33.400	57.354	92.185	45.281
1400	33.400	59.931	94.660	48.621
1500	33.400	62.324	96.964	51.961
1600	33.400	64.557	99.120	55.301
1700	33.400	66.651	101.145	58.641
1800	33.400	68.620	103.054	61.981
1900	33.400	70.481	104.860	65.321
2000	33.400	72.243	106.573	68.661
2100	33.400	73.917	108.203	72.001
2200	33.400	75.511	109.756	75.341
2300	33.400	77.032	111.241	78.681
2400	33.400	78.487	112.663	82.021
2500	33.400	79.882	114.026	85.361
2600	33.400	81.220	115.336	88.701
2700	33.400	82.507	116.597	92.041
2800	33.400	83.747	117.811	95.381
2900	33.400	84.942	118.983	98.721
3000	33.400	86.095	120.116	102.061
298.15	25.350	23.281	42.550	5.745

Uncertainties in Functions

300	0.100	0.150	0.200	0.020
1000	0.300	0.400	0.500	0.200
3000	4.000	4.000	5.000	9.000

Table III-71   THERMODYNAMIC FUNCTIONS at 1 atm: Ag(g)

T	$C_p{}^\circ$	$-\{G^\circ - H^\circ(0)\}/T$	S°	$H^\circ - H^\circ(0)$
$\overline{\text{K}}$	$\dots\dots$	J·(mol·K)$^{-1}$	$\dots\dots$	kJ·mol^{-1}
0	0.000	0.000	0.000	0.000
100	20.786	129.395	150.181	2.079
200	20.786	143.802	164.589	4.157
300	20.786	152.231	173.017	6.236
400	20.786	158.210	178.997	8.314
500	20.786	162.849	183.635	10.393
600	20.786	166.638	187.425	12.472
700	20.786	169.843	190.629	14.550
800	20.786	172.618	193.404	16.629
900	20.786	175.067	195.853	18.708
1000	20.786	177.257	198.043	20.786
1100	20.786	179.238	200.024	22.865
1200	20.786	181.046	201.833	24.943
1300	20.786	182.710	203.496	27.022
1400	20.786	184.251	205.037	29.101
1500	20.786	185.685	206.471	31.179
1600	20.786	187.026	207.812	33.258
1700	20.786	188.286	209.073	35.337
1800	20.786	189.474	210.261	37.415
1900	20.786	190.598	211.385	39.494
2000	20.786	191.665	212.451	41.572
2100	20.786	192.679	213.465	43.651
2200	20.786	193.646	214.432	45.730
2300	20.786	194.570	215.356	47.808
2400	20.786	195.454	216.240	49.887
2500	20.786	196.303	217.089	51.965
2600	20.786	197.118	217.904	54.044
2700	20.786	197.903	218.689	56.123
2800	20.786	198.658	219.445	58.201
2900	20.787	199.388	220.174	60.280
3000	20.792	200.093	220.879	62.360
3100	20.794	200.774	221.561	64.439
3200	20.798	201.434	222.221	66.519
3300	20.803	202.074	222.861	68.599
3400	20.811	202.694	223.483	70.680
3500	20.820	203.297	224.086	72.761
3600	20.831	203.883	224.673	74.844
3700	20.845	204.452	225.243	76.927
3800	20.863	205.007	225.800	79.013
3900	20.884	205.547	226.342	81.100
4000	20.909	206.073	226.871	83.190
298.15	20.786	152.102	172.888	6.197

Uncertainties in Functions

300	0.001	0.003	0.004	0.001
1000	0.001	0.003	0.004	0.001
4000	0.004	0.004	0.005	0.005

Table III-72   THERMODYNAMIC FUNCTIONS: Ti(cr & l)

$\dfrac{T}{K}$	$C_p°$	$-\{G°-H°(0)\}/T$	$S°$	$H°-H°(0)$
	........J·(mol·K)$^{-1}$........			kJ·mol^{-1}
0	0.000	0.000	0.000	0.000
100	14.310	2.639	8.229	0.559
200	22.300	8.819	21.204	2.477
300	25.095	14.640	30.875	4.870
400	26.380	19.663	38.291	7.451
500	27.349	24.006	44.281	10.137
600	28.411	27.819	49.360	12.925
700	29.511	31.221	53.822	15.821
800	30.456	34.301	57.828	18.822
900 *	31.002	37.120	61.452	21.899
1000	32.681	39.720	64.764	25.045
1100	39.222	42.148	68.146	28.597
1156 α	45.189	43.457	70.233	30.953
1156 β	27.975	43.457	73.520	34.753
1200	28.565	44.579	74.576	35.997
1300	29.952	46.977	76.917	38.922
1400	31.402	49.197	79.189	41.989
1500	32.916	51.271	81.407	45.204
1600	34.494	53.222	83.581	48.574
1700	36.136	55.071	85.721	52.105
1800	37.841	56.833	87.835	55.804
1900	39.611	58.520	89.928	59.676
1944 β	40.409	59.241	90.844	61.436
1944 l	46.800	59.241	98.354	76.036
2000	46.800	60.355	99.683	78.657
2100	46.800	62.282	101.967	83.337
2200	46.800	64.136	104.144	88.017
2300	46.800	65.921	106.224	92.697
2400	46.800	67.642	108.216	97.377
2500	46.800	69.304	110.126	102.057
2600	46.800	70.909	111.962	106.737
2700	46.800	72.463	113.728	111.417
2800	46.800	73.967	115.430	116.097
2900	46.800	75.425	117.072	120.777
3000	46.800	76.840	118.659	125.457
3100	46.800	78.214	120.194	130.137
3200	46.800	79.549	121.679	134.817
3300	46.800	80.848	123.120	139.497
3400	46.800	82.112	124.517	144.177
3500	46.800	83.343	125.873	148.857
3600	46.800	84.543	127.192	153.537
3700	46.800	85.713	128.474	158.217
3800	46.800	86.854	129.722	162.897
3900	46.800	87.969	130.938	167.577
4000	46.800	89.058	132.123	172.257
298.15	25.060	14.540	30.720	4.824

Uncertainties in Functions

300	0.080	0.070	0.100	0.015
1000	0.700	0.500	0.600	0.400
4000	7.000	5.000	6.000	14.000

* Rapid change in C_p above this temperature.

Table III-73 THERMODYNAMIC FUNCTIONS at 1 atm: Ti(g)

$\dfrac{T}{K}$	$C_p°$	$-\{G°-H°(0)\}/T$	$S°$	$\dfrac{H°-H°(0)}{kJ \cdot mol^{-1}}$
	$\ldots\ldots\ldots J \cdot (mol \cdot K)^{-1}\ldots\ldots\ldots$			
0	0.000	0.000	0.000	0.000
100	26.974	127.885	151.138	2.325
200	26.487	144.789	170.018	5.046
300	24.399	155.059	180.340	7.584
400	23.104	162.281	187.164	9.953
500	22.360	167.786	192.232	12.223
600	21.914	172.208	196.266	14.435
700	21.632	175.891	199.622	16.611
800	21.454	179.041	202.498	18.765
900	21.354	181.790	205.018	20.905
1000	21.323	184.228	207.266	23.038
1100	21.362	186.416	209.299	25.172
1200	21.474	188.402	211.162	27.313
1300	21.657	190.220	212.888	29.469
1400	21.910	191.897	214.502	31.647
1500	22.228	193.455	216.024	33.853
1600	22.604	194.911	217.470	36.094
1700	23.029	196.279	218.853	38.376
1800	23.497	197.570	220.182	40.702
1900	23.999	198.795	221.466	43.076
2000	24.528	199.959	222.711	45.502
2100	25.079	201.072	223.921	47.983
2200	25.647	202.137	225.100	50.519
2300	26.228	203.161	226.253	53.112
2400	26.818	204.147	227.382	55.765
2500	27.416	205.098	228.489	58.476
2600	28.019	206.019	229.576	61.248
2700	28.626	206.911	230.644	64.080
2800	29.235	207.777	231.696	66.973
2900	29.846	208.620	232.733	69.927
3000	30.457	209.441	233.755	72.942
3100	31.067	210.242	234.764	76.019
3200	31.675	211.024	235.760	79.156
3300	32.280	211.788	236.744	82.354
3400	32.879	212.536	237.716	85.612
3500	33.473	213.270	238.678	88.929
3600	34.058	213.989	239.629	92.306
3700	34.635	214.694	240.570	95.741
3800	35.201	215.388	241.501	99.232
3900	35.756	216.069	242.423	102.780
4000	36.297	216.739	243.335	106.383
298.15	24.430	154.902	180.189	7.539

Uncertainties in Functions

300	0.030	0.010	0.010	0.002
1000	0.100	0.030	0.050	0.020
4000	0.500	0.100	0.200	0.800

Table III-74   THERMODYNAMIC FUNCTIONS: TiO$_2$(cr & l)

$\dfrac{T}{K}$	$C_p°$	$-\{G°-H°(0)\}/T$	$S°$	$H°-H°(0)$
	........J·(mol·K)$^{-1}$........			kJ·mol^{-1}
0	0.000	0.000	0.000	0.000
100	18.860	3.286	10.232	0.695
200	49.290	11.872	31.065	3.837
300	55.245	21.688	50.961	8.782
400	61.457	31.176	67.809	14.653
500	64.967	39.954	81.928	20.987
600	67.414	47.980	94.000	27.612
700	69.358	55.323	104.543	34.454
800	71.035	62.072	113.916	41.475
900	72.557	68.310	122.371	48.656
1000	73.981	74.107	130.090	55.983
1100	75.343	79.524	137.206	63.450
1200	76.661	84.610	143.818	71.050
1300	77.950	89.405	150.006	78.781
1400	79.216	93.943	155.829	86.639
1500	80.466	98.254	161.337	94.624
1600	81.704	102.362	166.570	102.732
1700	82.932	106.287	171.560	110.964
1800	84.153	110.047	176.335	119.318
1900	85.367	113.657	180.917	127.794
2000	86.577	117.131	185.327	136.392
2100	87.783	120.480	189.580	145.110
2185 cr	88.806	123.237	193.083	152.615
2185 l	100.000	123.237	224.204	220.615
2200	100.000	123.927	224.889	222.115
2300	100.000	128.414	229.334	232.115
2400	100.000	132.709	233.590	242.115
2500	100.000	136.826	237.672	252.115
2600	100.000	140.781	241.594	262.115
2700	100.000	144.585	245.368	272.115
2800	100.000	148.250	249.005	282.115
2900	100.000	151.785	252.514	292.115
3000	100.000	155.199	255.904	302.115
3100	100.000	158.501	259.183	312.115
3200	100.000	161.697	262.358	322.115
3300	100.000	164.794	265.435	332.115
3400	100.000	167.798	268.420	342.115
3500	100.000	170.715	271.319	352.115
3600	100.000	173.549	274.136	362.115
3700	100.000	176.305	276.876	372.115
3800	100.000	178.986	279.543	382.115
3900	100.000	181.598	282.141	392.115
4000	100.000	184.144	284.672	402.115
298.15	55.080	21.507	50.620	8.680

Uncertainties in Functions

300	0.300	0.200	0.300	0.050
1000	1.000	0.600	1.000	0.500
4000	15.000	10.000	15.000	35.000

Table III-75   THERMODYNAMIC FUNCTIONS at 1 atm: $TiCl_4(g)$

$\dfrac{T}{K}$	$C_p°$	$-\{G°-H°(0)\}/T$	$S°$	$H°-H°(0)$
	........J·(mol·K)$^{-1}$.......			kJ·mol^{-1}
0	0.000	0.000	0.000	0.000
25	34.883	162.423	195.913	.837
50	48.485	186.583	223.964	1.869
75	59.567	202.821	245.900	3.231
100	66.834	215.929	264.086	4.816
150	77.673	237.064	293.337	8.441
200	85.791	254.164	316.856	12.538
250	91.543	268.734	336.656	16.980
300	95.532	281.510	353.721	21.663
350	98.323	292.916	368.670	26.514
400	100.320	303.230	381.936	31.483
450	101.783	312.648	393.841	36.537
500	102.879	321.315	404.624	41.655
600	104.376	336.819	423.525	52.024
700	105.317	350.388	439.690	62.512
800	105.944	362.451	453.797	73.077
900	106.382	373.309	466.302	83.694
1000	106.699	383.179	477.528	94.349
298.15	95.408	281.064	353.131	21.487

Uncertainties in Functions

300	1.000	4.000	4.000	0.500
1000	1.000	4.000	4.000	0.500

Table III-76 THERMODYNAMIC FUNCTIONS: U(cr & l)

$\dfrac{T}{K}$	$C_p°$	$-\{G°-H°(0)\}/T$	$S°$	$H°-H°(0)$
	.........J·(mol·K)$^{-1}$.......			kJ·mol^{-1}
0	0.000	0.000	0.000	0.000
100	22.240	9.750	22.740	1.299
200	25.830	20.810	39.500	3.738
300	27.703	28.987	50.371	6.415
400	29.699	35.401	58.613	9.285
500	31.952	40.747	65.475	12.364
600	34.652	45.381	71.531	15.690
700	37.864	49.519	77.108	19.312
800	41.614	53.301	82.402	23.281
900	45.916	56.821	87.547	27.653
942 α	47.889	58.239	89.685	29.622
942 β	42.400	58.239	92.636	32.402
1000	42.400	60.308	95.170	34.862
1049 β	42.400	61.984	97.198	36.939
1049 γ	38.300	61.984	101.707	41.669
1100	38.300	63.868	103.525	43.622
1200	38.300	67.314	106.858	47.452
1300	38.300	70.475	109.923	51.283
1400	38.300	73.396	112.762	55.113
1408 γ	38.300	73.620	112.980	55.419
1408 l	47.739	73.620	119.173	64.139
1500	47.912	76.508	122.200	68.539
1600	48.124	79.461	125.299	73.340
1700	48.355	82.244	128.223	78.164
1800	48.600	84.876	130.994	83.012
1900	48.858	87.374	133.629	87.885
2000	49.125	89.750	136.141	92.784
2100	49.401	92.016	138.545	97.710
2200	49.682	94.184	140.849	102.664
2300	49.969	96.261	143.064	107.646
2400	50.260	98.256	145.197	112.658
2500	50.555	100.175	147.255	117.699
2600	50.853	102.024	149.243	122.769
2700	51.154	103.809	151.168	127.869
2800	51.458	105.534	153.034	133.000
2900	51.763	107.203	154.845	138.161
3000	52.070	108.821	156.605	143.353
3100	52.379	110.390	158.318	148.575
3200	52.689	111.914	159.985	153.828
3300	53.000	113.395	161.611	159.113
3400	53.312	114.837	163.198	164.428
3500	53.625	116.241	164.748	169.775
3600	53.939	117.609	166.263	175.153
3700	54.254	118.945	167.745	180.563
3800	54.569	120.248	169.196	186.004
3900	54.885	121.521	170.618	191.477
4000	55.201	122.766	172.011	196.981
298.15	27.665	28.855	50.200	6.364

Uncertainties in Functions

300	0.100	0.150	0.200	0.020
1000	0.400	0.400	0.600	0.300
4000	8.000	5.000	5.000	13.000

Table III-77 THERMODYNAMIC FUNCTIONS at 1 atm: U(g)

$\dfrac{T}{K}$	$C_p°$	$-\{G°-H°(0)\}/T$	$S°$	$H°-H°(0)$
	........J·(mol·K)$^{-1}$........			kJ·mol^{-1}
0	0.000	0.000	0.000	0.000
100	20.861	154.830	175.624	2.079
200	22.371	169.317	190.462	4.229
300	23.707	178.016	199.827	6.543
400	23.950	184.368	206.700	8.933
500	23.701	189.387	212.021	11.317
600	23.402	193.529	216.315	13.672
700	23.258	197.048	219.909	16.003
800	23.349	200.104	223.018	18.331
900	23.686	202.806	225.785	20.681
1000	24.253	205.232	228.308	23.076
1100	25.019	207.438	230.654	25.539
1200	25.950	209.465	232.871	28.086
1300	27.015	211.348	234.989	30.734
1400	28.185	213.110	237.034	33.493
1500	29.436	214.771	239.021	36.374
1600	30.749	216.348	240.963	39.384
1700	32.106	217.852	242.868	42.527
1800	33.494	219.294	244.743	45.808
1900	34.903	220.682	246.593	49.229
2000	36.321	222.024	248.420	52.792
2100	37.736	223.324	250.227	56.496
2200	39.143	224.587	252.015	60.341
2300	40.532	225.819	253.787	64.327
2400	41.897	227.021	255.542	68.451
2500	43.223	228.196	257.280	72.709
2600	44.511	229.348	259.002	77.099
2700	45.748	230.478	260.706	81.615
2800	46.931	231.588	262.393	86.253
2900	48.050	232.679	264.061	91.006
3000	49.101	233.753	265.709	95.868
3100	50.085	234.810	267.337	100.832
3200	50.994	235.852	268.943	105.891
3300	51.826	236.879	270.526	111.037
3400	52.578	237.891	272.086	116.261
3500	53.247	238.890	273.620	121.555
3600	53.833	239.876	275.129	126.911
3700	54.341	240.849	276.611	132.321
3800	54.772	241.809	278.066	137.777
3900	55.128	242.757	279.494	143.273
4000	55.415	243.693	280.893	148.801
298.15	23.694	177.882	199.680	6.499

Uncertainties in Functions

300	0.300	0.050	0.100	0.020
1000	0.500	0.100	0.300	0.100
4000	2.000	0.700	1.000	5.000

Table III-78   THERMODYNAMIC FUNCTIONS: UO_2(cr & l)

$\frac{T}{K}$	$C_p°$	$-\{G°-H°(0)\}/T$	$S°$	$H°-H°(0)$
	........J·(mol·K)$^{-1}$........			kJ·mol^{-1}
0	0.000	0.000	0.000	0.000
100	28.580	11.806	26.376	1.457
200	51.550	26.115	53.990	5.575
300	63.885	39.431	77.424	11.398
400	73.430	51.481	97.333	18.341
500	77.295	62.385	114.191	25.903
600	79.131	72.241	128.464	33.734
700	80.188	81.171	140.748	41.704
800	80.988	89.304	151.509	49.764
900	81.782	96.758	161.093	57.901
1000	82.701	103.631	169.755	66.124
1100	83.818	110.008	177.688	74.448
1200	85.180	115.958	185.038	82.896
1300	86.815	121.539	191.919	91.493
1400	88.740	126.800	198.421	100.269
1500	90.969	131.783	204.617	109.252
1600	93.510	136.522	210.568	118.473
1700	96.370	141.048	216.321	127.964
1800	99.553	145.386	221.918	137.758
1900	103.063	149.559	227.393	147.886
2000	106.903	153.585	232.776	158.381
2100	111.073	157.483	238.091	169.277
2200	115.576	161.267	243.361	180.607
2300	120.414	164.950	248.604	192.404
2400	125.586	168.545	253.837	204.701
2500	131.094	172.061	259.074	217.532
2600	136.938	175.509	264.328	230.931
2700	143.119	178.896	269.611	244.931
2800	149.638	182.231	274.933	259.566
2900	156.494	185.520	280.303	274.870
3000	163.688	188.770	285.728	290.876
3100	171.220	191.986	291.217	307.619
3123 cr	173.000	192.721	292.490	311.577
3123 l	131.000	192.721	317.466	389.577
3200	131.000	195.761	320.656	399.664
3300	131.000	199.607	324.688	412.764
3400	131.000	203.344	328.598	425.864
3500	131.000	206.977	332.396	438.964
3600	131.000	210.513	336.086	452.064
3700	131.000	213.955	339.675	465.164
3800	131.000	217.310	343.169	478.264
3900	131.000	220.581	346.572	491.364
4000	131.000	223.772	349.888	504.464
298.15	63.600	39.197	77.030	11.280

Uncertainties in Functions

300	0.500	0.150	0.200	0.020
1000	1.000	0.800	1.000	0.600
4000	15.000	6.000	10.000	25.000

Table III-79 THERMODYNAMIC FUNCTIONS: $UO_3(cr)$

$\dfrac{T}{K}$	C_p°	$-\{G^\circ-H^\circ(0)\}/T$	S°	$H^\circ-H^\circ(0)$
	........J·(mol·K)$^{-1}$........			kJ·mol^{-1}
0	0.000	0.000	0.000	0.000
100	39.290	11.828	29.698	1.787
200	66.730	30.187	66.442	7.251
300	81.893	47.495	96.616	14.736
400	89.589	62.975	121.394	23.368
500	93.131	76.763	141.807	32.522
600	95.269	89.073	158.988	41.949
700	96.945	100.143	173.802	51.561
800	98.553	110.182	186.852	61.335
900	100.282	119.362	198.558	71.276
1000	102.230	127.823	209.222	81.399
1100	104.455	135.676	219.068	91.731
1200	106.988	143.012	228.263	102.300
1300	109.853	149.906	236.937	113.140
1400	113.062	156.420	245.193	124.283
1500	116.626	162.604	253.113	135.764
298.15	81.670	47.192	96.110	14.585
Uncertainties in Functions				
300	0.800	0.300	0.400	0.050
1000	2.000	1.500	2.000	1.000
1500	2.500	3.000	3.000	2.000

Table III-80 THERMODYNAMIC FUNCTIONS: $U_3O_8(cr)$

$\dfrac{T}{K}$	$C_p°$	$-\{G°-H°(0)\}/T$	$S°$	$H°-H°(0)$
	........J·(mol·K)$^{-1}$........			kJ·mol^{-1}
0	0.000	0.000	0.000	0.000
100	116.690	35.150	87.910	5.276
200	195.060	89.398	195.983	21.317
300	239.022	140.088	284.025	43.181
400	263.565	185.493	356.987	68.597
483 λ trs	289.500	219.469	408.597	91.349
483	271.992	219.469	408.597	91.349
500	273.000	226.060	418.023	95.981
600	278.656	262.357	468.303	123.568
700	284.000	294.945	511.662	151.702
800	289.141	324.472	549.923	180.361
900	294.131	351.462	584.268	209.526
1000	299.000	376.328	615.511	239.183
1100	303.761	399.394	644.233	269.322
1200	308.426	420.921	670.864	299.932
1300	312.999	441.114	695.733	331.004
1400	317.486	460.144	719.093	362.529
1500	321.887	478.149	741.148	394.498
1600	326.206	495.245	762.060	426.904
1700	330.443	511.530	781.964	459.737
1800	334.600	527.086	800.969	492.990
1900	338.677	541.984	819.170	526.654
2000	342.675	556.283	836.644	560.723
298.15	238.000	139.199	282.550	42.740

Uncertainties in Functions

300	0.500	0.300	0.500	0.100
1000	5.000	4.000	5.000	3.000
2000	15.000	8.000	9.000	20.000

Table III-81 THERMODYNAMIC FUNCTIONS: Th(cr & l)

$\dfrac{T}{K}$	C_p°	$-\{G^{\circ}-H^{\circ}(0)\}/T$	S°	$H^{\circ}-H^{\circ}(0)$
	........J·(mol·K)$^{-1}$........			kJ·mol^{-1}
0	0.000	0.000	0.000	0.000
100	22.690	10.848	24.814	1.393
200	25.260	22.441	41.541	3.820
300	26.245	30.664	51.992	6.399
400	27.084	36.995	59.656	9.065
500	27.953	42.160	65.793	11.816
600	28.834	46.541	70.967	14.656
700	29.720	50.359	75.478	17.583
800	30.609	53.755	79.505	20.600
900	31.500	56.823	83.162	23.705
1000	32.391	59.627	86.527	26.900
1100	33.284	62.216	89.656	30.183
1200	34.177	64.626	92.590	33.556
1300	35.070	66.885	95.361	37.019
1400	35.964	69.014	97.993	40.570
1500	36.858	71.030	100.504	44.212
1600	37.751	72.948	102.912	47.942
1650 α	38.198	73.874	104.080	49.841
1650 β	35.419	73.874	106.201	53.341
1700	36.017	74.840	107.268	55.127
1800	37.212	76.700	109.360	58.788
1900	38.407	78.473	111.404	62.569
2000	39.602	80.170	113.404	66.469
2023 β	39.877	80.550	113.859	67.383
2023 l	46.000	80.550	120.680	81.184
2100	46.000	82.053	122.399	84.725
2200	46.000	83.936	124.539	89.326
2300	46.000	85.746	126.583	93.926
2400	46.000	87.489	128.541	98.525
2500	46.000	89.169	130.419	103.126
2600	46.000	90.790	132.223	107.725
2700	46.000	92.357	133.959	112.326
2800	46.000	93.873	135.632	116.926
2900	46.000	95.341	137.246	121.525
3000	46.000	96.764	138.806	126.126
3100	46.000	98.145	140.314	130.725
3200	46.000	99.485	141.774	135.326
3300	46.000	100.788	143.190	139.926
3400	46.000	102.056	144.563	144.526
3500	46.000	103.289	145.897	149.125
3600	46.000	104.491	147.193	153.725
3700	46.000	105.662	148.453	158.326
3800	46.000	106.804	149.680	162.926
3900	46.000	107.919	150.874	167.526
4000	46.000	109.008	152.039	172.125
298.15	26.230	30.532	51.830	6.350

Uncertainties in Functions

300	0.050	0.300	0.500	0.050
1000	0.200	0.500	0.600	0.100
4000	4.000	6.000	4.000	20.000

Table III-82   THERMODYNAMIC FUNCTIONS at 1 atm: Th(g)

$\dfrac{T}{K}$	$C_p°$	$-\{G°-H°(0)\}/T$	$S°$	$H°-H°(0)$
	........J·(mol·K)$^{-1}$........			kJ·mol^{-1}
0	0.000	0.000	0.000	0.000
100	20.786	146.566	167.353	2.079
200	20.786	160.974	181.760	4.157
300	20.790	169.402	190.189	6.236
400	20.845	175.383	196.175	8.317
500	21.083	180.025	200.847	10.411
600	21.635	183.828	204.734	12.544
700	22.542	187.062	208.133	14.750
800	23.762	189.891	211.220	17.063
900	25.204	192.422	214.100	19.510
1000	26.769	194.728	216.836	22.108
1100	28.367	196.858	219.463	24.865
1200	29.930	198.848	221.999	27.781
1300	31.412	200.724	224.454	30.849
1400	32.790	202.504	226.833	34.060
1500	34.047	204.204	229.139	37.404
1600	35.183	205.832	231.374	40.866
1700	36.204	207.399	233.538	44.437
1800	37.119	208.909	235.634	48.105
1900	37.939	210.370	237.664	51.859
2000	38.673	211.784	239.630	55.691
2100	39.334	213.155	241.533	59.593
2200	39.932	214.487	243.378	63.558
2300	40.479	215.783	245.166	67.581
2400	40.980	217.043	246.900	71.655
2500	41.441	218.271	248.582	75.778
2600	41.872	219.469	250.217	79.945
2700	42.276	220.637	251.806	84.154
2800	42.659	221.778	253.351	88.403
2900	43.024	222.893	254.855	92.689
3000	43.373	223.983	256.320	97.011
3100	43.710	225.049	257.748	101.367
3200	44.032	226.093	259.142	105.756
3300	44.344	227.115	260.502	110.177
3400	44.648	228.117	261.831	114.628
3500	44.942	229.099	263.130	119.110
3600	45.226	230.062	264.401	123.620
3700	45.500	231.007	265.644	128.159
3800	45.765	231.934	266.862	132.724
3900	46.016	232.845	268.054	137.314
4000	46.255	233.740	269.223	141.929
298.15	20.789	169.274	190.060	6.197

Uncertainties in Functions

300	0.100	0.040	0.050	0.003
1000	0.500	0.100	0.200	0.100
4000	2.000	0.300	0.500	5.000

Table III-83   THERMODYNAMIC FUNCTIONS: Be(cr & l)

$\dfrac{T}{K}$	$C_p°$	$-\{G°-H°(0)\}/T$	$S°$	$H°-H°(0)$
	$\dots\dots\dots$ J·(mol·K)$^{-1}$ $\dots\dots\dots$			kJ·mol^{-1}
0	0.000	0.000	0.000	0.000
100	1.830	0.134	0.544	0.041
200	10.040	1.130	4.226	0.619
300	16.523	3.000	9.602	1.980
400	19.640	5.317	14.827	3.804
500	21.582	7.689	19.431	5.871
600	23.062	9.992	23.501	8.106
700	24.323	12.187	27.153	10.476
800	25.467	14.268	30.476	12.966
900	26.544	16.242	33.539	15.567
1000	27.579	18.115	36.389	18.274
1100	28.586	19.899	39.065	21.082
1200	29.575	21.603	41.595	23.990
1300	30.551	23.234	44.001	26.997
1400	31.517	24.800	46.300	30.100
1500	32.475	26.308	48.507	33.300
1550 α	32.953	27.041	49.580	34.936
1550 β	30.000	27.041	50.935	37.036
1560 β	30.000	27.195	51.128	37.336
1560 l	30.000	27.195	59.205	49.936
1600	30.000	28.005	59.964	51.136
1700	30.000	29.939	61.783	54.136
1800	30.000	31.756	63.498	57.136
1900	30.000	33.470	65.120	60.136
2000	30.000	35.091	66.659	63.136
2100	30.000	36.629	68.122	66.136
2200	30.000	38.093	69.518	69.136
2300	30.000	39.488	70.851	72.136
2400	30.000	40.822	72.128	75.136
2500	30.000	42.099	73.353	78.136
2600	30.000	43.324	74.530	81.136
2700	30.000	44.500	75.662	84.136
2800	30.000	45.633	76.753	87.136
2900	30.000	46.724	77.806	90.136
3000	30.000	47.777	78.823	93.136
3100	30.000	48.795	79.806	96.136
3200	30.000	49.779	80.759	99.136
3300	30.000	50.732	81.682	102.136
3400	30.000	51.655	82.577	105.136
3500	30.000	52.551	83.447	108.136
3600	30.000	53.421	84.292	111.136
3700	30.000	54.267	85.114	114.136
3800	30.000	55.089	85.914	117.136
3900	30.000	55.889	86.693	120.136
4000	30.000	56.669	87.453	123.136
298.15	16.443	2.960	9.500	1.950

Uncertainties in Functions

300	0.060	0.050	0.080	0.020
1000	0.100	0.100	0.200	0.100
4000	3.000	2.000	2.000	5.000

Table III-84   THERMODYNAMIC FUNCTIONS at 1 atm: Be(g)

$\dfrac{T}{K}$	C_p°	$-\{G^\circ - H^\circ(0)\}/T$	S°	$H^\circ - H^\circ(0)$
	.........J·(mol·K)$^{-1}$.......			kJ·mol^{-1}
0	0.000	0.000	0.000	0.000
100	20.786	92.673	113.459	2.079
200	20.786	107.080	127.867	4.157
300	20.786	115.509	136.295	6.236
400	20.786	121.488	142.275	8.314
500	20.786	126.127	146.913	10.393
600	20.786	129.916	150.703	12.472
700	20.786	133.121	153.907	14.550
800	20.786	135.896	156.682	16.629
900	20.786	138.345	159.131	18.708
1000	20.786	140.535	161.321	20.786
1100	20.786	142.516	163.302	22.865
1200	20.786	144.324	165.111	24.943
1300	20.786	145.988	166.774	27.022
1400	20.786	147.529	168.315	29.101
1500	20.786	148.963	169.749	31.179
1600	20.786	150.304	171.090	33.258
1700	20.786	151.564	172.350	35.337
1800	20.786	152.752	173.539	37.415
1900	20.786	153.876	174.662	39.494
2000	20.786	154.942	175.729	41.572
2100	20.786	155.957	176.743	43.651
2200	20.795	156.924	177.710	45.731
2300	20.801	157.848	178.635	47.811
2400	20.811	158.732	179.520	49.891
2500	20.825	159.581	180.370	51.973
2600	20.844	160.396	181.187	54.056
2700	20.870	161.181	181.974	56.142
2800	20.905	161.937	182.734	58.231
2900	20.950	162.667	183.468	60.323
3000	21.006	163.372	184.180	62.421
3100	21.075	164.055	184.869	64.525
3200	21.159	164.716	185.540	66.637
3300	21.259	165.357	186.192	68.757
3400	21.376	165.979	186.829	70.889
3500	21.512	166.584	187.450	73.033
3600	21.668	167.172	188.058	75.192
3700	21.843	167.744	188.654	77.367
3800	22.040	168.302	189.239	79.561
3900	22.258	168.847	189.815	81.776
4000	22.498	169.378	190.381	84.014
298.15	20.786	115.380	136.166	6.197

Uncertainties in Functions

300	0.001	0.002	0.003	0.001
1000	0.001	0.002	0.003	0.001
4000	0.004	0.003	0.004	0.005

Table III-85   THERMODYNAMIC FUNCTIONS: BeO(cr & l)

$\dfrac{T}{K}$	$C_p°$	$-\{G°-H°(0)\}/T$	$S°$	$H°-H°(0)$
	........J·(mol·K)$^{-1}$........			kJ·mol^{-1}
0	0.000	0.000	0.000	0.000
100	2.640	0.207	0.837	0.063
200	14.160	1.590	5.900	0.862
300	25.750	4.314	13.929	2.884
400	33.649	7.794	22.506	5.885
500	38.822	11.557	30.608	9.525
600	42.347	15.360	38.018	13.595
700	44.776	19.085	44.739	17.958
800	46.503	22.679	50.837	22.527
900	47.859	26.121	56.395	27.247
1000	49.153	29.407	61.504	32.097
1100	49.241	32.541	66.188	37.012
1200	49.744	35.527	70.492	41.958
1300	50.506	38.372	74.502	46.969
1400	51.424	41.089	78.278	52.065
1500	52.430	43.689	81.860	57.257
1600	53.475	46.182	85.277	62.552
1700	54.523	48.579	88.550	67.952
1800	55.551	50.887	91.696	73.456
1900	56.539	53.115	94.726	79.061
2000	57.472	55.269	97.650	84.762
2100	58.341	57.355	100.476	90.553
2200	59.136	59.378	103.208	96.428
2300	59.851	61.341	105.853	102.378
2373 α	60.319	62.739	107.731	106.764
2373 β	56.000	62.739	107.842	107.029
2400	56.000	63.250	108.476	108.541
2500	56.000	65.105	110.762	114.141
2600	56.000	66.904	112.958	119.741
2700	56.000	68.649	115.072	125.341
2800	56.000	70.344	117.108	130.941
2851 β	56.000	71.189	118.119	133.797
2851 1	84.000	71.189	148.284	219.797
2900	84.000	72.504	149.716	223.913
3000	84.000	75.125	152.563	232.313
3100	84.000	77.668	155.318	240.713
3200	84.000	80.137	157.984	249.113
3300	84.000	82.535	160.569	257.513
3400	84.000	84.867	163.077	265.913
3500	84.000	87.137	165.512	274.313
3600	84.000	89.347	167.878	282.713
3700	84.000	91.501	170.180	291.113
3800	84.000	93.601	172.420	299.513
3900	84.000	95.650	174.602	307.913
4000	84.000	97.650	176.729	316.313
298.15	25.565	4.255	13.770	2.837

Uncertainties in Functions

300	0.100	0.020	0.040	0.008
1000	0.300	0.150	0.300	0.100
4000	10.000	3.500	5.000	15.000

Table III-86   THERMODYNAMIC FUNCTIONS: Mg(cr & l)

T	$C_p°$	$-\{G°-H°(0)\}/T$	$S°$	$H°-H°(0)$
K	........J·(mol·K)$^{-1}$........			kJ·mol^{-1}
0	0.000	0.000	0.000	0.000
25	0.781	0.064	0.246	0.005
50	5.740	0.530	2.104	0.079
75	11.542	3.086	9.505	0.297
100	15.762	3.086	9.505	0.642
150	20.474	6.490	16.910	1.563
200	22.724	9.898	23.143	2.649
250	24.018	13.084	28.364	3.820
300	24.897	16.011	32.825	5.044
350	25.568	18.697	36.715	6.306
400	26.144	21.169	40.167	7.599
450	26.668	23.456	43.277	8.920
500	27.171	25.582	46.113	10.266
600	28.184	29.434	51.156	13.033
700	29.279	32.859	55.581	15.905
800	30.507	35.953	59.569	18.893
900	31.895	38.783	63.241	22.012
923 cr	32.238	39.403	64.050	22.749
923 l	34.309	39.403	73.234	31.226
1000	34.309	42.115	75.983	33.868
1100	34.309	45.345	79.253	37.299
1200	34.309	48.297	82.238	40.730
1300	34.309	51.015	84.984	44.161
1400	34.309	53.533	87.527	47.592
1500	34.309	55.879	89.894	51.022
1600	34.309	58.075	92.108	54.453
1700	34.309	60.139	94.188	57.884
1800	34.309	62.085	96.149	61.315
1900	34.309	63.927	98.004	64.746
2000	34.309	65.676	99.764	68.177
298.15	24.869	15.908	32.671	4.998

Uncertainties in Functions

300 cr	0.020	0.100	0.100	0.030
1000 l	0.200	0.150	0.150	0.100
2000	0.400	0.300	0.300	0.300

Table III-87   THERMODYNAMIC FUNCTIONS at 1 atm: Mg(g)

$\frac{T}{K}$	$C_p°$	$-\{G°-H°(0)\}/T$	$S°$	$H°-H°(0)$
	........J·(mol·K)$^{-1}$........			kJ·mol^{-1}
0	0.000	0.000	0.000	0.000
100	20.786	105.046	125.832	2.079
200	20.786	119.454	140.240	4.157
300	20.786	127.882	148.668	6.236
400	20.786	133.862	154.648	8.314
500	20.786	138.500	159.286	10.393
600	20.786	142.290	163.076	12.472
700	20.786	145.494	166.280	14.550
800	20.786	148.269	169.056	16.629
900	20.786	150.718	171.504	18.708
1000	20.786	152.908	173.694	20.786
1100	20.786	154.889	175.675	22.865
1200	20.786	156.698	177.484	24.943
1300	20.786	158.361	179.148	27.022
1400	20.786	159.902	180.688	29.101
1500	20.786	161.336	182.122	31.179
1600	20.786	162.677	183.464	33.258
1700	20.786	163.938	184.724	35.337
1800	20.786	165.126	185.912	37.415
1900	20.786	166.249	187.036	39.494
2000	20.786	167.316	188.102	41.572
2100	20.789	168.330	189.116	43.651
2200	20.795	169.297	190.084	45.731
2300	20.802	170.221	191.008	47.811
2400	20.812	171.106	191.894	49.892
2500	20.826	171.954	192.744	51.973
2600	20.846	172.770	193.561	54.057
2700	20.874	173.554	194.348	56.143
2800	20.909	174.311	195.108	58.232
2900	20.956	175.040	195.842	60.325
3000	21.013	175.746	196.554	62.424
3100	21.085	176.428	197.244	64.528
3200	21.171	177.089	197.915	66.641
3300	21.274	177.730	198.568	68.763
3400	21.396	178.353	199.205	70.897
3500	21.536	178.957	199.827	73.043
3600	21.696	179.545	200.436	75.205
3700	21.877	180.118	201.032	77.383
3800	22.080	180.676	201.619	79.581
3900	22.305	181.221	202.195	81.800
4000	22.553	181.752	202.763	84.043
298.15	20.786	127.753	148.539	6.197
Uncertainties in Functions				
300	0.001	0.003	0.003	0.001
1000	0.001	0.003	0.003	0.001
4000	0.003	0.010	0.010	0.020

Table III-88   THERMODYNAMIC FUNCTIONS: MgO(cr & l)

$\frac{T}{K}$	$C_p°$	$-\{G°-H°(0)\}/T$	$S°$	$H°-H°(0)$
	.........J·(mol·K)$^{-1}$........			kJ·mol^{-1}
0	0.000	0.000	0.000	0.000
100	8.030	0.580	2.590	0.201
200	26.690	4.225	14.100	1.975
300	37.384	9.747	27.181	5.230
400	42.769	15.594	38.772	9.271
500	45.560	21.242	48.644	13.701
600	47.301	26.533	57.114	18.349
700	48.527	31.440	64.503	23.144
800	49.464	35.990	71.046	28.045
900	50.224	40.216	76.917	33.031
1000	50.869	44.157	82.243	38.086
1100	51.436	47.844	87.119	43.202
1200	51.950	51.307	91.617	48.371
1300	52.430	54.570	95.794	53.591
1400	52.890	57.656	99.696	58.857
1500	53.341	60.582	103.361	64.168
1600	53.793	63.364	106.818	69.525
1700	54.255	66.018	110.093	74.927
1800	54.736	68.553	113.207	80.377
1900	55.243	70.983	116.180	85.875
2000	55.783	73.314	119.027	91.426
2100	56.364	75.557	121.763	97.033
2200	56.991	77.717	124.399	102.701
2300	57.674	79.803	126.947	108.433
2400	58.416	81.819	129.417	114.237
2500	59.227	83.771	131.818	120.119
2600	60.111	85.664	134.158	126.085
2700	61.075	87.502	136.444	132.144
2800	62.127	89.290	138.684	138.303
2900	63.271	91.032	140.884	144.572
3000	64.515	92.730	143.050	150.960
3100 cr	65.865	94.387	145.187	157.479
3100 l	84.000	94.387	170.026	234.479
3200	84.000	96.793	172.693	242.879
3300	84.000	99.132	175.277	251.279
3400	84.000	101.409	177.785	259.679
3500	84.000	103.626	180.220	268.079
3600	84.000	105.787	182.586	276.479
3700	84.000	107.893	184.888	284.879
3800	84.000	109.949	187.128	293.279
3900	84.000	111.956	189.310	301.679
4000	84.000	113.917	191.437	310.079
298.15	37.237	9.643	26.950	5.160
Uncertainties in Functions				
300	0.200	0.100	0.150	0.020
1000 cr	0.500	0.400	0.600	0.500
4000 l	10.000	5.000	10.000	20.000

Table III-89 THERMODYNAMIC FUNCTIONS: MgF_2(cr & l)

$\dfrac{T}{K}$	$C_p°$	$-\{G°-H°(0)\}/T$	$S°$	$H°-H°(0)$
	........J·(mol·K)$^{-1}$........			kJ·mol^{-1}
0	0.000	0.000	0.000	0.000
50	5.682	0.583	2.140	0.078
75	13.087	1.657	5.735	0.306
100	21.715	3.273	10.682	0.741
150	37.508	7.684	22.599	2.237
200	48.803	12.975	35.045	4.414
250	56.580	18.582	46.819	7.059
300	61.671	24.203	57.619	10.025
350	65.490	29.688	67.423	13.207
400	68.464	34.972	76.370	16.559
450	70.781	40.034	84.572	20.042
500	72.613	44.871	92.128	23.628
600	75.298	53.898	105.620	31.033
700	77.156	62.145	117.374	38.660
800	78.532	69.711	127.771	46.448
900	79.602	76.688	137.084	54.356
1000	80.472	83.156	145.517	62.361
1100	81.213	89.181	153.223	70.446
1200	81.853	94.817	160.317	78.600
1300	82.430	100.111	166.892	86.814
1400	82.953	105.103	173.020	95.084
1500	83.442	109.824	178.760	103.404
1536 cr	83.610	111.463	180.741	106.411
1536 l	94.895	111.463	218.952	165.103
1600	94.895	115.840	222.826	171.177
1700	94.895	122.305	228.579	180.666
1800	94.895	128.361	234.003	190.156
1900	94.895	134.057	239.133	199.645
2000	94.895	139.434	244.001	209.135
2100	94.895	144.524	248.631	218.624
2200	94.895	149.357	253.045	228.113
2300	94.895	153.958	257.264	237.603
2400	94.895	158.347	261.302	247.092
2500	94.895	162.543	265.176	256.582
2600	94.895	166.563	268.898	266.071
2700	94.895	170.420	272.479	275.561
2800	94.895	174.127	275.930	285.050
2900	94.895	177.695	279.260	294.540
3000	94.895	181.134	282.477	304.029
298.15	61.512	23.997	57.238	9.911
Uncertainties in Functions				
300	0.300	0.300	0.500	0.060
1000 cr	1.000	0.400	1.000	0.300
3000 l	5.000	2.500	4.500	8.000

Table III-90   THERMODYNAMIC FUNCTIONS: Ca(cr(α, β) & l)

$\dfrac{T}{K}$	$C_p°$	$-\{G°-H°(0)\}/T$	$S°$	$H°-H°(0)$
	.........J·(mol·K)$^{-1}$........			kJ·mol^{-1}
0	0.000	0.000	0.000	0.000
25	2.682	0.261	0.975	0.018
50	10.792	1.580	5.290	0.185
75	16.269	3.736	10.806	0.530
100	19.504	6.160	15.967	0.981
150	22.996	10.929	24.619	2.053
200	24.542	15.242	31.489	3.249
250	25.411	19.067	37.065	4.499
300	25.946	22.467	41.749	5.784
350	26.320	25.516	45.774	7.090
400	26.868	28.274	49.322	8.419
450	27.643	30.794	52.531	9.782
500	28.487	33.117	55.487	11.185
600	30.382	37.301	60.846	14.127
700	32.406	41.015	65.680	17.266
716 α	32.737	41.574	66.416	17.787
716 β	31.503	41.574	67.715	18.717
800	33.818	44.510	71.335	21.460
900	36.713	47.723	75.484	24.985
1000	39.706	50.701	79.507	28.806
1100	42.763	53.499	83.434	32.929
1115 β	43.226	53.905	84.017	33.574
1115 l	35.000	53.905	91.676	42.114
1200	35.000	56.673	94.247	45.089
1300	35.000	59.673	97.049	48.589
1400	35.000	62.436	99.642	52.089
1500	35.000	64.998	102.057	55.589
1600	35.000	67.385	104.316	59.089
1700	35.000	69.621	106.438	62.589
1800	35.000	71.722	108.438	66.089
1900	35.000	73.705	110.331	69.589
2000	35.000	75.582	112.126	73.089
2200	35.000	79.058	115.462	80.089
2400	35.000	82.220	118.507	87.089
2500	35.000	83.700	119.936	90.589
298.15	25.929	22.348	41.588	5.736

Uncertainties in Functions

300 cr, α	0.300	0.400	0.400	0.040
1000 cr, β	0.600	0.700	0.700	0.500
2500 l	6.000	3.000	4.000	6.000

Table III-91   THERMODYNAMIC FUNCTIONS at 1 atm: Ca(g)

$\dfrac{T}{K}$	$C_p°$	$-\{G°-H°(0)\}/T$	$S°$	$H°-H°(0)$
	.........J·(mol·K)$^{-1}$........			kJ·mol^{-1}
0	0.000	0.000	0.000	0.000
100	20.786	111.284	132.070	2.079
200	20.786	125.692	146.478	4.157
300	20.786	134.120	154.906	6.236
400	20.786	140.100	160.886	8.314
500	20.786	144.738	165.524	10.393
600	20.786	148.528	169.314	12.472
700	20.786	151.732	172.518	14.550
800	20.786	154.508	175.294	16.629
900	20.786	156.956	177.742	18.708
1000	20.786	159.146	179.932	20.786
1100	20.786	161.127	181.913	22.865
1200	20.786	162.936	183.722	24.943
1300	20.787	164.600	185.386	27.022
1400	20.789	166.140	186.926	29.101
1500	20.793	167.574	188.361	31.180
1600	20.802	168.916	189.703	33.260
1700	20.818	170.176	190.965	35.341
1800	20.845	171.364	192.155	37.424
1900	20.889	172.488	193.283	39.510
2000	20.953	173.555	194.356	41.602
2100	21.046	174.570	195.381	43.702
2200	21.174	175.539	196.363	45.813
2300	21.343	176.465	197.308	47.938
2400	21.561	177.352	198.220	50.083
2500	21.836	178.205	199.106	52.252
2600	22.173	179.025	199.969	54.452
2700	22.578	179.817	200.813	56.689
2800	23.056	180.582	201.642	58.970
2900	23.612	181.322	202.461	61.303
3000	24.248	182.040	203.272	63.695
3100	24.965	182.738	204.078	66.155
3200	25.764	183.417	204.883	68.691
3300	26.644	184.080	205.689	71.311
3400	27.601	184.728	206.499	74.022
3500	28.633	185.361	207.314	76.833
3600	29.734	185.982	208.136	79.751
3700	30.897	186.592	208.966	82.782
3800	32.116	187.192	209.806	85.933
3900	33.382	187.783	210.656	89.207
4000	34.685	188.366	211.518	92.610
298.15	20.786	133.991	154.778	6.197

Uncertainties in Functions

300	0.001	0.004	0.004	0.001
1000	0.001	0.004	0.004	0.001
4000	0.003	0.010	0.010	0.020

Table III-92   THERMODYNAMIC FUNCTIONS: CaO(cr)

$\frac{T}{K}$	$C_p°$	$-\{G°-H°(0)\}/T$	$S°$	$H°-H°(0)$
	........J·(mol·K)$^{-1}$.......			kJ·mol^{-1}
0	0.000	0.000	0.000	0.000
25	0.300	0.025	0.101	0.002
50	2.630	0.207	0.867	0.033
75	8.330	0.732	2.945	0.166
100	14.700	1.670	6.210	0.454
150	26.050	4.507	14.400	1.484
200	33.640	8.070	23.020	2.990
250	38.590	11.890	31.090	4.800
300	42.183	15.601	38.361	6.828
350	45.066	19.340	45.096	9.014
400	46.980	22.950	51.247	11.319
450	48.330	26.411	56.862	13.703
500	49.331	29.717	62.009	16.146
600	50.720	35.879	71.135	21.154
700	51.654	41.492	79.028	26.275
800	52.346	46.627	85.972	31.476
900	52.897	51.349	92.171	36.739
1000	53.360	55.716	97.768	42.053
1100	53.767	59.774	102.874	47.410
1200	54.134	63.564	107.568	52.805
1300	54.474	67.118	111.915	58.236
1400	54.795	70.464	115.963	63.699
1500	55.100	73.625	119.754	69.194
1600	55.395	76.621	123.320	74.719
1700	55.680	79.468	126.687	80.273
1800	55.959	82.180	129.877	85.855
1900	56.233	84.771	132.910	91.464
2000	56.502	87.251	135.801	97.101
2200	57.030	91.914	141.211	108.454
2400	57.549	96.232	146.196	119.913
2600	58.061	100.256	150.823	131.474
2800	58.568	104.024	155.144	143.137
298.15	42.049	15.460	38.100	6.750

Uncertainties in Functions

300	0.400	0.300	0.400	0.060
1000	0.500	0.700	1.000	0.400
2800	1.000	3.000	4.000	3.000

Table III-93 THERMODYNAMIC FUNCTIONS: Li(cr & l)

$\dfrac{T}{K}$	$C_p°$	$-\{G°-H°(0)\}/T$	$S°$	$H°-H°(0)$
	........J·(mol·K)$^{-1}$........			kJ·mol^{-1}
0	0.000	0.000	0.000	0.000
100	12.970	2.390	7.330	0.494
200	21.550	8.162	19.852	2.338
300	24.881	13.680	29.274	4.678
400	27.584	18.541	36.742	7.280
453.69 cr	29.769	20.910	40.347	8.819
453.69 l	30.375	20.910	46.960	11.819
500	30.071	23.461	49.897	13.218
600	29.584	28.335	55.334	16.199
700	29.248	32.525	59.867	19.140
800	29.017	36.191	63.756	22.052
900	28.870	39.447	67.165	24.946
1000	28.795	42.374	70.202	27.829
1100	28.785	45.030	72.946	30.707
1200	28.836	47.462	75.452	33.588
1300	28.945	49.706	77.764	36.476
1400	29.111	51.787	79.915	39.379
1500	29.334	53.730	81.931	42.300
1600	29.611	55.553	83.832	45.247
1700	29.942	57.270	85.637	48.224
1800	30.328	58.894	87.359	51.237
1900	30.767	60.436	89.011	54.292
2000	31.260	61.905	90.601	57.393
2100	31.806	63.308	92.139	60.546
2200	32.406	64.653	93.633	63.756
2300	33.058	65.944	95.087	67.028
2400	33.764	67.188	96.509	70.369
2500	34.522	68.389	97.902	73.783
2600	35.334	69.551	99.272	77.275
2700	36.198	70.677	100.622	80.851
2800	37.115	71.770	101.954	84.517
2900	38.084	72.833	103.274	88.276
3000	39.107	73.870	104.582	92.135
298.15	24.860	13.584	29.120	4.632
Uncertainties in Functions				
300	0.200	0.150	0.200	0.040
1000	0.300	0.400	0.500	0.200
3000	4.000	2.000	2.000	4.000

Table III-94   THERMODYNAMIC FUNCTIONS at 1 atm: Li(g)

$\dfrac{T}{K}$	C_p°	$-\{(G^\circ-H^\circ(0)\}/T$	S°	$H^\circ-H^\circ(0)$
	$\ldots\ldots\ldots$ J$\cdot$(mol$\cdot$K)$^{-1}$ $\ldots\ldots\ldots$			kJ$\cdot$mol^{-1}
0	0.000	0.000	0.000	0.000
100	20.786	95.179	115.965	2.079
200	20.786	109.587	130.373	4.157
300	20.786	118.015	138.801	6.236
400	20.786	123.995	144.781	8.314
500	20.786	128.633	149.419	10.393
600	20.786	132.423	153.209	12.472
700	20.786	135.627	156.413	14.550
800	20.786	138.403	159.189	16.629
900	20.786	140.851	161.637	18.708
1000	20.786	143.041	163.827	20.786
1100	20.786	145.022	165.808	22.865
1200	20.786	146.831	167.617	24.943
1300	20.786	148.494	169.281	27.022
1400	20.786	150.035	170.821	29.101
1500	20.789	151.469	172.255	31.180
1600	20.793	152.811	173.597	33.259
1700	20.799	154.071	174.858	35.338
1800	20.810	155.259	176.047	37.419
1900	20.826	156.383	177.173	39.500
2000	20.849	157.449	178.241	41.584
2100	20.882	158.464	179.259	43.671
2200	20.925	159.431	180.232	45.761
2300	20.980	160.356	181.163	47.856
2400	21.048	161.242	182.057	49.957
2500	21.131	162.092	182.918	52.066
2600	21.230	162.909	183.749	54.184
2700	21.344	163.696	184.552	56.313
2800	21.475	164.455	185.331	58.453
2900	21.621	165.188	186.087	60.608
3000	21.784	165.896	186.823	62.778
3100	21.961	166.583	187.540	64.965
3200	22.153	167.249	188.240	67.171
3300	22.359	167.896	188.925	69.396
3400	22.576	168.524	189.595	71.643
3500	22.804	169.135	190.253	73.912
3600	23.042	169.731	190.899	76.204
3700	23.288	170.312	191.534	78.521
3800	23.541	170.878	192.158	80.862
3900	23.799	171.432	192.773	83.229
4000	24.060	171.973	193.379	85.622
298.15	20.786	117.886	138.673	6.197

Uncertainties in Functions

300	0.001	0.009	0.010	0.001
1000	0.001	0.009	0.010	0.001
4000	0.004	0.010	0.020	0.005

Table III-95 THERMODYNAMIC FUNCTIONS at 1 atm: $Li_2(g)$

$\dfrac{T}{K}$	$C_p°$	$-\{G°-H°(0)\}/T$	$S°$	$H°-H°(0)$
	........J·(mol·K)$^{-1}$.......			kJ·mol^{-1}
0	0.000	0.000	0.000	0.000
100	30.583	130.923	160.306	2.938
200	34.380	151.770	182.797	6.205
300	36.125	164.640	197.114	9.742
400	36.985	174.131	207.636	13.402
500	37.514	181.693	215.950	17.128
600	37.906	187.992	222.825	20.900
700	38.243	193.397	228.694	24.708
800	38.567	198.136	233.822	28.549
900	38.894	202.360	238.383	32.422
1000	39.228	206.171	242.499	36.328
1100	39.555	209.647	246.253	40.267
1200	39.852	212.843	249.708	44.238
1300	40.086	215.803	252.908	48.235
1400	40.235	218.561	255.884	52.252
1500	40.284	221.143	258.663	56.279
1600	40.230	223.570	261.261	60.306
1700	40.089	225.860	263.697	64.323
1800	39.872	228.026	265.982	68.322
1900	39.606	230.081	268.131	72.296
2000	39.321	232.034	270.156	76.242
2100	39.043	233.896	272.067	80.160
2200	38.801	235.672	273.878	84.052
2300	38.616	237.371	275.598	87.923
2400	38.509	238.998	277.239	91.778
2500	38.494	240.560	278.811	95.627
2600	38.581	242.060	280.322	99.480
2700	38.776	243.504	281.781	103.347
2800	39.079	244.897	283.196	107.239
2900	39.488	246.241	284.575	111.167
3000	39.996	247.542	285.922	115.140
3100	40.595	248.801	287.243	119.169
3200	41.271	250.023	288.542	123.262
3300	42.011	251.209	289.823	127.426
3400	42.801	252.364	291.089	131.666
3500	43.623	253.488	292.342	135.988
3600	44.461	254.585	293.582	140.392
3700	45.300	255.655	294.812	144.881
3800	46.124	256.702	296.031	149.452
3900	46.918	257.726	297.240	154.105
4000	47.670	258.729	298.438	158.836
298.15	36.103	164.439	196.890	9.675

Uncertainties in Functions

300	0.300	0.020	0.030	0.010
1000	0.600	0.060	0.100	0.200
4000	2.000	0.300	0.500	1.000

Table III-96   THERMODYNAMIC FUNCTIONS: Na(cr & 1)

$\dfrac{T}{K}$	$C_p°$	$-\{G°-H°(0)\}/T$	$S°$	$H°-H°(0)$
	.........J·(mol·K)$^{-1}$........			kJ·mol^{-1}
0	0.000	0.000	0.000	0.000
100	22.460	10.186	23.656	1.347
200	26.000	21.529	40.539	3.802
300	28.261	29.767	51.475	6.512
371.01 cr	31.509	34.533	57.752	8.614
371.01 l	31.799	34.533	64.754	11.212
400	31.532	36.811	67.137	12.130
500	30.659	43.599	74.077	15.239
600	29.920	49.155	79.599	18.267
700	29.353	53.839	84.167	21.229
800	28.973	57.880	88.059	24.144
900	28.787	61.426	91.459	27.030
1000	28.799	64.583	94.491	29.908
1100	29.012	67.429	97.244	32.796
1200	29.427	70.021	99.785	35.717
1300	30.045	72.403	102.163	38.689
1400	30.866	74.610	104.419	41.733
1500	31.891	76.669	106.582	44.869
1600	33.120	78.605	108.678	48.118
1700	34.553	80.434	110.728	51.499
1800	36.190	82.174	112.748	55.035
1900	38.032	83.836	114.754	58.744
2000	40.078	85.431	116.755	62.648
2100	42.328	86.971	118.765	66.767
2200	44.784	88.462	120.790	71.121
2300	47.444	89.912	122.838	75.730
298.15	28.230	29.633	51.300	6.460

Uncertainties in Functions

300	0.200	0.150	0.200	0.020
1000	0.300	0.400	0.500	0.200
2300	2.000	1.000	1.000	2.000

Table III-97   THERMODYNAMIC FUNCTIONS at 1 atm: Na(g)

$\dfrac{T}{K}$	$C_p°$	$-\{G°-H°(0)\}/T$	$S°$	$H°-H°(0)$
	........J·(mol·K)$^{-1}$.......			kJ·mol^{-1}
0	0.000	0.000	0.000	0.000
100	20.786	110.115	130.901	2.079
200	20.786	124.523	145.309	4.157
300	20.786	132.951	153.737	6.236
400	20.786	138.931	159.717	8.314
500	20.786	143.569	164.355	10.393
600	20.786	147.359	168.145	12.472
700	20.786	150.563	171.349	14.550
800	20.786	153.339	174.125	16.629
900	20.786	155.787	176.573	18.708
1000	20.786	157.977	178.763	20.786
1100	20.786	159.958	180.744	22.865
1200	20.786	161.767	182.553	24.943
1300	20.786	163.431	184.217	27.022
1400	20.786	164.971	185.757	29.101
1500	20.786	166.405	187.191	31.179
1600	20.786	167.747	188.533	33.258
1700	20.789	169.007	189.793	35.337
1800	20.792	170.195	190.982	37.416
1900	20.797	171.319	192.106	39.495
2000	20.805	172.385	193.173	41.575
2100	20.816	173.399	194.188	43.657
2200	20.833	174.367	195.157	45.739
2300	20.855	175.291	196.083	47.823
2400	20.885	176.176	196.972	49.910
2500	20.923	177.025	197.825	52.000
2600	20.970	177.841	198.646	54.095
2700	21.027	178.626	199.439	56.195
2800	21.099	179.383	200.205	58.301
2900	21.180	180.114	200.947	60.415
3000	21.275	180.820	201.666	62.538
3100	21.383	181.504	202.366	64.671
3200	21.504	182.167	203.046	66.815
3300	21.639	182.810	203.710	68.972
3400	21.788	183.434	204.358	71.143
3500	21.951	184.041	204.992	73.330
3600	22.127	184.631	205.613	75.534
3700	22.317	185.207	206.222	77.756
3800	22.519	185.768	206.820	79.998
3900	22.734	186.315	207.407	82.260
4000	22.961	186.850	207.986	84.545
298.15	20.786	132.823	153.609	6.197

Uncertainties in Functions

300	0.001	0.002	0.003	0.001
1000	0.001	0.002	0.003	0.001
4000	0.004	0.005	0.010	0.005

Table III-98   THERMODYNAMIC FUNCTIONS at 1 atm: $Na_2(g)$

$\dfrac{T}{K}$	$C_p°$	$-\{G°-H°(0)\}/T$	$S°$	$H°-H°(0)$
	........$J \cdot (mol \cdot K)^{-1}$........			$kJ \cdot mol^{-1}$
0	0.000	0.000	0.000	0.000
100	34.747	158.988	190.312	3.132
200	36.921	181.539	215.256	6.743
300	37.583	195.459	230.368	10.473
400	37.982	205.608	241.237	14.252
500	38.326	213.615	249.750	18.067
600	38.666	220.240	256.768	21.917
700	38.998	225.896	262.753	25.800
800	39.275	230.836	267.980	29.715
900	39.424	235.226	272.616	33.651
1000	39.387	239.176	276.770	37.593
1100	39.133	242.767	280.513	41.521
1200	38.671	246.056	283.900	45.413
1300	38.030	249.087	286.971	49.249
1400	37.263	251.894	289.762	53.015
1500	36.412	254.505	292.305	56.700
1600	35.523	256.941	294.626	60.296
1700	34.641	259.221	296.754	63.805
1800	33.791	261.362	298.709	67.226
1900	33.002	263.375	300.515	70.565
2000	32.297	265.275	302.189	73.829
2100	31.691	267.070	303.750	77.028
2200	31.197	268.771	305.213	80.171
2300	30.826	270.386	306.591	83.272
2400	30.589	271.922	307.897	86.341
2500	30.491	273.386	309.143	89.394
2600	30.537	274.784	310.340	92.444
2700	30.730	276.123	311.495	95.506
2800	31.070	277.406	312.618	98.595
2900	31.556	278.639	313.717	101.725
3000	32.182	279.827	314.797	104.911
3100	32.943	280.972	315.864	108.166
3200	33.828	282.079	316.924	111.503
3300	34.826	283.151	317.980	114.935
3400	35.923	284.191	319.035	118.472
3500	37.104	285.201	320.093	122.123
3600	38.351	286.185	321.156	125.895
3700	39.646	287.145	322.224	129.794
3800	40.970	288.082	323.299	133.825
3900	42.303	288.999	324.381	137.989
4000	43.627	289.897	325.468	142.285
298.15	37.575	195.243	230.136	10.403

Uncertainties in Functions

300	0.100	0.020	0.030	0.010
1000	0.300	0.030	0.050	0.050
4000	1.500	0.300	0.500	1.500

Table III-99   THERMODYNAMIC FUNCTIONS: K(cr & l)

T	$C_p°$	$-\{G°-H°(0)\}/T$	$S°$	$H°-H°(0)$
$\overline{K}$	$\ldots\ldots\ldots$J$\cdot$(mol$\cdot$K)$^{-1}$ $\ldots\ldots\ldots$			kJ$\cdot$mol^{-1}
0	0.000	0.000	0.000	0.000
100	24.650	18.260	35.590	1.733
200	26.820	31.842	53.467	4.325
300	29.671	41.054	64.863	7.143
336.86 cr	32.130	43.855	68.422	8.276
336.86 l	32.129	43.855	75.313	10.597
400	31.552	49.266	80.784	12.607
500	30.741	56.293	87.734	15.720
600	30.158	62.012	93.283	18.763
700	29.851	66.818	97.905	21.761
800	29.838	70.958	101.887	24.743
900	30.130	74.594	105.415	27.739
1000	30.730	77.839	108.618	30.779
1100	31.643	80.773	111.587	33.895
1200	32.870	83.459	114.391	37.118
1300	34.411	85.942	117.081	40.480
1400	36.268	88.260	119.697	44.011
1500	38.440	90.442	122.271	47.744
1600	40.929	92.511	124.830	51.710
1700	43.734	94.488	127.394	55.940
1800	46.856	96.388	129.980	60.467
1900	50.295	98.225	132.605	65.322
2000	54.050	100.010	135.278	70.537
2100	58.123	101.754	138.013	76.143
2200	62.512	103.466	140.817	82.172
298.15	29.600	40.907	64.680	7.088

Uncertainties in Functions

300	0.100	0.150	0.200	0.020
1000	0.300	0.500	0.600	0.200
2200	3.000	1.000	1.500	2.000

Table III-100   THERMODYNAMIC FUNCTIONS at 1 atm: K(g)

$\dfrac{T}{K}$	$C_p°$	$-\{G°-H°(0)\}/T$	$S°$	$H°-H°(0)$
	$\ldots\ldots\ldots J\cdot(mol\cdot K)^{-1}\ldots\ldots\ldots$			$kJ\cdot mol^{-1}$
0	0.000	0.000	0.000	0.000
100	20.786	116.738	137.524	2.079
200	20.786	131.146	151.932	4.157
300	20.786	139.574	160.360	6.236
400	20.786	145.554	166.340	8.314
500	20.786	150.192	170.978	10.393
600	20.786	153.982	174.768	12.472
700	20.786	157.186	177.972	14.550
800	20.786	159.962	180.748	16.629
900	20.786	162.410	183.196	18.708
1000	20.786	164.600	185.386	20.786
1100	20.786	166.581	187.367	22.865
1200	20.786	168.390	189.176	24.943
1300	20.789	170.054	190.840	27.022
1400	20.793	171.594	192.381	29.101
1500	20.801	173.028	193.816	31.181
1600	20.814	174.370	195.158	33.262
1700	20.836	175.630	196.421	35.344
1800	20.868	176.819	197.613	37.429
1900	20.913	177.943	198.742	39.518
2000	20.973	179.010	199.816	41.612
2100	21.054	180.025	200.841	43.714
2200	21.154	180.994	201.823	45.824
2300	21.275	181.920	202.766	47.945
2400	21.420	182.808	203.674	50.080
2500	21.589	183.660	204.552	52.230
2600	21.782	184.480	205.403	54.398
2700	22.007	185.270	206.229	56.589
2800	22.253	186.033	207.034	58.802
2900	22.525	186.771	207.820	61.040
3000	22.822	187.486	208.588	63.307
3100	23.143	188.179	209.342	65.605
3200	23.488	188.851	210.082	67.937
3300	23.857	189.506	210.810	70.304
3400	24.248	190.143	211.528	72.709
3500	24.660	190.764	212.237	75.154
3600	25.092	191.370	212.938	77.642
3700	25.543	191.963	213.631	80.173
3800	26.012	192.542	214.319	82.751
3900	26.498	193.109	215.000	85.376
4000	26.998	193.665	215.678	88.051
298.15	20.786	139.445	160.232	6.197

Uncertainties in Functions

300	0.001	0.002	0.003	0.001
1000	0.001	0.002	0.003	0.001
4000	0.004	0.005	0.010	0.005

Table III-101 THERMODYNAMIC FUNCTIONS at 1 atm: $K_2(g)$

$\dfrac{T}{K}$	$C_p°$	$-\{G°-H°(0)\}/T$	$S°$	$H°-H°(0)$
	........J·(mol·K)$^{-1}$.......			kJ·mol^{-1}
0	0.000	0.000	0.000	0.000
100	36.495	175.679	208.870	3.319
200	37.541	199.408	234.574	7.033
300	37.990	213.849	249.885	10.811
400	38.394	224.294	260.869	14.630
500	38.793	232.501	269.481	18.490
600	39.067	239.273	276.581	22.385
700	39.056	245.044	282.607	26.294
800	38.685	250.072	287.802	30.184
900	37.983	254.521	292.320	34.019
1000	37.050	258.503	296.276	37.773
1100	35.989	262.099	299.758	41.425
1200	34.902	265.368	302.843	44.970
1300	33.871	268.358	305.596	48.408
1400	32.952	271.108	308.071	51.748
1500	32.201	273.648	310.318	55.005
1600	31.661	276.006	312.379	58.197
1700	31.353	278.202	314.288	61.346
1800	31.305	280.257	316.077	64.477
1900	31.536	282.187	317.775	67.617
2000	32.056	284.008	319.405	70.794
2100	32.862	285.731	320.987	74.038
2200	33.946	287.369	322.540	77.376
2300	35.285	288.932	324.078	80.836
2400	36.847	290.428	325.612	84.441
2500	38.592	291.866	327.151	88.212
2600	40.470	293.253	328.701	92.164
2700	42.427	294.595	330.265	96.309
2800	44.406	295.897	331.844	100.651
2900	46.350	297.164	333.436	105.189
3000	48.205	298.400	335.039	109.918
3100	49.922	299.608	336.648	114.826
3200	51.460	300.790	338.258	119.897
3300	52.789	301.950	339.863	125.112
3400	53.886	303.089	341.456	130.448
3500	54.738	304.207	343.031	135.881
3600	55.342	305.307	344.582	141.388
3700	55.702	306.389	346.104	146.942
3800	55.831	307.454	347.591	152.521
3900	55.745	308.502	349.041	158.102
4000	55.463	309.533	350.449	163.664
298.15	37.982	213.626	249.650	10.741

Uncertainties in Functions

300	0.500	0.050	0.100	0.030
1000	0.800	0.200	0.300	0.300
4000	2.000	1.000	2.000	2.500

Table III-102   THERMODYNAMIC FUNCTIONS: Rb(cr & l)

$\dfrac{T}{K}$	$C_p°$	$-\{(G°-H°(0)\}/T$	$S°$	$H°-H°(0)$
	$\cdots\cdots\cdots J\cdot(mol\cdot K)^{-1}\cdots\cdots\cdots$			$kJ\cdot mol^{-1}$
0	0.000	0.000	0.000	0.000
100	25.510	26.058	46.958	2.009
200	27.450	41.952	65.252	4.660
300	31.231	51.817	76.973	7.547
312.47 cr	32.383	52.847	78.268	7.943
312.47 l	31.801	52.847	85.283	10.135
400	30.822	60.828	92.999	12.868
500	30.484	67.972	99.831	15.930
600	30.439	73.759	105.383	18.974
700	30.524	78.620	110.080	22.022
800	30.709	82.814	114.167	25.083
900	31.012	86.503	117.800	28.167
1000	31.476	89.800	121.090	31.290
1100	32.151	92.784	124.120	34.470
1200	33.094	95.514	126.955	37.729
1300	34.366	98.037	129.652	41.099
1400	36.026	100.389	132.257	44.615
1500	38.140	102.599	134.812	48.320
1600	40.768	104.692	137.354	52.261
1700	43.976	106.688	139.919	56.493
1800	47.828	108.607	142.539	61.077
1900	52.387	110.463	145.243	66.082
2000	57.719	112.272	148.063	71.581
2100	63.887	114.046	151.024	77.654
298.15	31.060	51.662	76.780	7.489
Uncertainties in Functions				
300	0.100	0.200	0.300	0.020
1000	0.500	0.700	0.800	0.300
2100	5.000	2.000	2.000	3.000

Table III-103   THERMODYNAMIC FUNCTIONS at 1 atm: Rb(g)

$\dfrac{T}{K}$	$C_p{}^\circ$	$-\{G^\circ - H^\circ(0)\}/T$	S°	$H^\circ - H^\circ(0)$
	$\cdots\cdots\cdots$ J$\cdot$(mol$\cdot$K)$^{-1}$ $\cdots\cdots\cdots$			kJ$\cdot$mol^{-1}
0	0.000	0.000	0.000	0.000
100	20.786	126.492	147.278	2.079
200	20.786	140.899	161.686	4.157
300	20.786	149.328	170.114	6.236
400	20.786	155.307	176.094	8.314
500	20.786	159.946	180.732	10.393
600	20.786	163.735	184.522	12.472
700	20.786	166.940	187.726	14.550
800	20.786	169.715	190.501	16.629
900	20.786	172.164	192.950	18.708
1000	20.786	174.354	195.140	20.786
1100	20.786	176.335	197.121	22.865
1200	20.786	178.143	198.930	24.943
1300	20.790	179.807	200.594	27.022
1400	20.795	181.348	202.135	29.102
1500	20.805	182.782	203.570	31.182
1600	20.821	184.123	204.913	33.263
1700	20.847	185.384	206.176	35.346
1800	20.884	186.572	207.368	37.433
1900	20.941	187.697	208.499	39.524
2000	21.013	188.764	209.575	41.621
2100	21.107	189.780	210.602	43.727
2200	21.225	190.749	211.587	45.844
2300	21.368	191.675	212.533	47.973
2400	21.538	192.564	213.446	50.118
2500	21.740	193.417	214.330	52.282
2600	21.974	194.238	215.187	54.468
2700	22.236	195.029	216.021	56.678
2800	22.529	195.793	216.835	58.916
2900	22.867	196.533	217.632	61.188
3000	23.229	197.249	218.413	63.493
3100	23.624	197.944	219.181	65.835
3200	24.052	198.620	219.938	68.219
3300	24.513	199.277	220.685	70.647
3400	25.005	199.918	221.424	73.122
3500	25.528	200.543	222.157	75.649
3600	26.082	201.153	222.883	78.229
3700	26.665	201.750	223.606	80.866
3800	27.277	202.335	224.325	83.563
3900	27.914	202.908	225.042	86.322
4000	28.577	203.470	225.757	89.147
298.15	20.786	149.199	169.985	6.197
Uncertainties in Functions				
300	0.001	0.002	0.003	0.001
1000	0.001	0.002	0.003	0.001
4000	0.004	0.005	0.010	0.005

Table III-104   THERMODYNAMIC FUNCTIONS at 1 atm: $Rb_2(g)$

$\dfrac{T}{K}$	$C_p°$	$-\{G°-H°(0)\}/T$	$S°$	$H°-H°(0)$
	........J·(mol·K)$^{-1}$........			kJ·mol^{-1}
0	0.000	0.000	0.000	0.000
100	37.157	195.254	229.798	3.454
200	37.726	219.746	255.765	7.204
300	38.032	234.481	271.122	10.992
400	38.291	245.078	282.100	14.809
500	38.468	253.370	290.666	18.648
600	38.437	260.188	297.681	22.496
700	38.095	265.977	303.585	26.325
800	37.436	271.002	308.632	30.104
900	36.535	275.431	312.991	33.804
1000	35.493	279.381	316.787	37.406
1100	34.410	282.936	320.119	40.902
1200	33.369	286.160	323.068	44.290
1300	32.426	289.102	325.701	47.579
1400	31.623	291.803	328.074	50.780
1500	30.992	294.294	330.234	53.910
1600	30.543	296.603	332.219	56.985
1700	30.284	298.753	334.062	60.025
1800	30.214	300.763	335.790	63.048
1900	30.326	302.650	337.426	66.074
2000	30.610	304.428	338.988	69.120
2100	31.048	306.110	340.491	72.202
2200	31.623	307.706	341.948	75.334
2300	32.312	309.225	343.369	78.530
2400	33.093	310.677	344.760	81.800
2500	33.943	312.068	346.128	85.151
2600	34.838	313.404	347.477	88.590
2700	35.754	314.691	348.809	92.120
2800	36.670	315.933	350.126	95.741
2900	37.567	317.134	351.429	99.453
3000	38.426	318.299	352.717	103.254
3100	39.233	319.430	353.990	107.137
3200	39.974	320.530	355.248	111.098
3300	40.642	321.600	356.488	115.130
3400	41.227	322.645	357.710	119.224
3500	41.727	323.664	358.913	123.372
3600	42.139	324.659	360.095	127.567
3700	42.463	325.633	361.254	131.798
3800	42.700	326.585	362.390	136.057
3900	42.855	327.518	363.501	140.335
4000	42.930	328.431	364.587	144.625
298.15	38.027	234.254	270.887	10.922

Uncertainties in Functions

300	0.500	0.050	0.100	0.020
1000	0.600	0.500	1.000	0.200
4000	2.000	0.800	2.000	2.000

Table III-105   THERMODYNAMIC FUNCTIONS: Cs(cr & l)

$\dfrac{T}{K}$	$C_p{}^\circ$	$-\{G^\circ-H^\circ(0)\}/T$	S°	$H^\circ-H^\circ(0)$
	$\ldots\ldots\ldots$ J·(mol·K)$^{-1}$ $\ldots\ldots\ldots$			kJ·mol^{-1}
0	0.000	0.000	0.000	0.000
100	25.820	33.478	54.978	2.150
200	27.790	49.330	73.475	4.829
300	32.379	59.527	85.430	7.771
301.59 cr	32.525	59.664	85.601	7.822
301.59 l	32.635	59.664	92.551	9.918
400	32.024	68.938	101.708	13.108
500	30.955	76.226	108.740	16.257
600	30.001	82.124	114.296	19.303
700	29.361	87.056	118.868	22.268
800	29.115	91.282	122.768	25.188
900	29.304	94.975	126.204	28.106
1000	29.948	98.256	129.320	31.064
1100	31.059	101.213	132.223	34.111
1200	32.643	103.913	134.990	37.292
1300	34.707	106.408	137.681	40.655
1400	37.251	108.737	140.343	44.249
1500	40.278	110.933	143.014	48.122
1600	43.791	113.022	145.723	52.321
1700	47.788	115.027	148.495	56.896
1800	52.273	116.965	151.351	61.895
1900	57.244	118.852	154.308	67.367
2000	62.702	120.701	157.381	73.360
298.15	32.210	59.367	85.230	7.711

Uncertainties in Functions

300	0.200	0.300	0.400	0.020
1000	0.300	0.500	0.600	0.100
2000	3.000	2.000	2.000	2.000

Table III-106   THERMODYNAMIC FUNCTIONS at 1 atm: Cs(g)

$\dfrac{T}{K}$	C_p°	$-\{G^\circ - H^\circ(0)\}/T$	S°	$\dfrac{H^\circ - H^\circ(0)}{kJ \cdot mol^{-1}}$
	........$J \cdot (mol \cdot K)^{-1}$........			
0	0.000	0.000	0.000	0.000
100	20.786	131.998	152.784	2.079
200	20.786	146.406	167.192	4.157
300	20.786	154.834	175.620	6.236
400	20.786	160.814	181.600	8.314
500	20.786	165.452	186.238	10.393
600	20.786	169.242	190.028	12.472
700	20.786	172.446	193.232	14.550
800	20.786	175.222	196.008	16.629
900	20.786	177.670	198.456	18.708
1000	20.786	179.860	200.646	20.786
1100	20.787	181.841	202.627	22.865
1200	20.791	183.650	204.436	24.944
1300	20.798	185.313	206.101	27.023
1400	20.813	186.854	207.642	29.104
1500	20.842	188.288	209.079	31.187
1600	20.886	189.630	210.426	33.273
1700	20.952	190.891	211.694	35.365
1800	21.047	192.081	212.894	37.464
1900	21.176	193.206	214.035	39.575
2000	21.343	194.275	215.125	41.701
2100	21.554	195.293	216.172	43.845
2200	21.814	196.265	217.180	46.013
2300	22.123	197.196	218.157	48.210
2400	22.486	198.089	219.106	50.440
2500	22.899	198.948	220.032	52.709
2600	23.372	199.777	220.940	55.024
2700	23.891	200.577	221.831	57.386
2800	24.481	201.352	222.712	59.808
2900	25.108	202.104	223.582	62.287
3000	25.783	202.834	224.444	64.831
3100	26.505	203.545	225.301	67.445
3200	27.270	204.238	226.155	70.133
3300	28.075	204.915	227.006	72.900
3400	28.918	205.577	227.857	75.749
3500	29.794	206.226	228.708	78.685
3600	30.701	206.863	229.560	81.709
3700	31.635	207.488	230.413	84.826
3800	32.592	208.102	231.270	88.037
3900	33.568	208.707	232.129	91.345
4000	34.559	209.303	232.991	94.751
298.15	20.786	154.705	175.491	6.197

Uncertainties in Functions

300	0.001	0.002	0.003	0.001
1000	0.001	0.002	0.003	0.001
4000	0.004	0.005	0.010	0.005

Table III-107   THERMODYNAMIC FUNCTIONS at 1 atm: $Cs_2(g)$

$\dfrac{T}{K}$	C_p°	$-\{G^\circ - H^\circ(0)\}/T$	S°	$H^\circ - H^\circ(0)$
	....	J·(mol·K)$^{-1}$	....	kJ·mol^{-1}
0	0.000	0.000	0.000	0.000
100	37.402	208.024	243.292	3.527
200	37.872	232.914	269.380	7.293
300	38.263	247.810	284.808	11.099
400	38.719	258.506	295.877	14.948
500	39.118	266.880	304.563	18.842
600	39.186	273.774	311.708	22.761
700	38.761	279.634	317.723	26.662
800	37.886	284.724	322.846	30.498
900	36.717	289.210	327.243	34.230
1000	35.438	293.208	331.046	37.838
1100	34.195	296.802	334.364	41.319
1200	33.091	300.056	337.291	44.681
1300	32.197	303.023	339.903	47.944
1400	31.546	305.743	342.264	51.129
1500	31.154	308.251	344.425	54.262
1600	31.023	310.575	346.430	57.369
1700	31.137	312.740	348.313	60.475
1800	31.478	314.766	350.102	63.604
1900	32.018	316.671	351.817	66.777
2000	32.727	318.470	353.477	70.013
2100	33.569	320.176	355.094	73.327
2200	34.510	321.799	356.677	76.730
2300	35.514	323.350	358.233	80.231
2400	36.547	324.835	359.766	83.834
2500	37.578	326.263	361.279	87.541
2600	38.578	327.638	362.772	91.349
2700	39.523	328.967	364.246	95.254
2800	40.393	330.253	365.700	99.251
2900	41.173	331.500	367.131	103.330
3000	41.851	332.711	368.539	107.482
3100	42.420	333.889	369.920	111.697
3200	42.877	335.036	371.275	115.962
3300	43.222	336.155	372.600	120.268
3400	43.458	337.246	373.894	124.603
3500	43.590	338.311	375.156	128.957
3600	43.625	339.351	376.384	133.318
3700	43.570	340.369	377.579	137.679
3800	43.434	341.363	378.739	142.029
3900	43.227	342.336	379.865	146.363
4000	42.956	343.288	380.956	150.673
298.15	38.255	247.581	284.571	11.029

Uncertainties in Functions

300	0.400	0.050	0.100	0.020
1000	0.500	0.300	0.400	0.200
4000	1.000	0.800	1.000	1.500

RECOMENDED KEY VALUES FOR MONOTOMIC GASES AND AQUEOUS IONS

Table IV-1.   CODATA Recomended Key Values for Monatomic Gases at 101 325 Pa (1 atm) and 100 000 Pa (1 bar) in alphabetic order.[a]

Element (g)	$\Delta_f H^o$(298.15 K) $\overline{\text{kJ·mol}^{-1}}$	S^o(298.15 K) $\overline{\text{J·K}^{-1}\text{·mol}^{-1}}$ (at 1 atm)	$[H^o$(298.15 K) $-H^o(0)]$ $\overline{\text{kJ·mol}^{-1}}$	S^o(298.15 K) $\overline{\text{J·K}^{-1}\text{·mol}^{-1}}$ (at 1 bar)
Ag	284.9 ± 0.8	172.888 ± 0.004	6.197 ± 0.001	172.997 ± 0.004
Al	330.0 ± 4.0	164.445 ± 0.004	6.919 ± 0.001	164.554 ± 0.004
Ar	0	154.737 ± 0.003	6.197 ± 0.001	154.846 ± 0.003
B	565 ± 5	153.327 ± 0.015	6.316 ± 0.002	153.436 ± 0.015
Be	324 ± 5	136.166 ± 0.003	6.197 ± 0.001	136.275 ± 0.003
Br	111.87 ± 0.12	174.909 ± 0.004	6.197 ± 0.001	175.018 ± 0.004
C	716.68 ± 0.45	157.991 ± 0.003	6.536 ± 0.001	158.100 ± 0.003
Ca	177.8 ± 0.8	154.778 ± 0.004	6.197 ± 0.001	154.887 ± 0.004
Cd	111.80 ± 0.20	167.640 ± 0.004	6.197 ± 0.001	167.749 ± 0.004
Cl	121.301 ± 0.008	165.081 ± 0.004	6.272 ± 0.001	165.190 ± 0.004
Cs	76.5 ± 1.0	175.492 ± 0.003	6.197 ± 0.001	175.601 ± 0.003
Cu	337.4 ± 1.2	166.289 ± 0.004	6.197 ± 0.001	166.398 ± 0.004
F	79.38 ± 0.30	158.642 ± 0.004	6.518 ± 0.001	158.751 ± 0.004
Ge	372 ± 3	167.795 ± 0.005	7.398 ± 0.001	167.904 ± 0.005
H	217.998 ± 0.006	114.608 ± 0.002	6.197 ± 0.001	114.717 ± 0.002
He	0	126.044 ± 0.002	6.197 ± 0.001	126.153 ± 0.002
Hg	61.38 ± 0.04	174.862 ± 0.005	6.197 ± 0.001	174.971 ± 0.005
I	106.76 ± 0.04	180.678 ± 0.004	6.197 ± 0.001	180.787 ± 0.004
K	89.0 ± 0.8	160.232 ± 0.003	6.197 ± 0.001	160.341 ± 0.003
Kr	0	163.976 ± 0.003	6.197 ± 0.001	164.085 ± 0.003
Li	159.3 ± 1.0	138.673 ± 0.010	6.197 ± 0.001	138.782 ± 0.010
Mg	147.1 ± 0.8	148.539 ± 0.003	6.197 ± 0.001	148.648 ± 0.003
N	472.68 ± 0.40	153.192 ± 0.003	6.197 ± 0.001	153.301 ± 0.003
Na	107.5 ± 0.7	153.609 ± 0.003	6.197 ± 0.001	153.718 ± 0.003
Ne	0	146.219 ± 0.003	6.197 ± 0.001	146.328 ± 0.003
O	249.18 ± 0.10	160.950 ± 0.003	6.725 ± 0.001	161.059 ± 0.003
P	316.5 ± 1.0	163.090 ± 0.003	6.197 ± 0.001	163.199 ± 0.003
Pb	195.2 ± 0.8	175.266 ± 0.005	6.197 ± 0.001	175.375 ± 0.005
Rb	80.9 ± 0.8	169.985 ± 0.003	6.197 ± 0.001	170.094 ± 0.003
S	277.17 ± 0.15	167.720 ± 0.006	6.657 ± 0.001	167.829 ± 0.006
Si	450 ± 8	167.872 ± 0.004	7.550 ± 0.001	167.981 ± 0.004
Sn	301.2 ± 1.5	168.383 ± 0.004	6.215 ± 0.001	168.492 ± 0.004
Th	602 ± 6	190.06 ± 0.05	6.197 ± 0.003	190.17 ± 0.05
Ti	473 ± 3	180.189 ± 0.010	7.539 ± 0.002	180.298 ± 0.010
U	533 ± 8	199.68 ± 0.10	6.499 ± 0.020	199.79 ± 0.10
Xe	0	169.576 ± 0.003	6.197 ± 0.001	169.685 ± 0.003
Zn	130.40 ± 0.40	160.881 ± 0.004	6.197 ± 0.001	160.990 ± 0.004

[a] Only the values of S^o for gases are affected by the change in pressure.

Table IV-2.   CODATA Recomended Key Values for Aqueous Ions
at 101 325 Pa (1 atm) and 100 000 Pa (1 bar) in alphabetic order.[a]

Ion (aq)	$\Delta_f H^o(298.15\ \text{K})$ $\overline{\text{kJ·mol}^{-1}}$	$S^o(298.15\ \text{K})$ $\overline{\text{J·K}^{-1}\text{·mol}^{-1}}$
Ag^+	105.79 ± 0.08	73.45 ± 0.40
Al^{3+}	-538.4 ± 1.5	-325 ± 10
Br^-	-121.41 ± 0.15	82.55 ± 0.20
CHO_3^-	-689.93 ± 0.20	98.4 ± 0.5
CO_3^{2-}	-675.23 ± 0.25	-50.0 ± 1.0
Ca^{2+}	-543.0 ± 1.0	-56.2 ± 1.0
Cd^{2+}	-75.92 ± 0.60	-72.8 ± 1.5
Cl^-	-167.080 ± 0.10	56.60 ± 0.20
ClO_4^-	-128.10 ± 0.40	184.0 ± 1.5
Cs^+	-258.00 ± 0.50	132.1 ± 0.5
Cu^{2+}	64.9 ± 1.0	-98 ± 4
F^-	-335.35 ± 0.65	-13.8 ± 0.8
H^+	0	0
H_4N^+	-133.26 ± 0.25	111.17 ± 0.40
HO^-	-230.015 ± 0.040	-10.90 ± 0.20
HPO_4^{2-}	-1299.0 ± 1.5	-33.5 ± 1.5
$H_2PO_4^-$	-1302.6 ± 1.5	92.5 ± 1.5
HS^-	-16.3 ± 1.5	67 ± 5
HSO_4^-	-886.9 ± 1.0	131.7 ± 3.0
Hg^{2+}	170.21 ± 0.20	-36.19 ± 0.80
Hg_2^{2+}	166.87 ± 0.50	65.74 ± 0.80
I^-	-56.78 ± 0.05	106.45 ± 0.30
K^+	-252.14 ± 0.08	101.20 ± 0.20
Li^+	-278.47 ± 0.08	12.24 ± 0.15
Mg^{2+}	-467.0 ± 0.6	-137 ± 4
NO_3^-	-206.85 ± 0.40	146.70 ± 0.40
Na^+	-240.34 ± 0.06	58.45 ± 0.15
O_4S^{2-}	-909.34 ± 0.40	18.50 ± 0.40
O_2U^{2+}	-1019.0 ± 1.5	-98.2 ± 3.0
Pb^{2+}	0.92 ± 0.25	18.5 ± 1.0
Rb^+	-251.12 ± 0.10	121.75 ± 0.25
Sn^{2+}	-8.9 ± 1.0	-16.7 ± 4.0
Zn^{2+}	-153.39 ± 0.20	-109.8 ± 0.5

[a] All of the values for properties in this table are the same at 1 atm and at 1 bar.

KEY SUBSTANCE INDEX
TO TABLES AND NOTES